African Savannas

Global Narratives & Local Knowledge
of Environmental Change

African Savannas

Global Narratives & Local Knowledge
of Environmental Change

EDITED BY
Thomas J. Bassett
Donald Crummey

JAMES CURREY
OXFORD

HEINEMANN
PORTSMOUTH, NH

James Currey Ltd
73 Botley Road
Oxford
OX2 0BS

Heinemann
A division of Reed Elsevier Inc.
361 Hanover Street
Portsmouth NH 03801-3912

British Library Cataloguing in Publication Data
African savannas : global narratives & local knowledge of
    environmental change
    1. Savannas - Africa, Sub-Saharan 2. Agriculture -
    Environmental aspects - Africa, Sub-Saharan 3. Agricultural
    systems - Africa, Sub-Saharan 4. Land use, Rural - Africa,
    Sub-Saharan 5. Range management - Africa, Sub-Saharan
    I. Bassett, Thomas, J.   II. Crummey, Donald
    333.7'4'0967

ISBN 0-85255-424-9 (James Currey paper)
ISBN 0-85255-429-X (James Currey cloth)

Library of Congress Cataloging-in-Publication Data
available on request from the Library of Congress

ISBN 0-325-07128-4 (Heinemann paper)
ISBN 0-325-07127-6 (Heinemann cloth)

Typeset in 10.6/11.6 pt Bembo by Long House Publishing Services

Contents

List of Photographs   vii

List of Figures   ix

List of Tables   xi

Preface   xiii

Notes on Contributors   xv

PART ONE   Introduction   1

1

Contested Images, Contested Realities:

Environment and Society in African Savannas   1

     THOMAS J. BASSETT & DONALD CRUMMEY

PART TWO   Longue Durée   31

2

Changing Land Use & Environmental Fluctuations

in the African Savannas   31

     MARTIN WILLIAMS

PART THREE   Land Users & Landscapes   53

3

Fire in the Savanna

Environmental Change & Land Reform

in Northern Côte d'Ivoire   53

     THOMAS J. BASSETT, ZUELI KOLI BI

     & TIONA OUATTARA

4

Investing in Soil Quality

Farmer Responses to Land Scarcity in Southwestern Burkina Faso   72

     LESLIE C. GRAY

5
Farmer Tree-Planting in Wällo, Ethiopia
1937–1997   91
> DONALD CRUMMEY & ALEX WINTER-NELSON

6
The Wild Vegetation Cover of Western Burkina Faso
Colonial Policy & Post-Colonial Development   121
> MAHIR ŞAUL, JEAN-MARIE OUADBA
> & OUETIAN BOGNOUNOU

PART FOUR   Pastoral Ecologies   161

7
Rethinking Interdisciplinary Paradigms
& the Political Ecology of Pastoralism in East Africa   161
> PETER D. LITTLE

8
Ecological 'Crisis' & Resource Management Policy
in Zimbabwe's Communal Lands   178
> WILLIAM A. MUNRO

PART FIVE   Policy, Producers & Resources   205

9
Littering the Landscape
Environmental Policy in Northeast Ethiopia   205
> DESSALEGN RAHMATO

10
Social Differentiation, Farming Practices
& Environmental Change in Mozambique   225
> MERLE L. BOWEN, ARLINDO CHILUNDO
> & CESAR A. TIQUE

Select Bibliography   248
Index   264

List of Photographs

2.1 Badland topography in the floor of caldera 5 at
K'one volcano, Ethiopia 44
2.2 Hot springs near Fantale Volcano, Ethiopia 48
3.1 Kakoli hunter holding a grivet monkey 63
3.2 Kakoli hunter (*donzo*) in a savanna woodland 63
3.3 Mural outside the GEPRENAF office in Ferkéssédougou 66
3.4 Cattle in heavily grazed area invaded by *Annona senegalensis* 67
3.5 Burnt fallow showing grass regrowth and *Isolberlina doka*
seedlings unscathed by fire 67
4.1 Woman harvesting cotton 83
4.2 Terraced field with young *Acacia albadia* trees 83
4.3 Preparing a field with an ox-drawn plow 84
4.4 Market women selling agricultural products 84
4.5 Maize interspersed with karité trees 85
5.1 *Dägga* landscape, Borru Peasant Association 103
5.2 *Wayna Dägga* landscape, Hitacha Peasant Association, near Sulula 104
5.3 *Wayna Dägga* landscape, Gwobeya Peasant Association with
Lake Häyq in the background 104
5.4 Yemam Yuséf with his coffee bushes, Bädädo Peasant Association 105
5.5 Wärqenäsh Täfara with her eucalyptus grove, Qorké Peasant
Association 105
5.6 Zäwdé Gétahun with her banana tree, T'is Aba Lima Peasant
Association 106
5.7 Ahmäd Bäkurä with his eucalyptus grove, Gerba Peasant Association 106
6.1 (a,b,c) *Parkia biglobosa* (néré or dawadawa): the tree, the flower
and a woman carrying her basket of beans 132
6.2 *Butyrospermum paradoxum* (shea or karité). The caterpillar
(*situmu* in Jula, *kpiye* in Bobo) that thrives on the tree
in the rainy season is gathered and eaten 133

6.3 Mature mango trees, in Péni, in the process of being severely lopped in order to rejuvenate them and open the field under them to cropping 158

10.1 Sheet and gully erosion in a small private farm on the outskirts of Namialo 236

10.2 Gully erosion on a non-cotton-producing family farmer's field 236

10.3 Trenches leading to soil erosion and pools of stagnant water following heavy rains in a small private farmer's field prepared by a rented tractor 237

10.4 Pest destruction on the bean crop of a non-cotton-producing family farmer without access to insecticides 237

10.5 Mulch used by small private farmers to provide vegetation cover and protect the soil against erosion 238

10.6 Sugar-cane residue used as fertilizer in family gardens in Namialo 238

10.7 Wood chips used by family farmers to fertilize the soil for vegetables 238

10.8 Family farmers value the fertility of the soil around termite mounds for cultivating maize and other food crops 239

10.9 Multi-cropping of cassava, groundnuts, maize and beans, practised by family farmers to protect the soil and manage pests 239

10.10 A rented tractor prepares the cotton fields of a small private farmer 239

List of Figures

	Reference map	xvi
2.1	Key regions influenced by the Southern Oscillation	34
2.2	Time series representation of the seven data sets shown in Table 2.1	35
2.3	Landforms and soil-forming parent sediments in the north-western Gezira, Sudan	37
2.4a	Geological and geomorphic map of K'one volcanic complex	42
2.4b	Awash River Valley complex	42
2.5	Hypothetical closing stages in the Quaternary evolution of K'one volcanic complex	43
2.6	Plane table map of K'one complex 'A' (see Fig. 2.4)	44
2.7	Levelled stratigraphic transect west of Lake Besaka	46
2.8	Schematic reconstruction of the late Quaternary evolution of Lake Besaka	47
3.1	Fire regimes and tree/grass ratios	57
3.2	Fulbe cattle population growth, 1973–90	59
3.3	Land cover change, Katiali, Côte d'Ivoire	60
3.4	Monthly sales of commercial game, Kakoli, Côte d'Ivoire, 1981/82 and 1997	62
3.5	Top six animals marketed by per cent of total sales, Kakoli, Côte d'Ivoire, 1981/82 and 1997	62
3.6	Commercial game sales by species, Kakoli, 1981/82 and 1997	62
4.1	Map of study area in Burkina Faso	73
5.1	Map of Tähulädäré Wärädä, Wällo	92
6.1	Project area and research sites	122
6.2a	Land use in Sara-Hantiaye, 1952	143
6.2b	Land use in Sara-Hantiaye, 1981	143
6.2c	Land use in Sara-Hantiaye, 1952 and 1981	144
6.3a	Land use in Bare,1952 and 1981	145

6.3b	Land use in Bare,1952 and 1981	146
6.4a	Land use in Soumousso, 1951 and 1981	147
6.4b	Land use in Soumousso, 1952 and 1981	148
6.5a	Land use in Péni, 1956/7 and 1983	149
6.5b	Land use in Péni, 1956 and 1983	150
6.6a	Land use in Bougoula, 1956/7 and 1983	151
6.6b	Land use in Bougoula, 1956 and 1983	152
6.7a	Land use in Kankalaba, 1956/7 and 1983	153
6.7b	Land use in Kankalaba, 1956 and 1983	154
7.1	Map of Lower Jubba Region, Somalia	172
10.1	Map of administrative posts, Meconta District, Nampula, Mozambique	226

List of Tables

2.1	Details of data sets used to construct Figure 2.2	34
4.1	Population change in Sara, Dimikuy, and Dohoun	77
4.2	Hectares under land type, 1981 and 1993	78
4.3	Average age of field in cultivation (in years) by village and ethnic group	78
4.4	Correlations among socioeconomic variables and manure use, fertilizer use, tree density	86
4.5	Percentage of fields where manure was applied	86
4.6	Average manure application (in sacks/ha) by village and group for farmers applying manure to their fields	87
4.7	Average fertilizer application (in sacks/ha) by village and group	87
4.8	Number of fields on which farmers made anti-erosion improvements	88
4.9	Density of trees per hectare by village and ethnic group	89
5.1	Population density by district	97
5.2	Interview groups	98
5.3	Descriptive statistics of three peasant associations	100
5.4	Percentage of households with less than X hectare per person	100
5.5	Percentage of respondents engaged in specific farm activities	100
5.6	Stated uses for trees	101
5.7	Trees and bushes planted by informants	102
5.8	Perceived changes in number of trees, Wällo	103
5.9	Perceived changes in number of trees, Gondär	108
5.10	Potentially explanatory variables of tree density	109
5.11	Factors influencing tree planting	110
5.12	Factors influencing limited tree planting	111
5.13	Factors influencing the presence of market gardening in Gwobeya	115
5.14	Period of initial tree-planting	117
6.1	Land use in the selected sites	142
7.1	The basic elements of a political ecology approach	165

7.2   Comparison of rangeland classification, Marsabit District, Kenya   170
7.3   Distance of grazing migrations of livestock by season, Afmadow herders 174
7.4   Average cattle owned by Afmadow herders at different grazing zones,
      long wet season   176
9.1   Estimate of cost of environmental protection, 1974–93 (US$ m.)   216
9.2   Estimate of conservation work area, 1976–90 ('000 ha)   218
10.1 Land-use changes in Namialo, Nampula Province, 1964–97   233
10.2 Category of farmers and crops cultivated   240
10.3 Natural soil fertility matrix   241
10.4 Matrix for crop rotation variation in the Namialo area   242

Preface

This collection originated in a collaborative research project organized by the Center for African Studies at the University of Illinois and funded by the John D. and Catherine T. MacArthur Foundation. We are deeply indebted to both these institutions, to the center for its enthusiastic support, and to the foundation for its generous funding. The project, entitled 'The African Environment: Experience and Control', ran from 1993 to 1998, and linked researchers at the University of Illinois at Urbana-Champaign with research teams in Burkina Faso, Côte d'Ivoire, Ethiopia and Mozambique. Each of those national teams, in turn, included one member of the Illinois team. The Illinois team incorporated a wide range of disciplines – anthropology, climatology, economics, geography (including geo-morphology), history and political science – while the African teams covered many of these same disciplines and added expertise in agronomy, archæology, botany, and soil science. The project represented a major commitment to cross-disciplinary dialogue, and, drawing on much, but not all, of this range of expertise, this collection embodies an unusual breadth of perspectives and research techniques. Our dialogue took place in the field in Africa, in bars and coffee shops in Africa and Champaign-Urbana, in seminar rooms on both continents, and in an annual series of symposia held at the University of Illinois. While we never overcame primary disciplinary commitments, each of the chapters is nonetheless influenced by the inter-disciplinary milieu out of which it arose.

A number of primary commitments shaped project research. We sought savanna environments and in each country we focused on areas with annual rainfall regimes of about 1,000 millimeters per annum and characterized by the farming and herding activities of small-scale enterprise. Each research team adopted themes of national concern: the pervasiveness of a national market for agricultural goods in Burkina Faso; commercial cotton growing, hunting and livestock raising in Côte d'Ivoire; drought and famine vulnerability in Ethiopia; radical change in macro-economic strategies in Mozambique.

The chapters which follow originated in papers presented at a retrospective symposium, 'African Savannas: New Perspectives on the Environment and Social Change', which took place in Urbana in April 1998. They represent only a small proportion of the research carried out within the project's framework. The Introduction refers to accounts of some of that research which are appearing elsewhere.

We have acquired numerous and heavy debts, beyond those mentioned already. We have benefited in more ways than we can adequately express from all the colleagues who participated in the project. At the University of Illinois these were: Tom Bassett; Merle Bowen; Don Crummey; Leslie Gray; Don Johnson; Pete Lamb (now at the University of Oklahoma); Mahir Şaul; Joseph Otieno; Sosina Asfaw; Cesar Tique; and Alex Winter-Nelson. Mahir Şaul was the Illinois link to the team in Burkina Faso, which was organized through the Centre Nationale de Recherche Scientifique and included: Ouétian Bognounou, an ethno-botanist; Jean-Baptiste Kiethega (an archaeologist); Jean-Marie Ouadba (a bio-geographer and director of the National Center); and Christophe Dya Sanou (a geographer and soil scientist). Tom Bassett was the liaison with the Côte d'Ivoire team which was organized through the Institut de Géographie Tropicale at the Université de Cocody (Abidjan) and included: Koli Bi Zuéli (geographer); Maméri Camara (a soil scientist): Sinaly Coulibaly (a geographer); and Tiona Ouattara (an historian). Don Crummey was the link to the Ethiopian team, which was organized through the Institute of Ethiopian Studies at Addis Ababa University and included: Asnake Ali (an historian); Bahru Zewde (an historian and Director of the Institute); Belay Tegene (a physical geographer); Dessalegn Rahmato (a political scientist); Daniel Gemechu (a physical geographer); and Sebsebe Demissew (a biologist). Merle Bowen was the link to the Mozambique team which was organized through the Instituto Nacional de Investigação Agronómica (INIA) in the Ministério da Agricultura e Desenvolvimento Rural and through Eduardo Mondlane University. The Mozambique team included Paulo Zucula (then Vice Minister of Agriculture), Carlos Zandamela (then Director of INIA), Feliciano Mazuze (head of the Department of Land and Water at INIA), and Arlindo Chilundo (an historian). Finally, at the Center for African Studies, Sue Swisher managed our accounts and made our numerous arrangements with aplomb, and Paul Tiyambe Zeleza, who became director of the center in 1995, was unstinting in his support.

Thomas J. Bassett   *Donald Crummey*

Notes on Contributors

Thomas J. Bassett is Professor of Geography at the University of Illinois, Urbana-Champaign. He has conducted research for more than twenty years in northern Côte d'Ivoire on the social and agricultural history of cotton, land tenure systems, Fulbe pastoralism, and the political ecology of environmental change. He is the author of *The Peasant Cotton Revolution in West Africa: Côte d'Ivoire, 1880–1995* (Cambridge University Press, 2001), and co-editor (with D. Crummey) of *Land in African Agrarian Systems* (University of Wisconsin Press, 1993).

Ouétian Bognounou is *Chargé de Recherche* in the INERA Institute (Institut d'Environnement et de Recherche Agricole) of the Centre for Science and Technology (CNRST) in Ouagadougou. His research concerns floristics and eth-nobotanics. He possesses an extensive knowledge of the flora of Burkina Faso and the local uses of plants. Over the years he has contributed to numerous research projects, the training of young people, and the policy of national environmental protection in Burkina Faso.

Merle L. Bowen is Associate Professor of Political Science at the University of Illinois, Urbana-Champaign and author of *The State Against the Peasantry: Rural Struggles in Colonial and Postcolonial Mozambique* (University Press of Virginia, 2000). She is currently engaged in a transcontinental comparative study on land reform and rural communities in Brazil, Mozambique and South Africa.

Arlindo G. Chilundo is Assistant Professor of History and coordinator of the Land Studies Unit at Eduardo Mondlane University. He also is a Higher Education Program Coordinator at the Ministry of Higher Education, Science and Technology. Dr. Chilundo has published many articles on the social and economic history of Mozambique and is the author of *Os Camponeses e os Caminhos de Ferro e Estradas em Nampula 1900–1961* (CIEDIMA, 2001) [*Peasants and Rail and Road Transportation Systems in Nampula (1900–1961)*].

Donald Crummey is Professor of History at the University of Illinois, Urbana-Champaign. From 1984 to 1994 he was director of the university's Center for African Studies. He was principal investigator on the project, 'The African Environment: Experience and Control', funded, from 1993 to 1998 by the John T. and Catherine C. MacArthur Foundation. He is the author of *Land and Society in the Christian Kingdom of Ethiopia: From the Thirteenth to the Twentieth Century* (University of Illinois Press, James Currey and Addis Ababa University Press, 2000).

Leslie C. Gray is an Assistant Professor in Environmental Studies and Political Science at Santa Clara University. Her main research interests include land tenure, population growth and the environment in Burkina Faso and coping strategies and drought in Sudan. She has recently begun a new project on land tenure and eco-tourism in Trinidad and Tobago.

Zuéli Koli Bi is *Maitre Assistant* in the Tropical Geography Institute at the University of Cocody in Abidjan where he heads the program on natural landscape studies. He is a specialist in geographic information systems and remote sensing. His research focuses on the influence of physical geographic factors in land use patterns in the savanna region of Côte d'Ivoire.

William A. Munro teaches African politics and international development at Illinois Wesleyan University. He is also a Senior Research Fellow at the Centre for Social and Development Studies, University of Natal (Durban). His research and publications focus mainly on the relationship between agrarian change and state formation in southern Africa. He is the author of *The Moral Economy of the State: Conservation, Community Development and State-Making in Zimbabwe* (Ohio University Press, 1998).

Peter D. Little is Professor of Anthropology, University of Kentucky and has published widely on social change and development, African pastoralism, and political ecology. He is the author of *The Elusive Granary: Herder, Farmer, and State in Northern Kenya* (Cambridge University Press, 1992) and *Economy without State: Survival and Accumulation in Stateless Somalia* (forthcoming, James Currey and Indiana University Press).

Jean-Marie Ouadba is *Chargé de Recherche* in the INERA Institute (Institut d'Environnement et de Recherche Agricole) of the National Centre for Science and Technology (CNRST), Ouagadougou, Burkina Faso. He conducts research on terrestrial ecosystem dynamics, forest ecology, and Sahelian floodplains, is the regional coordinator of SAWEG (Sahelian West African Expert Group), and has for many years collaborated with the International Union of World Conservation (IUCN).

Tiona F. Ouattara is *Maitre de Recherche* in History at the University of Cocody, Abidjan. He is the author of *Histoire des Fohobélé de Côte d'Ivoire, Une population sénoufo inconnue* (Karthala, 1999). His current research focuses on the settlement history of Côte d'Ivoire's secondary cities.

Dessalegn Rahmato is the manager of the Forum for Social Studies, an independent policy research institution based in Addis Ababa and formerly a Senior Research Fellow in the Institute of Development Research at Addis Ababa University. He has published numerous works on land tenure and agrarian issues, food security, and environmental policy. He is the winner of the 1999 Prince Claus Award (given to individuals from the Third World considered to have made significant contributions to their societies) in recognition of 'his achievements in the field of research and development'. He is currently working on the subject of poverty and famine in Ethiopia.

Mahir Şaul teaches in the department of Anthropology, University of Illinois, Urbana-Champaign. He has published articles and chapters on the ethnography of Burkina Faso, farm household organization, traders, Islam, pre-colonial and colonial political organization. Most recently he has published with Patrick Royer *West African Challenge to Empire: Culture and History in the Volta-Bani Anticolonial War* (James Currey and Ohio University Press, 2001).

César A. Tique is the Head of the Land and Water Department at the National Institute of Agrarian Research in Maputo, Mozambique. He is involved in wide ranging research on land use planning and conservation in Mozambique. His most recent publication is 'Gender Aspects of Soil Conservation: Adaptation and Conservation Strategies of Peasant Farmers in Gondola District, Manica Province', in *Strategic Women Gainful Men: Gender, Land and Natural Resources in Different Rural Contexts in Mozambique* edited by Rachel Waterhouse and Carin Vijfhuizen (University of Eduardo Mondlane, 2001).

Martin Williams is Foundation Professor of Environmental Studies at the University of Adelaide, Australia. He has carried out extensive fieldwork in Australia, Africa, India and China, and is the author of over 180 research papers on landscape evolution, climatic change and prehistoric environments in Australia, Africa and India. He has co-authored or co-edited 11 books, including *The Sahara and the Nile* (Balkema, 1980); *A Land Between Two Niles: Quaternary Geology and Biology of the Central Sudan* (Balkema, 1982); *Interactions of Desertification and Climate* (Arnold, with UNEP and WMO, 1996); and *Quaternary Environments* (Arnold, 1998).

Alex Winter-Nelson is an Associate Professor in the Department of Agricultural and Consumer Economics and the Center for African Studies at the University of Illinois. His research focuses on the impacts of economic policy on resource management and incomes among rural households in Africa. He teaches courses on international economic development and environmental economics at the graduate and undergraduate levels.

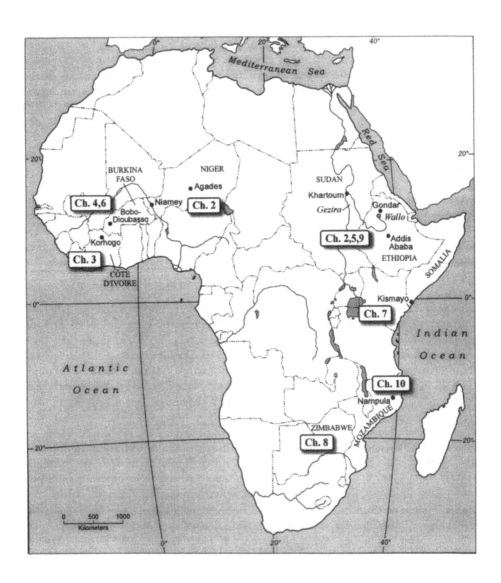

Reference Map

PART ONE

Introduction

1

Contested Images, Contested Realities
Environment & Society in African Savannas

THOMAS J. BASSETT & DONALD CRUMMEY

The environment is an increasingly influential framework within which both Africans and outsiders to the African continent interpret developments. Africa has entered the global debate under a number of rubrics – desertification, deforestation, drought and famine perhaps the most salient. This discourse on the environment emphasizes devastation, degradation, and human-ecological decline, reinforcing the negative images of the continent circulated by the mass media. It is similar, in this regard, to the apocalyptic character of development discourse with its emphasis on economic and political crises. We believe that images of chaos and destruction grossly misrepresent the history and geography of the relations between people and the environment in Africa.

This collection promotes a very different image, one of stress and struggle, to be sure, but one in which African farmers and herders appear as knowledgeable and responsible environmental managers, as actors rather than victims. Against the global and scientific discourses, our research establishes the importance of local knowledge.[1] It does so with reference to African savannas, which cover a significant proportion of Africa,[2] and have formed influential theaters within which the continent's peoples have forged their history. Savannas are significantly different from the desert edge, the canvas on which the 'desertification' narrative is typically painted.

As a whole, the collection represents an attempt to synthesize biophysical and historical processes. We include a chapter on geomorphology as a reminder that physical processes also have a history and that change in the twentieth century took place within a millennial and centennial framework of physical as well as cultural

---

[1] For a forceful account of the devaluing of the local and the practical in favor of a rationality with universalist claims, see James Scott, *Seeing Like a State: How certain schemes to improve the human condition have failed* (New Haven, CT: Yale University Press, 1998).

[2] Estimates vary from less than a half to two-thirds: R. J. Scholes and B. H. Walker, *An African Savanna: Synthesis of the Nylsvley study* (Cambridge: Cambridge University Press, 1993), p. 2; and M. E. Adams, 'Savanna environments' in W. A. Adams, A. S. Goudie and A. Orme (eds), *The Physical Geography of Africa* (Oxford: Oxford University Press, 1996), p. 196.

forces. While climate and soil are factors at work in all our cases, the case studies have paid particular attention to the interaction between farming and herding practices and the vegetated landscapes in which they take place.[3] In the first instance our attempts at synthesizing the ecological and social have been influenced by the approach of political ecology,[4] one which finds the dynamic of environmental change at the intersection of political-economic and ecological processes. More generally, the introduction of culture and our emphasis on dynamic processes taking place through time have placed us on the terrain marked out in recent decades by environmental history, and, already a generation earlier, by historical geography.[5]

When we started the research which led to this book, we did so with a set of concerns drawn from national and international agendas, which foregrounded environmental 'degradation'. We found those concerns subverted by our local and regional informants, who frequently described events from a perspective radically different from those perspectives which prevailed in national capitals and international fora. Local and regional perspectives opened to us processes of change and development of which national and international agencies are profoundly ignorant, and we became aware of a much greater gulf than any of the rifts which separate the natural from the social sciences – at times an *Alice in Wonderland* difference in perception between events as our informants described them, on the one hand, and the views of those same events, which prevailed elsewhere.

People and the Environment in Africa: Recent Contributions

In coming to this understanding we found ourselves in the company of others, most notably David Anderson and his collaborators in two separate collections, Richard

---

[3] Guyer and Richards insist on the importance of 'landscape' in understanding biodiversity in Africa: Jane Guyer and Paul Richards, 'The Invention of Biodiversity: Social Perspectives on the Management of Biological Variety in Africa,' *Africa*, LXVI, 1 (1996), pp. 1-13. This is a special issue of *Africa* dedicated to 'The Social Shaping of Biodiversity. Perspectives on the Management of Biological Variety in Africa'. Other contributors to this issue document a number of pertinent cases of landscape management.

[4] P. Blaikie, *The Political Economy of Soil Erosion* (New York: Longman, 1985); P. Blaikie, 'Changing Environments or Changing Views? A Political Ecology of Developing Countries,' *Geography*, 80, 348 (1995), pp. 203–14; R. Bryant and S. Bailey, *Third World Political Ecology* (London: Routledge, 1997); R. Peet and M. Watts, *Liberation Ecologies: Environment, development, social movements* (London: Routledge, 1996).

[5] See especially Donald Worster, *Dust Bowl. The Southern Plains in the 1930s* (Oxford/New York: Oxford University Press, 1979); and idem., 'Transformations of the Earth: Toward an Agroecological Perspective in History', *Journal of American History*, 76, 4 (1990), pp. 1087–1110; also William Cronon, *Nature's Metropolis. Chicago and the Great West* (New York: W. W. Norton, 1991); and with special reference to Africa, Robert Harms, *Games Against Nature: An Eco-cultural History of the Nunu of Equatorial Africa* (Cambridge: Cambridge University Press, 1987). Environmental history collections of value include: Donald Worster (ed.), *The Ends of the Earth* (Cambridge: Cambridge University Press, 1988); and Char Miller and Hal Rothman, *Out of the Woods. Essays in Environmental History* (Pittsburgh: University of Pittsburgh Press, [1997]). Classic works in historical geography include Carl Sauer, *Land and Life* (Berkeley; University of California Press, 1963); William L. Thomas (ed.), *Man's Role in Changing the Face of the Earth* (Chicago: University of Chicago Press, 1956); and Ralph H. Brown, *Historical Geography of the United States* (New York: Harcourt, Brace and World, 1948).

Grove and Douglas Johnson; James Fairhead and Melissa Leach; and Mary Tiffen, Michael Mortimore and Francis Gichuki.[6] The Anderson and Grove collection was inspired by the famine interventions of the 1980s. It established a history of environmental interventions in Africa, initiated by colonial governments and maintained into the era of independence by the succeeding African governments. These interventions rested overwhelmingly on European assumptions and only tenuously on empirical reality in Africa and were characterized by a language of crisis. Anderson and Grove concluded their introduction with the observation that:

> There exists an absolute necessity for scientists, social scientists, historians, development planners and the governments and agencies they advise, to begin to learn to speak to each other, and to become mutually aware of the full complexity of the social dynamics that operate in rural Africa. (p. 10)

Anderson and Johnson argued that the crisis atmosphere which surrounded drought and famine in northeast Africa in the 1980s ignored both long-term climatic and ecological processes and the long-term coping strategies of African societies.[7] Instead, governments and private agencies alike failed to distinguish between major crises and commonly occurring stress, and made illicit inferences about the role of indigenous livelihood practices in contributing to ecological stress, and, consequently, illicit and wasteful interventions. Moreover, they argue, injudicious intervention threatens to erode the capacity of African societies to deploy their historic strategies of survival.

Fairhead and Leach reinforced Anderson and Grove's account of the history of conservation in Africa with particular reference to the forest-savanna mosaic of the West African Republic of Guinea. They used historical documents, aerial photographs and the evidence of remote sensing to establish the radically divergent evolution of a national (and international) belief in degradation of the landscape through the continuing conversion of forest to savanna and of a local reality of *increasing* woody vegetation.[8] The explanation of actual landscape change the anthropologists Fairhead and Leach found in local, culturally informed agricultural practices. Tiffen, Mortimore and Gichuki also used historical documents together with aerial and ground photographs, the latter matched in pairs of the same landscape separated by upwards of 50 years, to establish another case resting on the Machakos District of Kenya – to Kenya's British rulers in the 1930s a paradigm of landscape degradation. They meticulously documented an increase in woody vegetation and a decrease in soil erosion, both associated with, and, indirectly, caused by, extraordinarily rapid

---

[6] David Anderson and Richard Grove (eds), *Conservation in Africa: Peoples, Policies and Practices* (Cambridge: Cambridge University Press, 1987); David Anderson and Douglas Johnson, *The Ecology of Survival. Case Studies from Northeast African History* (London/Boulder, CO: Lester Crook Academic Publishing and Westview Press, 1988); James Fairhead and Melissa Leach, *Misreading the African Landscape: Society and Ecology in a Forest-Savanna Mosaic* (Cambridge: Cambridge University Press, 1996); and Mary Tiffen, Michael Mortimore and Francis Gichuki (eds), *More People, Less Erosion: Environmental Recovery in Kenya* (Chichester, UK: John Wiley and Sons for the Overseas Development Institute, 1994).

[7] A point strongly seconded with reference to conservation biology by Guyer and Richards, 'The Invention of Biodiversity.'

[8] See also James Fairhead and Melissa Leach, 'Enriching the Landscape: Social History and the Management of Transition Ecology in the Forest-Savanna Mosaic of the Republic of Guinea', *Africa*, LXVI, 1 (1996), pp. 14–35.

population growth. Their principal explanatory framework was economic.[9]

All four of these books highlighted the radical scarcity of documentation about the nature and extent of environmental change in twentieth-century Africa and the filling of the resultant empirical void with inferences and explanations derived from pre-conceptions about environmental change in other, principally temperate, parts of the world. Anderson and Grove documented the existence of the latter mechanism as far back as the Cape Colony in the mid-nineteenth century; and Anderson has demonstrated how the American Dust Bowl helped lay the foundations for an interventionist program of soil conservation in Britain's African colonies.[10]

Still more recent contributions indicate the emergence of a literature on the environment in Africa, which incorporates the perspectives of local communities.[11] Much of the literature has reference to East and Southern Africa, where the ubiquitous alienation of colonial governments from their African subjects was intensified by European settler minorities. It demonstrates how counter-productive modern states in Africa – colonial and post-colonial – have been in their environmental interventions. A fundamental assumption of environmental history is that all landscapes are anthropogenic – shaped by human activity. This is particularly true of Africa, the only continent on which the entire range of human history has been played out, but twentieth-century governments have refused to see this reality or its implications for landscape management. European colonial views of the African landscape were dichotomous. One tendency depopulated the African landscape and re-constructed it as a tropical Eden, pristine and wild.[12] Another tendency recognized the human presence in Africa but, once again separating people from the landscapes which they had shaped, construed its role in almost entirely negative terms. The post-colonial governments of Africa took over the environmental attitudes of their predecessors, the pejorative colonial interpretation of indigenous land-use practices now understood within a framework of tradition versus modernity, with the state seeing itself as an agent firmly committed on the side of modernity.[13] We resume this discussion below in the section on state and environment in twentieth-century Africa.

[9] Geographer Michael Mortimore has subsequently published a general, insightful survey of environment and development issues in Africa: *Roots in the African Dust. Sustaining the Drylands* (Cambridge: Cambridge University Press, 1998). See also M. Leach and Robin Mearns (eds), *The Lie of the Land: Challenging Received Wisdom on the African Environment* (Oxford/Portsmouth, NH: James Currey/Heinemann, 1996).

[10] Richard Grove, 'Early themes in African conservation: the Cape in the nineteenth century', Chapter 1 in Anderson and Grove, *Conservation in Africa*; and David Anderson, 'Depression, Dust Bowl, Demography and Drought: The Colonial State and Soil Conservation in East Africa during the 1930s', *African Affairs*, 83 (1984), pp. 321–45.

[11] A valuable collection, which embodies a good deal of this literature, is Gregory Maddox, James Giblin and Isaria Kimambo (eds), *Custodians of the Land. Ecology and Culture in the History of Tanzania* (London/Nairobi/Dar es Salaam/Athens, OH: James Currey/Mkuki na Nyota/E.A.E.P./Ohio University Press, 1996).

[12] Terence Ranger, *Voices from the Rocks. Nature, Culture and History in the Matopos Hills of Zimbabwe* (Oxford/Bloomington, IN: James Currey/Indiana University Press, 1999), p. 61. See also David Anderson and Richard Grove, 'Introduction: The scramble for Eden: past, present and future in African conservation', in Anderson and Grove, *Conservation in Africa*.

[13] James Scott, *Seeing Like a State*, shows how the contemporary state is wedded to modernity and how this militates against its taking local knowledge into account.

Savanna Environments: Equilibrium or Disequilibrium?

If all landscapes are anthropogenic, humans do not cause all change within them. Climate, soils, topography and latitude combine to create conditions more or less favorable to distinct plant communities, which have their own dynamics in which human intervention is but one of many forces at work.

Africa's savannas have been the subject of much speculation and misunderstanding in the literature on the continent and its peoples. Ancient ideas about the influence of climate on health, character, and intellect were prevalent in the writings of European travelers and armchair geographers during the eighteenth and nineteenth centuries.[14] The interior savanna was considered to be 'superior' in many ways to the 'poisonous airs' of the swamps of Africa's coastal forests which were believed to bring 'fever' and death to Europeans.[15] Its higher elevation and relative dryness were assumed to bring better health, stronger character, and advanced civilizations. Reports of thriving commercial centers in West Africa and a Christian kingdom of the fabled Prester John in Ethiopia contributed to the imagined grandeur of the unknown interior.[16] Such ideas held sway despite contradictory evidence. Although disease claimed the lives of three-quarters of his men during his first exploration of the interior of West Africa in the late eighteenth century, Mungo Park still considered the savanna to be a hospitable environment.[17]

During the twentieth century, under European colonial rule, new views of the savanna emerged which stressed human impact on the environment as opposed to environmental influence on society. Colonial foresters and agronomists commonly viewed savannas as degraded landscapes that had been more wooded in the past. They folded these 'derived savannas' into the narrative of 'desiccation,' which, as we shall see, intimately linked rainfall decline with deforestation.[18] Colonial scientists knew that the Sahara had had a more humid past. Some viewed its drying up as a result of destructive activities by African farmers and pastoralists, activities which equally had produced savanna out of forest. This view was vigorously put forward in the debate about colonial forestry and agricultural policies, and continues to inform discussions on Africa's savannas.

Put simply, tropical savannas can be defined as mixed grass-tree plant com-

---

[14] Clarence Glacken traces the idea of environmental influences on society to the Greek physician Hippocrates, specifically to his assumed work, *Airs, Waters, and Places*: C. Glacken, *Traces on the Rhodian Shore* (Berkeley: University of California Press, 1967), pp. 80–8.

[15] Philip Curtin, *The Image of Africa: British Ideas and Action, 1780–1850* (Madison: University of Wisconsin Press, 1964), 2 Vols, pp. 77–80.

[16] *Ibid.*, pp. 25, 77, 180, 257, 353.

[17] *Ibid.*, p. 181.

[18] J. Fairhead and M. Leach, 'Dessication and Domination: Science and Struggles over Environment and Development in Colonial Guinea', *J. of African History* 41 (2000), pp. 35–54. The desiccation narrative may be traced to the classical Greek writer Theophrastus: R. Grove 'A Historical Review of Institutional and Conservationist Responses to Fears of Artificially Induced Global Climate Change: The Deforestation-desiccation Discourse in Europe and the Colonial Context 1500–1940', in Y. Chatelin and C. Bonneuil (eds), *Les Sciences Hors d'Occident au XXe Siècle*, Vol 3, *Nature et Environnement* (Paris, Orstom, 1995), pp.155–74. See also M. van Beusekom, 'From Underpopulation to Overpopulation: French Perceptions of Population, Environment, and Agricultural Development in French Soudan (Mali), 1900–1960', *Environmental History* IV(2) (1999), pp. 198–219.

munities, or biomes, of varying structure and composition.[19] Classification schemes abound in which tree density and height are the primary criteria for categorizing such subtypes as open savanna woodland, dense shrub savanna and the like.[20] Biogeographers and ecologists have produced a voluminous literature seeking to explain the determinants of the varying tree-grass ratios characteristic of savannas.[21] In recent years, the literature has been strongly influenced by debates in ecology over the importance of disturbance, instability, and non-equilibrium processes.[22] There are currently two competing ecological positions – the equilibrium and the non-equilibrium – which seek to describe the dynamics of environmental change. The equilibrium view argues that savannas contain relatively stable mixtures of trees and grasses whose balance is largely determined by climate, especially the amount of rainfall and the duration of the dry season. It points to the fact that, at the continental scale, tree density increases with rainfall as one moves towards the equator.[23] At the landscape level, it explains plant distributions by variations in landforms and soil conditions (e.g. soil catena);[24] and, at the field scale, by reference to niche specialization and competitive exclusion.[25]

The dynamics of this equilibrium model come from perturbations, both bio-physical and human-induced. Contemporary equilibrium thinking considers periodic disturbances such as wild fire, drought, herbivory and uneven rainfall to be fundamental in shaping the form and distribution of savanna environments. It recognizes that some of these disturbances – burning, farming, livestock raising, hunting, tree cutting – are human in origin. Overall, the role of fire in limiting the expansion of woody species is so critical that savannas are viewed by many as 'fire-climaxes'.[26] In the absence of fire, it is assumed that savannas would evolve towards dense deciduous forests following a linear and predictable plant succession pathway.[27]

Recent ecologists have challenged this equilibrium view of savanna ecosystems,

[19] Scholes and Walker, *An African Savanna*, p. 3.

[20] Adams, 'Savanna environments'; and G. Riou, *Savanes, l'herbe, l'arbre et l'homme en terres tropicales* (Paris: Masson Armand Colin, 1995).

[21] B. J. Huntley and B. H. Walker (eds), *Ecology of Tropical Savannas* (Berlin: Springer-Verlag, 1982); M. Cole, *The Savannas: Biogeography and biobotany* (New York: Academic Press, 1986); and O. T. Solbrig, E. Medina and J. F. Silva (eds), *Biodiversity and Savanna Ecosystem Processes*, Ecological Studies V. 121 (Berlin: Springer, 1996).

[22] D. B. Botkin, *Discordant Harmonies: A New Ecology for the Twenty-first Century* (New York: Oxford University Press, 1990); and K. S. Zimmerer and K.R. Young (eds), *Nature's Geography: New Lessons for Conservation in Developing Countries* (Madison: University of Wisconsin Press, 1998).

[23] F. Bourlière and M. Hadley, 'Present-day savannas: An overview', in F. Bourlière (ed.), *Ecosystems of the World 13: Tropical Savannas* (Amsterdam: Elsevier, 1983), p. 4.

[24] J.-M. Avenard, 'La savane, conditions et mécanismes de la dégradation des paysages', in Jean-François Richard (ed.), *La Dégradation des Paysages en Afrique de l'Ouest* (Dakar: Presses Universitaires de Dakar, 1990), pp. 55–76; and J.-C. Menaut and J. Cesar, 'The structure and dynamics of a West African savanna', in Huntley and Walker (eds), *Ecology of Tropical Savannas*, pp. 80–100.

[25] An example of niche specialization is the view that trees and grasses compete for water at different levels of the soil horizon: Scholes and Walker, *An African Savanna*, pp. 222–7; and Bourlière and Hadley, 'Present day savannas', p. 4.

[26] D. Gillon, 'The fire problem in tropical savannas', in Bourlière, *Ecosystems of the World*, p. 617.

[27] E. Adjanohoun, *Végétation des savanes et des rochers découvertes en Côte d'Ivoire Centrale* (Mémoire ORSTOM 7) (Paris: ORSTOM, 1964).

arguing that tree-grass mixtures are inherently unstable.[28] According to non-equilibrium thinking, tree-grass relationships are asymmetrical which results in trees being favored under certain conditions and grasses under others. Recent thinking further challenges the idea of niche specialization by pointing out that, in soils less than one meter in depth, tree and grass roots compete for the same water and nutrients and that classic plant succession models fail to predict the establishment sequence of herbaceous plants following disturbance.[29] In place of homogeneous belts of savanna types following rainfall and geomorphological gradients, the new savanna ecology emphasizes discontinuity, patchiness and disequilibrium. Patchiness is characteristic at multiple scales where nutrient concentration takes place, due to both natural and anthropogenic causes.

At the landscape scale, non-equilibrium thinking attributes tree-grass ratios principally to water and nutrient availability (especially nitrogen and phosphorous), and secondly to fire and herbivory.[30] Since water is dependent on fluctuating rainfall which is external to the system, disequilibrium ecologists now view savanna ecology as event-driven. Scholes and Walker's sixteen-year study of savanna ecology at Nylsvley, South Africa, leads them to conclude that 'change in savannas is seldom gradual and continuous. Rather, it occurs in fits and starts, linked to specific events.'[31]

In summary, a major shift in thinking has taken place in the savanna ecology literature over the past two decades. The equilibrium view of savannas as essentially stable ecosystems has receded as the disequilibrium view emphasizing instability and spatial variability of environmental change has become more prominent. This conceptual turn is partly scale-dependent. At the continental scale, the broad pattern of climatic control (rainfall amount, seasonality) seems to hold. At the regional, landscape and plant scales, tree-grass dynamics show much greater variation and uncertainty. Rather than a suite of savanna types following a humid-arid gradient subject to landform position, African savannas appear as patchwork mosaics in which tree-grass ratios and productivity are determined first by water and nutrient supply, and secondly by fire and herbivory. Human modification of savanna environments through cultivation, hunting, livestock raising, wood harvesting and settlement contributes fundamentally to the mosaic pattern of savanna landscapes. This changing theoretical view of the human role in shaping savannas has very practical implications, as we shall see – not least with respect to policy towards African pastoral societies, or to the use of fire by farmers and herders.

[28] D. G. Sprugel, 'Disturbance, Equilibrium, and Environmental Variability; What is 'Natural' Vegetation in a Changing Environment?' *Biological Conservation*, 58 (1991), pp. 1–18; and I. Scoones, *Living with Uncertainty: New directions in pastoral development in Africa* (London: Intermediate Technology Publications, 1995).

[29] Scholes and Walker, *An African Savanna*, pp. 211–13, 227–9; and J.-C. Menaut, 'The vegetation of African savannas', in Bourlière, *Ecosystems of the World*, pp. 109–49.

[30] P. G. H. Frost, J-C. Menaut, B. H. Walker, E. Medina, O. T. Solbrig and M. Swift, 'Responses of Savannas to Stress and Disturbance', *Biology International Special Issue 10* (1986); A.J. Belsky, 'Spatial and temporal landscape patterns in arid and semi-arid African savannas', in L. Hansson, L. Fahrig, and G. Merriam (eds), *Mosaic Landscapes and Ecological Processes* (London: Chapman & Hall, 1995), pp.31–56; Cole, *The Savannas*, pp. 30, 42.

[31] Scholes and Walker, *An African Savanna*, p. 262.

Understanding Environmental Change in Africa:
The Challenge of Master Narratives

In the understanding of the dynamic interactions between people and the environment in Africa in general, and in its savannas in particular, perception plays an important role. Perception, in turn, is shaped by a host of factors, which acquire coherence and meaning through their organization into larger frames of reference and interpretation, into 'master narratives', a mode of structuring reality the importance of which post-modernist discourse theory has established.[32] Discourses are defined by Barnes and Duncan as 'frameworks that embrace particular combinations of narratives, concepts, ideologies, and signifying practices, each relevant to a particular realm of social action'.[33] Narratives have differing scales of reference and a host of sub-narratives, many of them diverging from one another. Many of these narratives are narratives of 'degradation', or of what William Cronon has called 'declension'.[34] One such narrative, for example, holds that two forces born in the eighteenth century have dominated the relations between society and the environment in Western Europe and North America – enlightenment thought and capitalist economic practice. The Enlightenment gave rise to a particular mode of rationalism which took a cold and distancing view of nature. This form of rationalism joined hands with capitalism, which was driven by ceaseless accumulation and a strictly utilitarian approach to nature. The combination, according to this narrative, is the principal explanation of the global destruction of species and ecosystems at a pace unprecedented in human history.[35]

A second master narrative, of profound influence in thinking about the environment, globally and in Africa, originated with Robert Malthus and holds that

[32] Some pertinent geographical interpretations, uses, and critiques of discourse theory are: J. Crush, 'Imagining development' in J. Crush (ed.), *Power of Development* (London: Routledge, 1995), pp. 1-26; Peet and Watts (eds), *Liberation Ecologies*; M. Gandy, 'Crumbling Land: The Postmodernity Debate and the Analysis of Environmental Problems', *Progress in Human Geography*, 20, 1 (1996), pp. 23-40; and D. Demeritt, 'Social Theory and the Reconstruction of Science in Geography', *Transactions of the Institute of British Geographers*, 21, 3 (1996), pp. 484-503. For an environmental historian's response to post-modernism, see William Cronon, 'A Place for Stories: Nature, History, and Narrative', *The Journal of American History* (1992), pp. 1347-76. Allan Hoben discusses the influence of master narratives in 'The Cultural Construction of Environmental Policy: Paradigms and Politics in Ethiopia', Chapter 11 in Leach and Mearns, *Lie of the Land*. See also the introduction by Leach and Mearns to this collection, 'Environmental change and policy: challenging received wisdom in Africa.' Hoben draws on E. Roe, '"Development Narratives" or Making the Best of Blueprint Development', *World Development*, XIX, 4 (1991), pp. 287-300. Mortimore describes the globally derived character of thinking about the African environment: *Roots in the African Dust*, Chapter 2.

[33] T. Barnes and J. Duncan (eds), *Writing Worlds: Discourse, Text and Metaphor in the Representation of Landscape* (London: Routledge, 1992), p. 8.

[34] Cronon, 'A Place for Stories', pp. 1363-5.

[35] Donald Worster's introduction to *The Ends of the Earth* sets forth such a narrative with characteristic elegance. Richard Grove argues that trends in Enlightenment thought in the same period are equally an influential source of modern environmentalism: Richard Grove, *Green Imperialism. Colonial Expansion, Tropical Island Edens and the Origins of Environmentalism, 1600-1860* (Cambridge: Cambridge University Press, 1995), Chapter 7. Simon Schama makes a related point in *Landscape and Memory* (New York: A.A. Knopf, 1995).

population increase inevitably leads to environmental degradation.[36] Population growth outstrips the production of food, leading to ever more intensive and destructive uses of natural resources. Like all narratives it is a mode of structuring reality which brings with it cause and effect ready-made. Where there is population growth there must be environmental degradation; where environmental degradation is perceived it must be explained by population growth. The population of Africa is growing rapidly, therefore Africa faces unprecedented environmental crises.[37] More specific terms around which narratives of African environmental degradation continue to be woven are the aforementioned 'desertification' and 'deforestation'.[38]

The United Nations Convention to Combat Desertification was adopted on 17 June 1994. Six years later, 52 different African countries had acceded to it. The convention incorporates a steady stream of UN thinking on deforestation which attributes it significantly to human activity. A 1977 UN Conference on Desertification defined desertification as 'the dimunition of or destruction of the biological potential of the land, and can lead ultimately to desert-like conditions'; and singled out the 'overexploitation' of land by farmers, herders, and woodcutters as the main cause of degradation.[39] A 1992 UN report concluded its definition of desertification with the phrase, 'resulting mainly from adverse human impact'.[40] The accent placed on human as opposed to biophysical dynamics was mimicked by NGOs, scholars, and popular writers alike. Lloyd Timberlake's popular account stressed four human causes of desertification: 'overcultivation, overgrazing, deforestation, and poor irrigation'.[41] A UNEP fact sheet echoes Timberlake, stating that:

> Desertification is caused by climate variability and human activities. In the past, drylands recovered easily following long droughts and dry periods. Under modern conditions, however, they tend to lose their biological and economic productivity quickly unless they are sustainably managed. Today drylands on every continent are being degraded by overcultivation, overgrazing, deforestation, and poor irrigation practices. Such overexploitation is generally caused by economic and social pressure, ignorance, war, and drought.[42]

---

[36] Michael Glantz (ed.), *Drought Follows the Plow* (Cambridge: Cambridge University Press, 1994) adopts a modified Malthusian approach. The Malthusian line of interpretation is critically evaluated in B. L. Turner, Goran Hyden and Robert W. Kates (eds), *Population Growth and Agricultural Change in Africa* (Gainesville: University Press of Florida, 1993), and in H. B. S. Kandeh and Paul Richards, 'Rural People as Conservationists: Querying Neo-Malthusian Assumptions about Biodiversity in Sierra Leone', *Africa*, LXVI, 1 (1996), pp. 90–103.

[37] For a fine account of how colonial thinking about population was politically driven, switching back and forth between over- and under-population views, see van Beusekom, 'From Underpopulation to Overpopulation'.

[38] Both McCann and Mortimore provide good introductions to the problematic nature of these concepts as applied to Africa: James McCann, *Green Land, Brown Land, Black Land. An Environmental History of Africa, 1800-1900* (Portsmouth, NH/Oxford: Heinemann/James Currey, 1999); Mortimore, *Roots in the African Dust*.

[39] United Nations Environment Programme, *Report of the United Nations Conference on Desertification, 29 August–9 September 1977* (Nairobi: UNEP, 1977), p. 3.

[40] UNEP, *Report of the Executive Director. Status of Desertification and Implementation of the United Nations Plan of Action to Combat Desertification* (Nairobi: UNEP, 1992), p. v.

[41] L. Timberlake, *Africa in Crisis: The Causes, the Cures of Environmental Bankruptcy* (London: Earthscan, 1985), p. 62.

[42] 'Fact Sheet 11 on the Convention to Combat Desertification', http://www.unccd.int/publicinfo/

The UN declares desertification to be at its worst in Africa where up to two-thirds of the continent is either desert or arid lands.

Ideas about desertification and deforestation are tightly interconnected. The modern tie was established as early as the eighteenth century when European naturalists inferred from the deforestation of tropical islands like Barbados and St Helena that removing dense woody vegetation generated both drought and soil erosion.[43] Sometimes these processes were seen as closely linked, sometimes as divergent in cases where de-vegetation caused soil erosion in the absence of drought. Nevertheless, a link between deforestation and drought, which empirically seems observable at the level of an island, was then projected to the continental scale. This thinking wound its way from British imperial India into South Africa and on into the twentieth century where it received considerable amplification from the great drought of 1973 in the African Sahel. However, that event also boosted research into global climate, which established that the principal engine of climate change at the continental and sub-continental levels was the interaction between the atmosphere and the surface of the ocean and implied that the removal of vegetation would have only local effects on climate.

Although a causal link between tree felling and drought has been rendered highly problematic in the instance of the African Sahel, the two concepts continue to flourish in separate, but still intertwined, existences. The United Nations thinking on desertification is posited on the belief that, at the global level, deserts are growing.[44] The Sahara is the world's largest desert and what is true globally must be true of it. This case rests on a shaky empirical basis. Swedish researchers have persuasively shown that areas believed to be overtaken by desertification in the 1970s were covered with savanna vegetation with the return of normal rainfall in the 1980s.[45] Their data indicate that the desert border, the Sahel, oscillates in a north-south direction and that this movement can largely be explained by inter-annual rainfall variation.[46] Michael Mortimore sums up the debate on desertification, arguing that the case for significant human causation 'has rested, to an undesirable extent, on fragmentary evidence, unsystematic field observation and hypothetical arguments'.[47]

Deforestation is a global narrative parallel to desertification. It is particularly focused on the Amazon, but has plenty of proponents in Africa. As Fairhead and Leach point out, the French brought a belief in African deforestation with them in

---

[42] (cont.) factsheets/showFS.php?number=11, accessed 10 November 2000.

[43] Grove, *Green Imperialism*, pp. 108–25 and Chapter 6.

[44] For a discussion of the issues involved, see Jeremy Swift, 'Desertification: narratives, winners and losers', Chapter 4 in Leach and Mearns, *Lie of the Land*; McCann, *Green Land*, Chapter 4; and Mortimore, *Roots in the African Dust*, pp. 17-25.

[45] U. Helldén, 'Desertification—Time for an Assessment?' *Ambio*, 20, 8 (1991), pp. 372–83.

[46] C. Tucker and H. D. Newcomb, 'Expansion and Contraction of the Sahara Desert from 1980 to 1990', *Science*, 253(1991), pp. 299-301. Andrew Warren rejects as inaccurate the maps in the UN's *World Atlas of Desertification* which show large areas in southern Mali as undergoing desertification: A. Warren, 'Desertification', in Adams, Goudie and Orme, *The Physical Geography of Africa*, pp. 351–2.

[47] M. Mortimore, *Adapting to Drought: Farmers, Famines and Desertification in West Africa* (Cambridge: Cambridge University Press, 1989), p. 25.

the creation of their late nineteenth-century empire.[48] Anxiety about deforestation was widely diffused throughout colonial Africa by the 1930s.[49] By the later decades of the colonial era British and French foresters, led by the Frenchman Auguste Aubréville, had developed an elaborate, and still influential, narrative pertaining to West Africa.

Aubréville, a botanist who later became the French Inspector General of Water and Forests of the Colonies, believed that the African landscape was more extensively forested in the recent past. Drawing upon the equilibrium thinking which links plant communities with climatic regimes, he convinced himself that the rain forest area of West Africa had declined from its former extent.[50] He argued forcefully, and at times melodramatically, that human activities were the driving force behind this forest loss. Farmers and herders were the main villains, who with axe and fire steadily degraded forested lands. He further hypothesized that deforestation produced feedbacks such as increased aridity ('dessication') and soil degradation, which ultimately made the land less habitable.[51]

The Guinea savanna of West Africa was one such example. Aubréville believed that this tree-grass savanna used to be covered by relatively dense open forests (*les forêts claires*) dominated by single species like *Isoberlina doka* or *Daniella oliveri*. 'There is no doubt in our mind,' he wrote, 'that the soudanian savannas were in the past covered by an open but closed forest formation where grasses were only found in a few natural clearings.'[52] He held African farmers and herders responsible for the transformation of these forested lands into savannas, a process he called 'savannaization.' Open forests, the supposed climax vegetation, were being replaced by shrub and grass savannas. Wooded savannas (*les savanes boisés*) he considered to be 'vestiges of the primitive forest vegetation, sometimes little modified, which used to cover vast areas of the guinean climatic zone'.[53] In the 1930s, even these scattered remnants were disappearing 'limb by limb' under the repeated 'attacks' of fire and land clearance to the point where 'the definitive destruction of the forest flora, the denudation of the soil and its serious consequences are imperceptibly reducing the habitability of tropical countries leading them into a desert state'.[54] At a general level, this image of an increasingly degraded wooded savanna giving way to a grass savanna and ultimately desert-like conditions persists in the minds of contemporary

---

[48] Fairhead and Leach, *Misreading the African Landscape*, p. 25.

[49] Lord Hailey, *An African Survey. A Study of Problems Arising in Africa South of the Sahara* (London: Oxford University Press, 1938), Chapter XIV, 'Forests'. James Fairhead and Melissa Leach, 'Statistics, policy and power', Chapter 8 of their *Reframing Deforestation Global Analyses and Local Realities: Studies in West Africa* (London: Routledge, 1998), discusses the development from the beginning of the twentieth century of a pan-West African community of colonial foresters.

[50] A. Aubréville, *Climats, forêts et désertification de l'Afrique tropicale* (Paris: Société d'Edition de Géographie Maritime et Coloniale, 1949), p. 318. See also *idem, La Forêt Coloniale: Les forêts de l'Afrique occidentale française,* Annales de l'Académie des Sciences Coloniales, IX (Paris: Société d'Éditions Géographiques, Maritimes, et Coloniales, 1938).

[51] For a scientific, historical, institutional perspective on the idea of dessication in West Africa, see Fairhead and Leach, 'Dessication and Domination'.

[52] Aubréville, *Forêt Coloniale*, p. 58. The paradoxical, 'open but closed', refers to a landscape with a *closed* canopy of trees, but an *open* undergrowth.

[53] *Ibid.*, p. 48.

[54] *Ibid.*, p. 62.

environmental planners, even though Fairhead and Leach have shown how inadequate Aubréville's understanding of human–ecological dynamics is, and despite substantial skepticism in the scientific community.[55]

As Crummey and Winter-Nelson point out in Chapter 5, a belief in deforestation has acquired great efficacy in Ethiopia, where it influences the activities of government and NGOs alike, although lacking any more substance than Aubréville's views of West African ecological history.[56] They are supported by McCann, who has deconstructed the Ethiopian narrative showing that it originated in sketchy estimates with essentially no empirical basis.[57] The study of land-cover change in northern Côte d'Ivoire by Bassett, Koli Bi and Ouattara contradicts the claim made in Ivorian environmental policy documents that desert-like conditions are spreading into the savanna region.

Environmental thinking and its narratives about Africa are promoted by a host of organizations, governmental and non-governmental alike, many of them international in reach. These organizations include bilateral and multilateral aid agencies such as USAID, the World Bank, the United Nations Environmental Program, as well as international NGOs like the International Red Cross, the International Union for the Conservation of Nature and the World Wildlife Fund. African governments play roles subordinate to these external agencies to whom they are beholden for funding and expertise.[58] Moreover, most African experts trained in environmentally related sciences were trained abroad and imbued, with their technical training, with the broader framework of meaning within which their mentors placed it. Whether that training was in the capitalist or the socialist world, its framework was self-consciously 'modern', isolated natural processes from human influence, interpreting the latter as inevitably 'degrading,' and arrogated to science the role of defining reality.[59] It allowed no place for 'local', 'practical' knowledge,

[55] For Fairhead and Leach, see *Misreading the African Landscape*. For summaries of the general debate about the purported process of desertification, see Mortimore, *Roots*; Swift 'Desertification;' and van Beusekom, 'From Underpopulation to Overpopulation'. The original sources include L. D. Stamp, 'The Southern Margin of the Sahara: Comments on Some Recent Studies on the Question of Dessication in West Africa', *Geographical Review*, XXX, 2 (1940), pp. 297–300; B. Jones, 'Dessication and the West African Colonies', *Geographical Journal*, XCI, 5 (1938), pp. 401–23; and E. P. Stebbing, 'The Encroaching Sahara: The Threat to the West African Colonies', *Geographical Journal*, DXXXV, 5 (1935), pp. 506–24.

[56] Hoben, 'Cultural construction;' Al Gore, *Earth in the Balance: Ecology and the human spirit* (Boston, MA: Houghton Mifflin, 1992).; and W. Ellis, 'Africa's Stricken Sahel', *National Geographic Magazine*, 172, 2 (1987), pp. 140–79.

[57] McCann, *Green Land*, Chapter 5. The basic erroneous data live on, most recently having been incorporated into the *National Action Programme to Combat Desertification* (Addis Ababa: Federal Democratic Republic of Ethiopia, Environmental Protection Agency, November 1998), p. 28; and into Shibru Tedla and Kifle Lemma, *Environmental Management in Ethiopia: Have the National Conservation Plans Worked?* (Addis Ababa: Organization for Social Science Research in Eastern and Southern Africa, 1998), pp. 9 and 11.

[58] For perceptive insights into the practical application of internationally derived environmental thinking in one African country, see James Keeley and Ian Scoones, 'Knowledge, Power and Politics: the Environmental Policy-making Process in Ethiopia', *Journal of Modern African Studies*, XXXVIII, 1 (2000), pp. 89–120.

[59] For a recent account by African experts incorporating a degradationist viewpoint, see M. A. Mohamed Salih and Shibru Tedla (eds), *Environmental Planning, Policies and Politics in Eastern and*

devaluing the latter in favor of the findings of 'science'. Our collection, by contrast, asserts the importance of the local and practical and insists that environmental change in Africa cannot be inferred, it must be established and measured.

Globally derived inferences about environmental change in Africa, ostensibly resting on science, lead to the kinds of misreadings of African landscape history documented by Fairhead and Leach. A number of younger scholars have found landscape stories similar to that of Fairhead and Leach with respect to the eighteenth- and nineteenth-century Ethiopian highlands, colonial Eritrea, and dryland Namibia.[60] One of the principal implications of this literature is the fundamental need for the kinds of empirical research into the nature and extent of landscape change in Africa presented in this collection.

State and Environment in Twentieth-Century Africa

In creating their African empires at the end of the nineteenth century, Britain and France brought the environmental baggage picked up during their earlier experiences in Asia, the Indian Ocean and the Caribbean, baggage which entailed the belief that the relationship of their 'native' subjects with the environment was destructive.[61] African hunting, livestock raising and farming practices were denigrated, and colonial administrations adopted an array of policies to save African soils, forests, rangelands and animals from 'further degradation'.[62] Colonial conservation also reflected a keen self-interest in the botanical, biological and mineral

---

[59] (cont.) *Southern Africa* (London/New York: Macmillan/St Martin's Press, 1999).

[60] See especially Anderson and Grove, *Conservation in Africa*; Fairhead and Leach, *Misreading*; and Tiffen et al., *More People, Less Erosion.* See also Alfons Rittler, *Land Use, Forests and the Landscape of Ethiopia, 1699–1865. An enquiry into the historical geography of central-northern Ethiopia* (Berne: University of Berne, 1997), Soil Conservation Research Programme, Ethiopia, Research Report 38; Pauline Boerma, 'Seeing the Wood for the Trees: Deforestation in the Central Highlands of Eritrea since 1890', unpublished Ph.D. dissertation, University of Oxford, 1999; and Richard Frederick Rohde, 'Nature, Cattle Thieves and Various Other Midnight Robbers. Images of People, Place and Landscape in Damaraland, Namibia', unpublished Ph.D. dissertation, University of Edinburgh, 1997.

[61] See here, in addition to Grove, *Green Imperialism*; Fairhead and Leach, *Reframing Deforestation*, pp. 164–81.

[62] McCann, *Green Land,* provides a survey of the influence of the global narratives of desertification, deforestation and erosion on European colonial conservation policies in twentieth-century Africa. See also Pamela A. Maack, '"We Don't Want Terraces!" Protest and Identity under the Uluguru land usage scheme', Chapter 6 in Maddox, Giblin and Kimambo, *Custodians of the Land*; Thomas Spear, 'Struggles for the land. The political and moral economies of land on Mount Meru', Chapter 9 in *ibid.*; and C. A. Conte, 'The Forest Becomes Desert: Forest Use and Environmental Change in Tanzania's West Usambara Mountains', *Land Degradation and Development*, X (1999), pp. 291–309. For a recent discussion of 'deforestation' in West Africa, see Fairhead and Leach, *Reframing Deforestation*. Some chapters in Anderson and Grove retain their pertinence to these issues: Chapter 4, Richard Bell, 'Conservation with a human face: conflict and reconciliation in African land use planning'; Chapter 5, Katherine Homewood and W. A. Rodgers, 'Pastoralism, conservation and the overgrazing controversy;' and Chapter 11, Andrew Millington, 'Environmental degradation, soil conservation and agricultural policies in Sierra Leone, 1895–1984.'

resources, with which Africa's people had long established relationships.[63] In South Africa, by the middle of the nineteenth century, the influence of British experience in India was expressing itself in concerns for watershed protection. By the 1880s the first game reserves were established.[64] Some twenty years later, in 1907, the Society for the Preservation of the Wild Fauna of the Empire was founded. State interventions which followed included the creation of an array of game reserves, and, in the 1940s and 1950s, full-blown national parks, which channelled the interests of European environmentalists, primarily in Africa's megafauna.[65] The first of the national parks was proclaimed in 1926 in the Matopos Hills of Southern Rhodesia.[66] In every instance these reserves and parks were directed to the exclusion of local peoples, whose historic patterns of land use had created the landscapes which colonial rulers now deemed worthy of protection, and this exclusion continues to the present.[67]

The speed with which the colonial regimes separated their African subjects from natural resources was related, in part, to the pressure of white settlers. This was particularly dramatic in South Africa where alienation of land throughout the nineteenth century meant that, with the enactment of the Native Land Act of 1913, Africans were in secure possession of rather less than 13 per cent of the land in the Union. A similar, if not quite so radical, exclusion of Africans from the land occurred in Southern Rhodesia starting in the 1890s.[68] A policy of native reserves was adopted in Kenya in 1904 and was implemented in such a way as to restrict severely the land available to African farmers.

Africans were also separated from access to grazing, forest and game resources. District boundaries and native reserves constricted the movement of pastoralists,

---

[63] Christophe Bonneuil, 'Entre science et empire, entre botanique et agronomie: Auguste Chevalier, savant colonial', in Patrick Petitjean (ed.), *Les Sciences Hors d'Occident au XXe Siècle*, Vol. 2, *Les Sciences Coloniales: Figures et Institutions* (Paris: ORSTOM, 1996), pp. 15–36.

[64] William Beinart and William Coates, *Environment and History. The taming of nature in the USA and South Africa* (London and New York: Routledge, 1995); and Roderick Neumann, *Imposing Wilderness. Struggles over Livelihoods and Nature Preservation in Africa* (Berkeley: University of California Press, 1998). The creation of Yellowstone Park in 1872 provided British and French interest groups (e.g., big game hunters, foresters) with a model that ultimately involved the displacement of local land users to the periphery of newly established protected areas. See also Anne Bergeret, 'Les Forestiers Coloniaux Français: Une doctrine et des politiques qui n'ont cessé de "rejeter de souche"', in Chatelin and Bonneuil, *Les Sciences Hors d'Occident au XXe Siècle*, pp. 59–74; and S. Stevens, 'The Legacy of Yellowstone', in S. Stevens (ed.), *Conservation through Survival: Indigenous Peoples and Protected Areas* (Washington, DC: Island Press, 1997), pp. 13–32.

[65] Beinart and Coates, *Environment and History*. See also E. Steinhart, 'Hunters, Poachers and Gamekeepers: Towards a Social History of Hunting in Colonial Kenya', *Journal of African History*, 30, 2 (1989), 247–64; John M. MacKenzie, 'Chivalry, social Darwinism and ritualised killing: the hunting ethos in Central Africa up to 1914', Chapter 2 in Anderson and Grove, *Conservation in Africa*; and John McCracken, 'Colonialism, capitalism and ecological crisis in Malawi: a reassessment', Chapter 3 in *ibid*.

[66] Ranger, *Voices from the Rocks*, pp. 62, 96–7.

[67] David Turton, 'The Mursi and National Park development in the Lower Omo River', Chapter 8 in Anderson and Grove, *Conservation in Africa*; Dan Brockington and Katherine Homewood, 'Wildlife, pastoralists and science', Chapter 5 in Leach and Mearns, *Lie of the Land*; and Ethiopian 'National Plan to Combat Desertification', pp. 33–6.

[68] Terence Ranger, *Revolt in Southern Rhodesia, 1896–97; A study in African resistance* (Evanston, IL; Northwestern University Press, 1967).

denying them access to historic sources of water and grazing.[69] In their colony of Tanganyika the Germans alienated land and reserved forests by 1910 in support of white settlement in the Meru and Usambara mountains.[70] In Kenya forest reservation began as early as 1911. The pace of intervention quickened in the 1920s with the establishment of colonial forest services and the increasing reservation of forest lands. In 1923 the French organized their African forest service under a decree of 18 July 1923, and in 1924 the British established an Imperial Forestry Institute in Oxford, the same year in which the Belgians began to implement a forest reserve policy in the Congo. The amount of land reserved to forest varied greatly from colony to colony with a high of around 7 per cent of Nyasaland's surface area. By the mid-1930s only Côte d'Ivoire in French Africa had a complete forest service, but forest reserves had, nonetheless, been established throughout the French territories.[71]

The 1930s also saw the foundations laid for what, in the following decade, became aggressive programs of soil conservation in British Africa. A growing interest in conservation was reinforced by the American Dust Bowl which heightened consciousness around the world of the dangers of careless soil management and gave rise to a bundle of strategies, which, however well adapted to temperate North America, were of dubious relevance to tropical Africa.[72] In the 1940s vigorous campaigns promoting terracing, bunding, crop rotation and cattle culling were imposed in Britain's colonies, where they gave rise to movements of protest and resistance which helped create a mass base for budding movements of African nationalism.[73]

Independence in eastern and southern Africa was won by movements which owed at least some of their rural support to opposition to colonial conservation policies. This made the governments of independent Africa, initially at least in the 1960s, politically less disposed to intervention. But they took over, with little modification, the expert, modernist thinking which had occasioned the interventionism of colonial officials. Before long they were again intervening in the relations between African farmers and herders and the natural resources on which their livelihoods depended. Unlike the 1930s, the impact was no longer concentrated on East, Central and Southern Africa. Indeed, the triggering event – drought and ensuing famine – took place in 1973 in the West African Sahel, bringing with it notions about desertification and the relentless advance of the Sahara. The 1973 Sahelian drought also introduced new actors, Western non-governmental

---

[69] For a few of many examples, see Beverley Gartrell, 'Prelude to disaster: the case of Karamoja', and Neal Sobania, 'Pastoralist migration and colonial policy: a case study from northern Kenya', in Anderson and Johnson, *Ecology of Survival*, pp. 193–217 and 219–39.

[70] Neumann, *Imposing Wilderness*.

[71] Guéhi Jonas Ibo, 'La Politique Coloniale de Protection de la Nature en Côte d'Ivoire, (1900–1958)', *Revue Française d'Histoire d'Outre-Mer*, LXXX (1993), No. 298, pp. 83–104; Aubréville, *La Forêt Coloniale*, pp. 191–201; and Lord Hailey, *African Survey*, pp. 984–1024.

[72] Worster, *Dust Bowl*; Anderson, 'Depression, Dust Bowl, Demography and Drought.'

[73] For one example, see Maack, '"We Don't Want Terraces"'. See also Carl Rosberg and John Nottingham, *The Myth of Mau Mau. Nationalism in Kenya* (New York: Praeger, 1966), pp. 163–6, 170, 237, 328; and D. A. Low and Alison Smith, *History of East Africa* (Oxford: Clarendon Press, 3 vols., 1963, 1968, 1976), III, pp. 118–19 and 515–16, where it is construed as 'opposition to agricultural improvement'; Ranger, *Voices from the Rocks*, Chapters 5 and 6.

organizations. NGOs were motivated by a variety of concerns ranging from famine relief to economic development to classical conservation. The Club du Sahel was formed following the 1973–4 drought to co-ordinate aid donor activities in the Sahelian countries. Part of the Organization for Economic Co-operation and Development system, it considers itself 'a forum for informal exchange and brain-storming between partners from the North and the South, both public and private' with regard to development aid.[74]

As Western governments increasingly channeled resources through these NGOs they became, from the standpoint of African governments, powerful and influential. The narratives which shaped their interventions were a mix of the old and the new. Thinking about drought and famine in the 1970s was rooted in the long-standing desiccation paradigm attributing drought primarily to human activities, which, in the case of the Sahel, primarily meant the UN mantra of 'over-cultivation, overgrazing, and deforestation.'[75] But it was also an expression of a new environmentalism, an outgrowth of the 1960s and the aftermath of Rachel Carson's *Silent Spring*.[76] One response was the revival of Stebbing's idea of planting forest belts to arrest the Sahara's advance.[77]

Drought returned with even greater intensity in 1984 and triggered in Ethiopia the worst famine in twentieth-century Africa. As relief agencies cast about for explanations they found them ready-made in the founding assumptions of the Soil Conservation Research Project (SCRP), set up in 1981, under the leadership of the Swiss, Hans Hurni, and the Ethiopian Highlands Reclamation Study (EHRS), which, funded by the World Bank, began its work in 1983.[78] The explanation was

---

[74] Robert Berg, 'Foreign aid in Africa: here's the answer – is it relevant to the question?' in R. J. Berg and J.S.Whitaker (eds), *Strategies for African Development* (Berkeley: University of California Press, 1986), p. 511. The Club du Sahel's mission is taken from the organization's website at http://www.oecd.org/sah/about/, accessed 19 January 2001.

[75] Timberlake, *Africa in Crisis*, p. 62. More nuanced and historical treatments are provided by Richard H. Grove, 'A historical review of institutional and conservationist responses to fears of artificially induced global climate change', in Chatelin and Bonneuil, *Les Sciences Hors d'Occident au XXe Siècle*, Vol 3, pp 155–69; and van Beusekom, 'From Underpopulation to Overpopulation', pp. 201–7. The linkage of deforestation leading to degradation and, eventually, famine, appears in a report sponsored by a leading NGO, the International Union for the Conservation of Nature: IUCN, *Ethiopia. National Conservation Strategy. Phase I Report. Prepared for the Government of the Peoples Republic of Ethiopia with the assistance of IUCN. Based on the work of Adrian Wood and Michael Ståhl* (Gland, Switzerland: IUCN, March 1990); for the point at issue see p. 6.

[76] Rachel Carson, *Silent Spring With an Introduction by Vice President Al Gore* (Boston/New York: Houghton Mifflin Co., 1962, re-issue 1994).

[77] Erik Eckholm and Lester R. Brown, *Spreading Deserts – The Hand of Man*, World Watch Paper 13 (1977), p. 33; Jeffrey A. Gritzner, *The West African Sahel: Human agency and environmental change*, University of Chicago Geography Research Paper No. 226 (1988). See Stebbing, 'The Encroaching Sahara', p. 508.

[78] For an informed review of institutions, environmental thinking and environmental policy-making in Ethiopia, see Keeley and Scoones, 'Knowledge, power and politics.' The principal findings and recommendations of the EHRS are presented in Michael Constable and Deryke Belshaw, 'The Ethiopian Highlands Reclamation Study: Major findings and recommendations', in People's Democratic Republic of Ethiopia, Office of the National Committee for Central Planning, *Towards a Food and Nutrition Strategy for Ethiopia. The Proceedings of the National Workshop on Food Shortages for Ethiopia held at Alemaya University of Agriculture, 8–12 December 1986* (Addis Ababa, December 1989),

that Ethiopian farmers had mined their landscape to exhaustion, denuding it of trees and generating soil erosion at a rate approaching the apocalyptic. The answer was a host of government interventions: excluding farmers and stock raisers from much of the hillside pastureland which hitherto they had used; the planting of millions of trees, mostly eucalyptus, on those hillsides; enforced terracing and bunding; and the resettlement of hundreds of thousands of farmers and their families under conditions of duress.[79] In the promotion of official conservation, and, to a lesser extent, in resettlement the Ethiopian government was enthusiastically supported by NGOs like the Red Cross and Catholic Relief Services. At no point was the master narrative of farmer-generated environmental destruction questioned, yet, as both Dessalegn and Crummey and Winter-Nelson show in Chapters 9 and 6 respectively, it rested on particularly slim foundations.

Similar narratives, interventions, and data gaps exist in the literature on livestock raising and environmental degradation in Africa. Chapter 7 by Little and Chapter 8 by Munro in this volume show that livestock management policies in eastern and southern Africa have been consistently premised on equilibrium-based notions of fixed carrying capacity and climax vegetation communities, and on a caricature of Africans as destroyers of the environment. After reviewing the range of interventions and policies that have historically reduced pastoralists' access to rangelands, both conclude that state policies that reduce herd mobility actually increase the 'overstocking' for which governments blame herders and account for much of the recorded range degradation.[80] Indeed, Little goes so far as to argue that pastoralists contribute positively to biodiversity, and separating them from historic grazing lands leads not to 'natural regeneration,' but to a botanically impoverished landscape.[81]

---

[78] (cont.) pp. 142–79. Of the multitudinous studies commissioned as part of the EHRS particularly germane to the present argument is Yeraswork Admassie, Mulugetta Abebe and Markos Ezra, *Ethiopian Highlands Reclamation Study. Report on the sociological survey and sociological considerations in preparing a development strategy* (Addis Ababa: Land Use Planning and Regulatory Department, Ministry of Agriculture, December 1983). The SCRP published its results in a series of annual reports starting with volumes 1–4 published together in 1984: Provisional Military Government of Socialist Ethiopia, Ministry of Agriculture, Soil and Water Conservation Department, Soil Conservation Research Project, *Volumes 1–4. Compilation of Phase I Progress Reports (Years 1981, 1982, and 1983)* (Berne: University of Berne in association with The United Nations University, Tokyo, June 1984). Volumes 5–9 (with the omission of a volume 7) were published in 1986, 1988 and 1991.

[79] For immediate responses to the famine, see: Kurt Jansson, Michael Harris and Angela Penrose, *The Ethiopian Famine* (London: Zed Press, 1987); and Dawit Wolde Giorgis, *Red Tears. War, Famine and Revolution in Ethiopia* (Trenton, NJ: Red Sea Press, 1989). Conservation measures undertaken by the Ethiopian government are summarized and criticized in: Shibru and Kifle, *Environmental Management in Ethiopia*; and in Yeraswork Admassie, *Twenty Years to Nowhere. Property Rights, Land and Conservation in Ethiopia* (Lawrenceville, NJ: The Red Sea Press, 1997).

[80] See also M. Niamer-Fuller, *Managing Mobility in African Rangelands: The Legitimization of Transhumance* (London: Intermediate Technology Publications, 1999); and G. Oba, N. C. Stenseth and W. J. Lusigi, 'New Perspectives on Sustainable Grazing Management in Arid Zones of sub-Saharan Africa', *BioScience* 50, 1 (2000), pp. 35–51.

[81] Peter Little, 'Pastoralism, Biodiversity, and the Shaping of the Savanna Landscapes in East Africa', *Africa*, LXVI, 1 (1996), pp. 37–51. The implications of disequilibrium ecology for pastoral management have carried over into discussions of land tenure reform: T. J. Bassett and D. Crummey (eds), *Land in African Agrarian Systems* (Madison: University of Wisconsin Press, 1993); and C. Toulmin and J. Quan (eds), *Evolving Land Rights, Policy, and Tenure in Africa* (London: DFID/IIED/ NRI, 2000).

Both authors argue that the environmental data are highly suspect, reflecting in part a disconnection between social science and range ecology research. They also underscore the need to consider non-equilibrium ecological processes in any assessment of rangeland degradation, and point to the need for interdisciplinary research methods that will improve our understanding of the nature and direction of environmental change in Africa's savannas.

Shaping the Landscape: Innovation, Investment and Adaptation

The contributors to this volume demonstrate that African farmers and herders modify landscapes in ways far more subtle and nuanced than is allowed for in any of the master narratives governing environmental policy in Africa. Our contributors are particularly reticent about 'degradation', which, as we have argued elsewhere, is always contextual – always judged from a particular viewpoint and its attendant values.[82] Rarely, if ever, did our informants volunteer this term or an equivalent. What our research has confirmed is the strength and resilience of local knowledge, which reveals deep understanding of natural processes, the basis for a broad array of strategies for exploiting natural resources, the impact of which is well understood.[83] We have been particularly struck by the innovation and adaptation to change which attends those strategies, and by the ways in which African land users are continually investing in natural assets.[84] Given the continental narratives of deforestation, it was of particular interest to us to find farmers in Burkina Faso, Côte d'Ivoire and Ethiopia to have adopted tree-planting as one of their investment strategies. In Burkina Faso the area under cereal cultivation has been under steady expansion for decades, yet there has not been a fully proportionate loss in green land cover. Şaul's

---

[82] D. L. Johnson, S. H. Ambrose, T. J. Bassett, M. L. Bowen, D. E. Crummey, J. S. Isaacson, D. N. Johnson, P. Lamb, M. Saul, and A. E. Winter-Nelson, 'Meaning of Environmental Terms', *Journal of Environmental Quality*, XXVI, 3 (May–June, 1997), pp. 581–9.

[83] Classic examples of this approach are William E. Allan, *The African Husbandman* (New York: Barnes & Noble Inc., 1965); Marvin Miracle, *Agriculture in the Congo Basin; Tradition and change in African rural economies* (Madison: University of Wisconsin Press, 1967); Robert McC. Netting, *Hill Farmers of Nigeria: Cultural Ecology of the Kofyar of the Jos Plateau* (Seattle: University of Washington Press, 1968); and Paul Richards, *Indigenous Agricultural Revolution* (London: Hutchinson, 1985). This is also a major theme of Michael Mortimore's work, summarized in *Roots in the African Dust*; and of Fairhead and Leach, *Misreading*. See also Jane Guyer, 'Diversity at Different Levels: Farm and Community in Western Nigeria', *Africa*, LXVI, 1 (1996), pp. 71–89. Dr Belay Tegene, in research which was part of the MacArthur collaborative project, found a wealth of practical knowledge about soils and about their advantages and disadvantages under the actual range of climatic conditions which Ethiopian farmers face. Their conservation practices were rooted not in theory but in hard-earned experience. These techniques, he believes, are essential components of a successful national strategy for soil conservation. Belay Tegene, 'Indigenous Soil Knowledge and Fertility Management Practices of the South Wällo Highlands', *Journal of Ethiopian Studies*, XXXI, 1 (1998), pp. 123–58; and *idem.*, 'Potential and Limitations of an Indigenous Structural Soil Conservation Technology of Welo, Ethiopia', *Eastern Africa Social Science Research Review*, XIV, 1 (1998), pp. 1–18.

[84] But one of many examples of innovation and adaptation may be found in David Anderson, 'Cultivating pastoralists: ecology and economy among the Il Chamus of Baringo: 1840–1980', in Anderson and Johnson, *Ecology of Survival*, pp. 241–60.

research reveals the answer in booming investment in fruit trees. In neighboring northern Côte d'Ivoire, Bassett, Koli Bi and Ouattara document a parallel expansion of cashew and mango plantations. In famine-afflicted Ethiopia, where the government promotes the notion of farmer-driven deforestation, Crummey and Winter-Nelson argue that individual farmers are planting trees and bushes to meet their own, and market-driven, demands for fuel, building materials and stimulants. National accounts of loss of tree cover simply do not take this planting adequately into account.

In contrast to these locally based strategies of afforestation, stands the dreary story of government-imposed afforestation in Ethiopia, which Dessalegn recounts in Chapter 9.[85] Joseph Otieno has found a parallel story of imposed afforestation this time by corporations in southwestern Kenya where the heightened demand for fuel wood to cure tobacco has led to increased tree cover.[86] Farmers have planted these trees to meet their contractual obligations. Nevertheless, smallholder farmers perceive this expanding tree cover in negative terms. Many of the trees are eucalyptus and farmers blame them for reduced crop yields and lower stream levels.[87] Otieno concludes that 'Many policy makers assume that increased tree cover, irrespective of species results in an enhanced environment'. The value of increased tree cover is clearly a subjective judgement linked to the resource management goals of the land manager, a point that Bassett, Koli Bi and Ouattara also make in Chapter 3.

Farmers invest in soil as well as in trees. Leslie Gray's study of southwest Burkina Faso in Chapter 4 shows how cultivators invest in soil improvement. Land rights are uncertain in the area and both local people and land-hungry immigrants have adopted soil conservation and improvement techniques as a means of establishing rights and countering their uncertainty. The techniques include applying manure, constructing anti-erosion devices and preserving and planting trees.

Sara Berry reminds us, in the words of the West African slogan, that 'No condition is permanent'.[88] Contrary to the image of profound conservatism held by urban Africans, foreign experts, and their own governments alike, African farmers and herders have long experience of change which reaches them from many corners. From the earliest times in Africa people have been developing and adopting new technology. At the turn of the millennium, so far as farmers are concerned, that means new trees and crops, plows, and manufactured fertilizers and pesticides and herbicides. In the twentieth century national markets, linked to the

---

[85] For Dessalegn, see also 'Environmentalism and Conservation in Wällo Before the Revolution', *Journal of Ethiopian Studies*, XXXI, 1 (1998), pp. 43–86. See also Bahru Zewde, 'Forests and Forest Management in Wällo in Historical Perspective', *Journal of Ethiopian Studies*, XXXI, 1 (1998), pp. 87–121.

[86] Joseph Otieno, 'Tobacco Under Contract: Agricultural Development and Environmental Change in Kuria District, Western Kenya', unpublished Ph.D. dissertation, University of Illinois at Urbana-Champaign, 1998.

[87] Eucalyptus, as Crummey and Winter-Nelson point out in Chapter 5, is overwhelmingly the preferred tree of Ethiopian farmers. Their longer experience of eucalyptus compared with the farmers of Western Kenya and the local roots of its adoption have mitigated the tree's negative potential.

[88] Sara Berry, *No Condition is Permanent: the Social Dynamics of Agrarian Change in sub-Saharan Africa* (Madison: University of Wisconsin Press, 1993).

global market, have provided both opportunities – increasing demand for tropical commodities, food, fruit and fuel – and constraints, not least of which have been uncertain and depressed prices. Changes in property and in the access to natural resources have equally constrained land users and enabled them to meet their own needs and the needs of the market more satisfactorily. The distribution of property has never been stagnant, as the chapters which follow remind us once again, thanks to land re-distribution in Ethiopia, immigrant farmers in Burkina Faso, national policy encouraging production of beef and cotton in Côte d'Ivoire, and the staking out of private ranches in the rangelands of northern Kenya. Land users throughout Africa have equally had to cope with changes in the legal framework for holding property, some of these changes being quite fundamental as the nationalization of land in Mozambique and Ethiopia and its subsequent privatization in Mozambique demonstrated, some of them more subtle, as is the case with our Ivorian and Burkinabè farmers.

Both the actual distribution of property and the legal framework within which it is held have implications for soil management and cropping strategies. Land users have also had to cope with change in their access to natural resources, in some cases directly as a result of government policies – the enclosure of hillsides and the consequent contraction of grazing land in Ethiopia from the 1980s onwards is but one of many examples – and sometimes slightly more indirectly as is the case with the constant encroachment on rangeland in Kenya. Government policy may further affect natural resources through social policies like those promoting the sedentarization of East African pastoralists. Finally, population growth has been and continues to be at a high level in Africa, and, as Ester Boserup has demonstrated, can lead to innovation every bit as much as it leads to deterioration.[89]

Much environmental thinking in Africa embodies Malthusian assumptions which, at their most reductionist, hold that Africans are blindly driven by force of population growth to degrade the environment, or where it recognizes Africans to have conscious agency, that agency is perceived as narrowly constrained by inherited cultural ideas and practices.[90] Against these views, we argue that population growth

---

[89] Ester Boserup, *The Conditions of Agricultural Change. The Economics of Agrarian Change under Population Pressure* (New York: Aldine Publishing Co., 1965). For a positive relationship between population growth and indigenous conservation in Africa, see Kandeh and Richards, 'Rural People as Conservationists.'

[90] The argument of this chapter has been stated elsewhere and by others, as early as 1988 in the case of Johnson and Anderson: see the 'Introduction' to *The Ecology of Survival*. That it should still have to be made is a sad reflection on much environmental thinking in Africa. For an example of 'blind drive' thinking, see P. Stähli, 'Changes in settlement and land use in Simen, Ethiopia, especially from 1954 to 1975', in B. Messerli and K. Aerni (eds), *Simen Mountains-Ethiopia*, Vol. 1, *Cartography and its Application for Geographical and Ecological Problems* (Bern: Geographisches Institut der Universität Bern, 1978). A classic example is the idea that East African pastoralists are motivated by cultural values to accumulate livestock, and that this 'cattle complex' is economically irrational and environmentally destructive. See Munro, Chapter 8 in this book, and R. McC. Netting, *Cultural Ecology* (Menlo Park, CA: Cummingh's Publishing Co., 1977), pp. 40–56. A second example is the idea that the 'target income mentalities' of African farmers lead them to reduce their effort to produce crops for the market once their income objectives have been realized. See John C. de Wilde, *Experiences with Agricultural Development in Tropical Africa*, Vol. 1, *The Synthesis* (Baltimore, MD: Johns Hopkins University Press, 1967), p. 66.

does encourage innovation, and that we should be directing our attention to the relative success or failure of these innovations in meeting the goals of the innovators. Against the image of agricultural conservatism, we posit a rural Africa dynamic and innovating, constrained much more by the vagaries of climate and by the actual choices available in the national and global markets than by any *a priori* reluctance to enter them, and characterized by deep knowledge of the biophysical processes which provide so many of the resources on which rural livelihoods depend.

Methodology

The following chapters draw on a wide range of methodologies. They are directed to: (i) measuring landscape change; (ii) recovering the pertinent social and cultural past; and (iii) understanding contemporary political-ecological dynamics. The understanding of contemporary dynamics poses two very different challenges: a grasp of events and processes making for landscape change at the local and regional level in Africa; and a critical evaluation of the environmental narratives driving the formulation and implementation of policy at the national and international levels.

Measuring landscape change. We believe that one of the principal challenges to understanding human-environmental interactions in Africa is measuring landscape change. To be sure, this will always be done within frustrating ranges of precision, but some measurement is a giant step ahead from the current abyss of ignorance and a necessary pre-condition for the development of responsible environmental policies. Williams, in Chapter 2 in this collection, demonstrates the importance of geomorphology.[91] Geomorphology, along with climatology, speaks to the pace of landscape change, and both are essential, and sometimes the only, tools available for understanding change at the scale of centuries and millennia.[92] Williams also demonstrates their pertinence to an understanding of change over periods as short as decades. He points out the need for distinguishing between change occurring at very different time scales and how often the failure to do so has led to serious misunderstanding. Geomorphology is also pertinent to an understanding of soil erosion, since the rate of soil *reconstitution* in Africa is still very poorly understood.

[91] For an earlier statement in this vein see John A. J. Gowlett, 'Human adaptation and long-term climatic change in northeast Africa: an archaeological perspective', in Anderson and Johnson, *Ecology of Survival*, pp. 27–45. A more extended statement of Williams' views may be found in Martin A. J. Williams and Robert C. Balling Jr., *Interactions of Desertification and Climate* (London: Arnold for the World Meteorological Organisation and the United Nations Environmental Program, 1996). Michael Glantz has repeatedly demonstrated how climatology can illuminate African environmental problems. See particularly the introduction to *Drought Follows the Plow*; but see also Michael H. Glantz, 'Drought and economic development in sub-Saharan Africa', in Michael H. Glantz (ed.), *Drought and Hunger in Africa: Denying Famine a Future* (Cambridge: Cambridge University Press, 1987), pp. 37–58.

[92] S. E. Nicholson, 'The Methodology of Historical Climate Reconstruction and its Application to Africa', *Journal of African History*, XX, 1 (1979), pp. 31–49. Mortimore demonstrates how important a grasp of climate variability is for an understanding of the strategies of African farmers and herders: *Roots in the African Dust*, Chapter 5.

Other tools – matched photographs, aerial photographs, remote sensing, vegetation transects and the changing incidence of wildlife – also permit some degree of measurement of landscape change through time. The expertise involved in deploying these techniques virtually mandates inter-disciplinary work if their value is to be fully realized. Matched photographs are used by replicating historic landscape photographs and juxtaposing present image against past. The methodology is not a new one, even in its application to Africa,[93] but it has been little used. It holds out a great deal of promise, even if, like all methodologies, it brings with it its own limitations. Most recently Tiffen, Mortimore and Gichuki have demonstrated its value, and a pair of dissertations have used it to good effect.[94]

Aerial photographs and remote sensing are additional powerful tools which have been used too rarely by social scientists. There is an aerial photographic record in Africa going back at least to the 1950s and satellites have been recording the African landscape for 30 years now.[95] Together they contain a vast amount of information about landscape change in Africa. Even with their varying degrees of resolution, they can establish the extent of landscape change between two or more points in time, and, given the inferential basis of most judgments about landscape change in Africa, they ought to play a larger role in discussions and exercise more influence on policy-making processes than they now do.

Aerial photography and remote sensing are of value in establishing changes in vegetation at the landscape scale.[96] At the patch scale, the technique of vegetation transects is useful to record vegetation parameters (species composition, height, density) along a specified distance.[97] Integrating information across the different scales from patch to landscape can yield valuable results.[98]

[93] J. R. Hastings and R. H. Turner, *The Changing Mile. An Ecological Study of Vegetation Change with Time in the Lower Mile of an Arid and Semiarid Region* (Tucson, AR: University of Arizona Press, 1965); and H. L. Shantz and B. L. Turner, *Photographic Documentation of Vegetational Changes in Africa over a Third of a Century* (Tucson: University of Arizona, College of Agriculture, Report 169, 1958).

[94] Tiffen *et al.*, *More People, Less Erosion*;. Rohde, 'Nature, Cattle Thieves and Various Other Midnight Robbers;' and Boerma, 'Seeing the Wood for the Trees'. See also Donald Crummey, 'Deforestation in Wällo: Process or Illusion?', *Journal of Ethiopian Studies*, XXXVI, 1 (1998), pp. 1–41.

[95] An impressive example of the use of aerial photographs, integrated with other kinds of information, is Holly T. Dublin, 'Dynamics of the Serengeti-Mara Woodlands. An Historical Perspective', *Forest and Conservation History*, XXXV (1991), pp. 169–78. Fairhead and Leach integrate the findings of aerial photographs and remote sensing to persuasive effect in *Misreading the African Landscape*. Tiffen *et al*, *More People, Less Erosion*, also uses aerial photography. See also Halvor Wøien, *Woody Plant Cover and Farming Compound Distribution on the Mafud Escarpment, Ethiopia. An aerial photo interpretation of changes 1957–1986* (Trondheim: University of Trondheim, Centre for Environment and Development, 1995). For the use of remote sensing in the study of land degradation in northwestern Ghana, see K. Nsiah-Gyabah, *Environmental Degradation and Desertification in Ghana: A study of the Upper West Region* (Aldershot: Avebury Studies in Green Research, 1994), pp. 105–37.

[96] John Innes, 'Measuring environmental change', in D. L. Peterson and V. T. Parker (eds), *Ecological Scale: Theory and Applications* (New York: Columbia University Press, 1998), pp. 429–57.

[97] Matthew D. Turner, 'The interacton of grazing history with rainfall and its influence on annual rangeland dynamics in the Sahel', in Zimmerer and Young (eds), *Nature's Geography*, pp. 237–61.

[98] K. Poiani, B. Richter, M. Anderson and H. Richter, 'Biodiversity Conservation at Multiple Scales: Functional Sites, Landscapes, and Networks', *BioScience*, L, 2 (2000), pp. 133–46; John A. Wiens, 'Landscape mosaics and ecological theory', in Hansson *et al.* (eds), *Mosaic Landscapes*, pp. 1–26; and R. Reid, R. Kruska, N. Muthui, A. Taye, S. Wotton,C. Wilson and W. Mulatu, 'Land-use and

In and of themselves remote sensing, aerial photography, matched photography and vegetation transects tell us nothing about the social dynamics of landscape change. These dynamics must be ascertained separately and directly with the techniques of social science and historical inquiry. They can not be inferred directly from the landscape. The challenge to all environmental research in Africa is to demonstrate the interaction of multi-scale biophysical and social processes in the explanation of environmental change. To do so requires confronting the important epistemological and methodological differences in how natural science and social science knowledge is generated.[99] To assist in this task Piers Blaikie provides a useful heuristic model, or 'chain of explanation', in which he conceptualizes the inter-action of multi-scale social processes involved in the dynamics of soil erosion.[100] Fully to answer the challenge of understanding environmental change dynamics requires intensive field study and hybrid research methodologies.[101]

Recovering the Past. All the persuasive accounts of landscape change in Africa draw on the documentary record, published and unpublished, and on the oral testimonies of the actors involved. The recent essays in African environmental history demonstrate this most convincingly.[102] In the chapters which follow Crummey and Winter-Nelson rely on the testimonies of elderly informants; Bassett, Koli Bi and Ouattara conducted semi-structured interviews with individual men and women stratified by age, income, and ethnicity; Dessalegn uses sources from the Wällo provincial archives; while Munro, Şaul and his colleagues, and Bowen and her colleagues draw upon both oral and written historical sources and field observations to reconstruct their land-use histories.

A number of the chapters rest on 'hybrid' research methods that allowed the collection of information from a diversity of actors. Participant observation, key informant interviews, and informal discussions with different groups and individuals formed an important part of this research strategy, as did 'landscape walks' in which local people shared their understanding of the influence of fire, grazing pressure, new farming technologies, prices, and government policies on soil and vegetation

---

[98 (cont.)] Land-cover Dynamics in Response to Changes in Climatic, Biological and Socio-political Forces: The Case of Southwestern Ethiopia', *Landscape Ecology*, 15 (2000), pp. 339–55.

[99] P. Blaikie, 'Political Ecology in the 1990s. An Evolving View of Nature and Society', *CASID Distinguished Speaker Series, No. 13*. Michigan State University, East Lansing (1994); and *idem*, 'Changing Environments or Changing Views?'

[100] P. Blaikie, 'Explanation and Policy in Land Degradation and Rehabilitation for Developing Countries', *Land Degradation and Rehabilitation*, 1 (1989), pp. 23–37.

[101] S. Batterbury, T. Forsyth and K. Thomson, 'Environmental Transformations in Developing Countries: Hybrid Research and Democratic Policy', *The Geographical Journal*, CLXIII, pp. 126–32; and Bassett and Koli Bi, 'Environmental Discourses'.

[102] See the articles in Maddox *et al.*, *Guardians of the Land*; in Anderson and Grove, *Conservation in Africa*; and the contributions to the special African issue of *Environmental History*, IV, 2 (1998) – van Beusekom, 'From Underpopulation to Overpopulation', pp. 198–219; Gregory Maddox, 'Africa and Environmental History', pp. 162–7; Christopher A. Conte, 'Colonial Science and Ecological Change: Tanzania's Mlalo Basin, 1888–1946', pp. 220–44; and Sarah T. Phillips, 'Lessons from the Dust Bowl: Dryland Agriculture and Soil Erosion in the United States and South Africa, 1900–1950', pp. 245–66. Harms, *Games against Nature*, draws heavily on oral tradition and works at a much greater time depth than the foregoing contributions.

changes. Government officials, extension agents, representatives of aid donor organizations and NGOs are important sources of information for perceptions of environmental change and its causes. Survey research methods provided quantitative information complementary to these qualitative approaches. The chapters on Ethiopia and Côte d'Ivoire drew on information gained from questionnaires administered to samples of the rural communities representative of location, occupation, ethnicity, and economic standing.

Discourse Analysis. Discourse analysis of environment and development texts provided insights into the origins of key ideas and words like desertification, and their relationship to a nexus of power relations, policies, and interventions. We were particularly struck by the continuity of such conceptual frameworks as blaming land users, of spatial images like the creeping desert, and of authoritative positions established on extraordinarily weak foundations. Each of the countries studied by the MacArthur-funded research project possessed what Richard Peet and Michael Watts call a 'regional discursive formation' in which 'certain modes of thought, logics, themes, styles of expression, and typical metaphors run through the discursive history of a region'.[103] Discourse analysis was a valuable tool in allowing the deconstruction of such terms as 'savannaization,' 'deforestation', and the like.

The Argument of the Book

Five critical themes concerning the environment and social change in Africa run through the chapters which follow. These themes speak directly to the formation of environmental policy. Our contributors demonstrate that current thinking about the environment in Africa rests on the shakiest of empirical foundations. They further show that much of this thinking about the environment in Africa has its roots in the European colonial period and is dominated by global narratives of degradation. By contrast, they argue that scholars interested in African environmental change and policy-makers concerned to shape environmental policy must themselves be shaped by local knowledge. Further, for policy to have any predictable impact, it must take fully into account inevitably conflicting social, political and economic forces. Finally, they assert that environmental policy in Africa must rest on a foundation of empirically established change.

Environmental Data Gaps. A recurring theme of our contributors is the inadequate base on which thinking about the environment rests. The case of Ethiopia is particularly pertinent. Three of the chapters – by Crummey and Winter-Nelson, Dessalegn, and Williams – challenge the conventional view that recent population growth and stagnant technology have resulted in widespread deforestation and soil erosion, which, in turn, have caused famine. Williams endorses McCann's conclusions that some parts of the highlands have been without trees for a very long time (if they ever had them), while other parts changed from farmland to canopy forest

---

[103] R. Peet and M. Watts, ' Liberation Ecology: Development, Sustainability, and Environment in an Age of Market Triumphalism', in Peet and Watts, *Liberation Ecologies.*

in the space of a century. Crummey and Winter-Nelson examine the case of Wällo Province where devastating famines occurred in the early 1970s and mid-1980s. They use matched historical photographs and informant accounts to argue that tree cover has actually expanded in Wällo as a result of farmer initiatives. They join Williams in rejecting the catastrophic view which emphasizes recent and progressive deforestation in Ethiopia. Dessalegn takes up the narrative of catastrophic soil erosion, summarizing a scientific literature that undermines the studies on which the narrative was based. This research questions the validity of extrapolating soil loss and run-off rates obtained from small experimental test plots to farmers' fields in the first place and beyond that to the entire Ethiopian highlands.[104]

The chapters on land-use and land-cover change in Côte d'Ivoire and Burkina Faso provide complementary landscape histories that challenge neo-Malthusian visions of population-induced environmental degradation. In Côte d'Ivoire, Bassett, Koli Bi and Ouattara argue that grazing pressure, expanding cropland, and less aggressive fires have produced an increasingly wooded landscape in the north central savanna region, thereby rejecting the West African savanna degradation narrative that paints a picture of tree loss, desiccation, and desertification. In the case of Burkina Faso, the authors argue that the expansion of such commercial opportunities as cotton, grain farming, and fruit orchards, rather than population growth, has led to a decline in fallow vegetation. The West African and Ethiopian case studies illustrate the dynamics of technological innovation, new markets, and changing policy frameworks which have shaped land use and land cover over the past half-century.

Much of the existing work on the environment in Africa compounds the inadequacy of empirical data with a failure to engage the epistemological and methodological challenges of measuring and interpreting data on environmental change. Peter Little demonstrates this in the case of a long-term ecological monitoring project conducted in Marsabit, Kenya, where he links inconsistent and contradictory claims of land degradation to contrasting commitments to equilibrium and non-equilibrium theoretical positions. Martin Williams underlines the difficulty of distinguishing between human and non-human causes of environmental change. Taken together, the case studies demonstrate that land-use and land-cover changes have been important over the past 50–60 years, and that the nature and direction of these changes contrast sharply with the apocalyptic images found in the environmental crisis narratives.

Challenging Dominant Narratives. The classic narrative, that appears and reappears in the development and policy texts, is one of environmental crisis. A rapidly diminishing resource base portends disaster for future generations. As Dessalegn and Bassett and his colleagues demonstrate, the narrative begins with an apocalyptic statement describing an environmental catastrophe. Whether it is the loss of billions of tons of top soil each year in the Ethiopian highlands or the devastation of savanna woodlands wrought by bush fires set by farmers and herders, the narrative defines the problem as environmental, not social or political. It gives little attention to the

---

[104] M. Stocking, 'Soil erosion: breaking new ground', in Leach and Mearns, *The Lie of the Land*, pp.140–54. In studies that were part of the MacArthur-funded research project Belay Tegene has further challenged some of the assumptions of the catastrophic school. See the articles referred to above, in Note 83.

social dimensions of land use, or to the local cultural-ecological and larger political circumstances which frame natural resource management.

William Munro shows how policy-makers in colonial and post-colonial Zimbabwe have viewed rangeland degradation as a technical problem that could be fixed through land-use and livestock-management controls. Conservation interventions, in his words, focused on 'what farmers *should* be doing rather than on why they were doing what they did'. One of the key ecological concepts of the bureaucratic land managers is 'carrying capacity' which they use to establish 'appropriate' population-resource relationships. Little and Munro show that such models fail to consider the goals or rationale of herd owners and hence the role of cattle in indigenous farming systems, and the relationship between livestock management and the environment. They agree with other contributors that environmental crisis narratives impede our understanding of environmental change dynamics, establishing instead a basis for external intervention, control and appropriation.

Leslie Gray challenges yet another dominant narrative, one which holds that 'land titling leads to land improvement,' in her study of how immigrant farmers in southwestern Burkina Faso are driven by the insecurity of their land holdings to invest in soil improvement.

Local Knowledge. Underlying this collection is the conviction that local, not 'scientific', knowledge is, in the first instance, the key to a proper understanding of the African landscape. [105] To be sure, science, both social and natural, has much to offer in the study of Africa's environmental problems. But those problems must be the ones which Africa's farmers and herders confront in their daily lives. Both colonial and post-colonial thinking about nature-society relations in Africa has too often defined human-environment issues without reference to local goals, let alone knowledge. The implication is both epistemological and practical. A social problem cannot be defined without reference to livelihoods and values, and, too often, the values deployed by governments on the African landscape have been metropolitan values. The obvious need to engage and mobilize the interests of local people in addressing environmental issues raises tensions and contradictions. The reality is that 'learning from and with rural people ... has changed less than the rhetoric',[106] and this suggests that there is a politics of representing local knowledge (and knowledge production in general) which hinges on its denigration and/or silencing.[107]

---

[105] Much of the literature on local knowledge is concerned with its utility for 'development' and natural resource management. A good example of this applied orientation is D. M. Warren, L. J. Slikkerveer and D. Brokensha (eds), *The Cultural Dimension of Development: Indigenous knowledge systems* (London: Intermediate Technology Publications, 1995). For a critique of the concepts of 'local' and 'participatory development', see P. Peters, ''Who's Local Here?' The Politics of Participation in Development', *Cultural Survival Quarterly*, (Fall 1994), pp. 22–5; and P. Little, 'The link between local participation and improved conservation: a review of issues and experiences', in D. Western and R. M. Wright (eds), *Natural Connections: Perspectives in Community-based Conservation* (Washington, DC: The Island Press, 1994), pp. 347–72. See also A. Gupta, *Postcolonial Development: Agriculture in the Making of Modern India* (Durham, NC: Duke University Press, 1998), pp. 168–83.

[106] R. Chambers and P. Richards 'Preface', in Warren *et al.*, *The Cultural Dimension of Development*, p. xiii.

[107] Michael Goldman offers a compelling example of how both expert and local knowledge was suppressed by the World Bank and the Laotian government in the course of developing project

A recurring theme of this book is that Westerners and Western-trained African elites have typically held disparaging views of local environmental knowledge, both of which entail practical experience. This negative view springs from at least three sources. The first is the plain ignorance of non-locals who fail to spend time with farmers, herders, and hunters to learn how and why they use resources the way they do. As a result, non-locals are unable to interpret the landscape or fathom its political, cultural and ecological history. They are faced with what James Scott terms 'illegible' landscapes and knowledge systems.[108] Unable to make sense of the local, yet compelled by scientific and state bureaucratic conventions to simplify and generalize, their classic response has been to draw on global narratives to fill the void. These 'totalizing models'[109] invariably contain pejorative views of local knowledge (for example, overgrazing the commons) which serve as foils to the purported authoritative knowledge of developers and bureaucrats. Thus, the representation of indigenous peoples, practices, and views as 'traditional,' 'irrational,' 'primitive' – in short, all that is not 'modern' – is central to the discourse on both development and the environment,[110] which portrays local resource management practices such as burning savanna grasses as environmentally destructive. On the other hand, the literature, especially on development, contains stunning contradictions where developers promote the expansion of palm oil or coffee plantations under the banner of 'comparative advantage', thereby condoning tropical forest clearance.[111]

A second reason scientists and developers regularly disparage local knowledge is because their analytical models are unable to accommodate the diversity of local situations. Disequilibrium environments, hybrid landscapes, overlapping rights to the same parcel of land – such diversity and uncertainty are disconcerting to modelers whose equations require simplifying assumptions and universalistic claims. Most development professionals belittle local knowledge to mask their own analytical shortcomings. The fact that local knowledge and experiences are constantly evolving and vary by age, ethnicity, gender, and economic standing further muddies the terrain of investigation. In the end, developers and policy-makers take the most expedient road to regulate people and resources. Rather than building upon the strengths of local knowledge and practices, they devise new rules, institutions, and spaces of conservation and development, and increasingly do so in the name of 'participatory development' without a hint of the irony entailed in these new modes and spaces of regulation.

[107] (cont.) planning documents for the controversial Nam Theun 2 dam project. M. Goldman, 'The Birth of a Discipline: Producing Authoritative Green Knowledge, World Bank Style', *Ethnography*, 2, 2 (2001).

[108] Scott, *Seeing Like a State*, pp. 18, 24, 29, 39.

[109] V. Broche-Due, 'Producing nature and poverty in Africa: an introduction', in V. Broche-Due and R. A. Schroeder, *Producing Nature and Poverty in Africa* (Stockholm: Nordiska Afrikainstitutet, 2000).

[110] See Stuart Hall, 'The West and the rest: discourse and power', in S. Hall *et al.*, *Modernity: An introduction to modern societies* (Oxford: Blackwell, 1996); and Gupta, *Postcolonial Development*, pp. 168–83.

[111] For additional examples of negative views of local knowledge and practices and their centrality to the 'development' discourse, see J. Ferguson, *The Anti-Politics Machine: 'Development', Depoliticization, and Bureaucratic Power in Lesotho* (Minneapolis: University of Minnesota Press, 1990); and A. Escobar, *Encountering Development: The Making and Unmaking of the Third World* (Princeton, NJ: Princeton University Press, 1995).

Third, local knowledge is devalued by being ignored or silenced in development texts. This is the *tabula rasa* view of history popularized by Walter Rostow in his influential book, *The Stages of Economic Growth*.[112] History begins in the developing world only after European contact. The spaces of conservation and development appear to be inhabited by people without history or culture.[113] The West has nothing to learn from 'the Rest' of the world. This blank slate perspective effectively reduces the 'indigenous' and 'local' to a vacuous 'tradition', which is simplistically opposed to the 'modern'. In place of complex cultures and human–environment interactions, modernization theorists give us blank spaces. This discourse masks the fact that the local is never entirely local. Its hybrid character is shaped through interactions with other cultures, polities and economic systems. For example, peasants commonly combine 'customary' farming techniques such as intercropping with chemical fertilizers, pesticides, and crop varieties introduced during the colonial period, or more recently in the context of a development project. The integration of these new technologies into 'indigenous' farming systems presents new possibilities and constraints that invariably modify local agricultural calendars and practices. Neither indigenous nor purely modern, the farming system mixes and matches from a field of options and evolves into distinctive forms. A number of the chapters in this collection also show that local knowledge and practices are typically differentiated by social (ethnicity, age, gender, occupation, economic status) and geographical (population density, rainfall, elevation, regional uneven development) criteria. It is important to recognize this hybrid and differentiated quality of local environmental knowledge in our investigations of environmental change dynamics which build upon the perception of resource users.

In summary, the papers in this volume seek to advance our understanding of the value of local knowledge in providing sorely needed information about and insights into the nature and direction of environmental change. In the process, it also helps to illuminate the varied interests and actors who have a stake in belittling local knowledge. The politics of this power/knowledge struggle vary from place to place and over time. In contrast to the simplifying narratives, a diversity of landscapes and livelihoods stand out to undermine the claims of authoritative knowledge implicit in environment and development discourses. The case studies collected here demonstrate that the links between social and environmental change do not conform to the global narratives which obscure more than they clarify the experiences of African rural peoples with their environments. We argue that the devaluation of local knowledge has a political basis in which various agents (the state, aid donors, urban-based elites, NGOs) seek to control local peoples and resources for varied ends (taxation, access to rural resources, aid funds). We contend that the interpretation of landscapes entails a consideration of political and ecological processes and the discourses and interventions that give them form and meanings. One of the objectives of this book is to point to the contrasting interpretations of environmental change in Africa, and to argue that these contested landscapes reflect struggles over the control and appropriation of people and resources.

[112] W. W. Rostow *The Stages of Economic Growth: A Non-Communist Manifesto* (Cambridge: Cambridge University Press, 1960).
[113] This idea is well developed by Eric R. Wolf in his *Europe and the People Without History* (Berkeley: University of California Press, 1982).

Political Ecological Analysis. Raymond Bryant and Sinead Bailey state well the *political* premise of political ecology, that 'environmental change is not a neutral process amenable to technical management. Rather, it has political sources, conditions and ramifications that impinge on existing socio-economic inequalities and processes.'[114] However, the approach itself emerged from the intellectual traditions of *political* economy and cultural *ecology*. In so doing, it went beyond the geographers' traditional approach of examining the impact of human activities on the environment to the exclusion of analyses of the social and political processes surrounding resource use. However, as Peter Little argues, the political ecology approach has tended to emphasize the social and political dimensions of resource access and management at the expense of the physical ecological dimension. By contrast, this collection strives to bridge the natural and social scientific dimensions by means of a range of analytical and methodological strategies, thereby recovering some of the original promise of political ecology. In the first instance, as Williams illustrates, some environmental changes are not induced by humans. Nevertheless, most of the contributions explore human influences on the environment and attempt to link social and environmental processes.

Our contributors highlight the importance of a range of social forces, some of them in conflict with each other, and of state intrusion into rural economies in shaping patterns of resource access and management. For example, Munro and Little show how livestock raising strategies and range condition in Kenya, Somalia and Zimbabwe are linked to power, patronage, and accumulation in rural and peri-urban areas. Leslie Gray, connecting the social and political variables with biophysical measures of agricultural intensification, shows that poor households are less capable of improving soil quality than their more prosperous neighbors. The Ethiopian studies demonstrate how government intervention blocking local access to hillsides and watersheds has exacerbated ecological conditions elsewhere and placed great strain on agricultural production. Fully to realize the potential of political ecology requires a concerted effort to bridge the natural and social sciences and their distinctive epistemologies and methods.[115]

Environmental Change and Policy. This volume is timely in light of the extraordinary amount of environmental planning taking place in Africa today. Most countries have either completed or are in the process of formulating environmental action plans. Whether imposed by international aid donors as a condition for receiving further aid or inspired by concerns of business, NGOs and grassroots actors, environmental planning is a current preoccupation of African governments. Of course, environmental planning is not new. Many of the papers point to striking continuities between colonial-era conservation policies and those of the post-independence period. To repeat, these continuities include: (i) the prevalence of environmental crisis narratives; (ii) glaring environmental data gaps; (iii) the dismissal of indigenous resource management strategies; (iv) the unmitigated failure of state-imposed conservation schemes; and (v) the resistance of land users to attempts by the

---

[114] Bryant and Bailey, *Third World Political Ecology*, p. 28.
[115] For a fuller discussion of the epistemological and methodological challenges of political ecology, see P. Blaikie, 'A Review of Political Ecology: Issues, Epistemology and Analytical Narratives', *Zeitschrift für Wirtschaftsgeographie*, 43 (1999), pp. 131–47.

state to control their access to resources. We believe that improved knowledge of environmental change is critical to any serious effort to engage in environmental planning. Our multi-scale and multi-method approach centered on resource users located in webs of social and ecological relations has produced very different images of the African savanna from those found in most environment and development texts. We argue that environmental policy must be based on a more subtle and more rigorous approach to research, one which engages the physical geographic as much as the social origins of change, and which respects both the autonomy of ecological processes and the value of the time-proven methods of environmental management which African societies have developed.

Conclusion

This is, in the first instance, a book of environmental scholarship. As such, one of its important implications is directed to future research in this area. At its most basic level it affirms the value of scholarship organized around environmental issues. Environmental questions are valuable ones around which to organize scholarship on Africa, since they demand the integration of what have often been distinct domains of economy, culture, politics and society. And they yield rich information about how Africans have constructed and continue to construct their worlds.

The collection reinforces the value of collaborative research, reflecting the richness of the findings which collaboration makes possible and the constraints which current sources of knowledge place on the individual researcher. In particular, it emphasizes, in ways which previous collections have not, the importance of collaboration across the divide between the social and the physical sciences. Disciplinary boundaries are collapsing everywhere, but there is still an enormous divide between the practices of, for example, ecology and history or of climatology and sociology. Social scientists have far too often ignored the natural forces which shape life and livelihood in Africa, alas, to much the same extent that ecological research proceeds without reference to the millennial shaping hand of human management.

Our collection also reinforces the importance of empirically establishing the nature and degree of landscape change in Africa as a foundation for sound policy and effective intervention. The inadequate empirical base on which environmental policy and intervention is founded in Africa would not be tolerated in Western Europe, Japan or the United States. The degree of urgency which accompanies so many calls for intervention is far too often directly proportional to the ignorance out of which it arises. Outsiders have been constructing Africa according to their own will for far too long. The projections of environmentalists are no more justified by the loftiness of their motives than were the projections of the imperialists at the end of the nineteenth century. Africa possesses the human resources, knowledge, and values necessary to a sound foundation of environmental intervention and management. Science and technology can help achieve the goals of environmental planning, but those goals must arise from and be embedded in the local communities which continue to manage the continent's natural legacy.

PART TWO
Longue Durée

2

Changing Land Use & Environmental Fluctuations in the African Savanna

MARTIN WILLIAMS

Introduction

The African savannas were the birthplace of humanity. The cultural and biological importance of the African savannas to early human origins is well-documented.[1] Less well known is their possibly paramount role in modern human survival during the apparent near demise of the entire human population some 71,000 years ago. The catastrophic eruption of Toba volcano in northern Sumatra at this time appears to have caused a dramatic and sustained reduction in global temperature and possible failure of the tropical summer monsoon.[2] Dense clouds of ash particles were erupted high into the atmosphere and rapidly distributed around the globe, blocking incoming solar radiation. The Toba eruption was of an order of magnitude bigger than the eruption of Krakatoa and blanketed all of peninsular India in ash, no doubt destroying plants and animals in the process. Evidence from mitochondrial DNA suggests that small groups of humans survived in Africa, eventually moving out to re-occupy Asia and Africa.[3] The African savannas encompass a wide range of ecosystems and a rich and diverse flora and fauna,[4] and this ecological diversity was doubtless a powerful reason for the survival of modern humans in Africa at a time of widespread famines elsewhere.

Climatic variability is a key characteristic of the African savanna environments, and one to which plants, animals and human societies have long adapted. During

[1] E. S. Vrba, G. H Denton, T. C. Partridge and L. H. Burckle (eds), *Paleoclimate and Evolution, with Emphasis on Human Origins* (New Haven, CT: Yale University Press, 1995).

[2] G. A. Zielinski, P. A. Mayewski, L. D. Meeker, S. Whitlow, and M. S. Twickler, 'Potential Atmospheric Impact of the Toba Mega-eruption – 71,000 Years Ago', *Geophysical Research Letters*, 23, 8 (1996), pp. 837–40.

[3] S. H. Ambrose, 'Late Pleistocene Human Population Bottlenecks, Volcanic Winter, and Differentiation of Modern Humans', *Journal of Human Evolution*, 34 (1998), pp. 623–51.

[4] R. H. V. Bell, 'The effect of soil nutrient availability on community structure in African ecosystems', in B. J. Huntley and B. H. Walker (eds), *Ecology of Tropical Savannas* (Berlin: Springer-Verlag, 1982), pp. 193–216.

the last 25,000 years, for example, the climate, in what are now the seasonally wet tropical savannas of Africa, has ranged from cold, dry and windy to warm and wet.[5] How did the hunter-gatherers and early pastoral and farming societies of the African savanna respond to the vicissitudes of an ever-changing climate? The question is not entirely rhetorical, but serves to highlight two important themes in African history. First, the environment (of which we are all a part) has fluctuated in the past and will continue to fluctuate in the future, irrespective of any human impact. Second, by their very interaction with their ever-changing habitat, human societies create further change, some of it reversible and some of it not, at least at the scale of human generations.

Interactions between human societies and the environment are complex and hard to unravel. Separating cause from effect can be problematic. In particular, it is often technically difficult to distinguish between environmental variations caused, directly or indirectly, by human interventions and those stemming from other, non-human causes. Even a cursory reading of the burgeoning literature on land degradation, desertification, deforestation, drought, and famine in Africa demonstrates the need for clarity of thought and rigorous evaluation of unproven assumptions, however popular or widely accepted they may be.[6]

My aim in this chapter is to use a series of seven case studies drawn from different parts of the African savanna in an attempt to answer the following question: 'How might we identify recent environmental changes caused by human activities from those caused by other factors?' A related but more subtle question is: 'How might we distinguish recent environmental changes which are directly attributable to human activities from those changes which at first blush seem to be anthropogenic but which in fact have nothing to do with human actions?'

The seven examples are drawn from a range of North African savanna environments in Ethiopia, Niger, Somalia and Sudan with which I am directly familiar. Three examples relate to accelerated soil erosion in Ethiopia and Niger, all of which have quite different causes and repercussions. One example deals with the depositional legacy of the Nile in the Sudan, concluding that salinity in cotton fields along the lower White Nile has more to do with the alluvial history of the Nile than with current climate and land use. The remaining examples deal with water use in Somalia, the curious case of a lake in Ethiopia that was rising during the height of the 1964–76 droughts, and a more general analysis of historic floods and droughts in Ethiopia and the Sudan. These case studies are designed to show that careful fieldwork and appropriately eclectic forms of analysis can allow us to isolate the degree to which recent land-use change is a response to, and a cause of, environmental fluctuations in the African savannas. In order to set the stage for the

---

[5] F. Gasse, 'Water resources variability in tropical and subtropical Africa in the past', in *Water Resources Variability in Africa during the XXth Century*, IAHS Pub. No. 252 (1998), pp. 97–105; and *idem.*, 'Hydrological Changes in the African Tropics since the Last Glacial Maximum', *Quaternary Science Reviews*, 19 (2000), pp. 189–211.

[6] M. Leach and R. Mearns (eds), *The Lie of the Land. Challenging Received Wisdom on the African Environment* (Oxford: James Currey, 1996); J. C. McCann, *Green Land, Brown Land, Black Land: An Environmental History of Africa, 1800–1990* (Portsmouth, NH/Oxford: Heinemann/James Currey, 1999), especially Chapter 5, 'A tale of two forests: narratives of deforestation in Ethiopia, 1840-1996', pp. 79–107.

wider discussion of environmental changes in these diverse regions, it seems appropriate to begin with climatic variability and historic floods and droughts.

Historic floods and droughts in Ethiopia, the Sudan and the Sahel

The persistence and severity of the drought that first became evident in the Sahel region of West Africa in 1968 eventually alerted world attention to the importance of rainfall failure as a potent cause of soil loss and ecological decline. (The incidence of similar prolonged and severe droughts in 1913 and 1939-41 was overshadowed by the tragic impact of the two World Wars and so went largely unnoticed.) The Sahel drought rapidly spread to encompass all of the southern margin of the Sahara, a belt of land extending for nearly 5,000 km from the Atlantic to the Red Sea and including over twenty African countries and many hundreds of millions of people. This drought re-kindled both scientific and political interest in the causes and consequences of rainfall and river flow variations and land degradation in the African savannas. Arising from this came a renewal of interest in desertification, culminating in the signing of an international convention in 1994.

The concept of desertification is elusive and inherently ambiguous.[7] As a result, past attempts to define it have often resulted in studied vagueness. Some recent definitions are founded on unproven assumptions about the ultimate causes of desertification, stating as given what has yet to be proved. For example, the first edition of the *World Atlas of Desertification* begs the question by defining it as 'land degradation in arid, semi-arid and dry sub-humid areas resulting *mainly* from adverse human impact'.[8] With the emphasis essentially on 'adverse human impact,' this definition allows little role for climatic desiccation or a series of severe droughts as significant contributors to dryland degradation.

As a belated scientific and political response to this natural catastrophe, the definition of desertification was subtly altered at the June 1992 Earth Summit Conference on Environment and Development held at Rio de Janeiro. The role of climate was now given explicit recognition, and desertification was redefined as 'land degradation in arid, semi-arid and dry sub-humid areas resulting from various factors, including climatic variations and human activities'.[9]

An additional factor contributing to this small but significant change of emphasis was the growing scientific recognition of the part played by El Niño Southern Oscillation (ENSO) events in modulating the global incidence of major floods and droughts.[10] Increasing attention was now being paid to sea surface temperature

[7] M. Mainguet, *Desertification: Natural Background and Human Mismanagement* (Berlin: Springer-Verlag, 2nd edn, 1994); D. S. G. Thomas and N. J. Middleton, *Desertification: Exploding the Myth* (Chichester: Wiley, 1994); M. A. J. Williams and R. C. Balling, Jr., *Interactions of Desertification and Climate* (London: Arnold with UNEP and WMO, 1996); and UNEP, *World Atlas of Desertification*, with editorial commentary by N.Middleton and D.S.G.Thomas (London: Arnold, 1997).

[8] UNEP, *World Atlas of Desertification*, with editorial commentary by N. Middleton and D. S. G. Thomas (London: Edward Arnold, 1st edn,1992), vii (emphasis added).

[9] UNCED, *Earth Summit Agenda 21: Programme of Action for Sustainable Development* (New York: United Nations Environment Programme, 1992).

[10] E. M. Rasmusson and J. M. Wallace, 'Meteorological Aspects of the El Niño/Southern Oscillation',

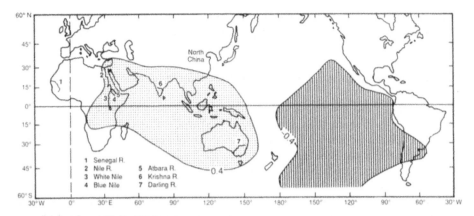

Figure 2.1 Key regions influenced by the Southern Oscillation

Note: During an ENSO event (negative SOI) the stippled area shows surface atmospheric pressure above normal, and the hatched area shows pressure below normal. During an anti-ENSO event the converse is true. The map shows correlations of annual pressure anomalies with those in Jakarta, greater than 0.4 (stippling) and −0.4. (After Whetton *et al.* 'Rainfall and river flow variability.)

Table 2.1 Details of data sets used to construct Figure 2.2

Data set	Record length	Mean length	Std dev.	Units
SOI	1870–1986	0.18	0.79	–
North China				
Rainfall Index	1870–1979	3.07	0.44	–
Indian Rainfall	1871–1985	992.2	82.6	mm
Krishna Discharge	1901–60	57.1	14.5	10^6M1
Senegal Discharge	1904–64	24.0	7.8	10^6M1
Nile Discharge	1871–1984	84.3	12.2	10^6M1
Darling Discharge	1886–1984	36.3	10.7	$(M1)^{1/4}$

Note: The means and standard deviations were calculated for the reference period 1911–60. To allow for strong positive skewness, Darling annual flow (June to May) was corrected using a fourth-root transform.
(After Whetton *et al.*, 'Rainfall and riverflow variability'.)

(SST) anomalies and their possible utility in forecasting future rainfall and river flow.[11]

The River Nile is the lifeblood of Egypt and the Sudan.[12] Over the last hundred years, fluctuations in Nile discharge have been attributed to a variety of causes, including variations in sunspot activity and sea surface temperature anomalies. An

[10] (cont.) *Science*, 222 (1983), pp. 1195-1202; and H. F. Diaz and V. Markgraf (eds), *El Niño: Historical and Paleoclimatic Aspects of the Southern Oscillation* (Cambridge: Cambridge University Press, 1992).

[11] H. J. Simpson, M. A Cane, S. K. Lin, S. E. Zebiak, and A. L. Herczeg, 'Forecasting Annual Discharge of River Murray, Australia, from a Geophysical Model of ENSO', *Journal of Climate*, 6, 5 (1993), pp. 386–90.

[12] F. A. Hassan, 'Historical Nile Floods and their Implications for Climatic Change', *Science*, 212 (1981), pp. 1142–5.

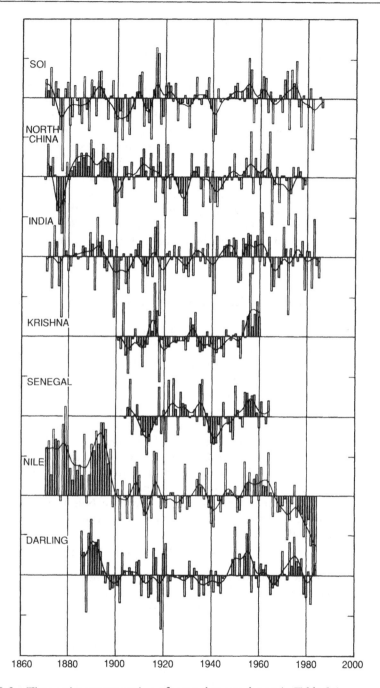

Figure 2.2 Time series representation of seven data sets shown in Table 2.1

Note: Each series has been standardized with respect to the means and standard deviations given in Table 1, and each interval on the vertical axis represents two standard deviations. A smoothed version of the time series is shown by the solid curved line. (After Whetton *et al.* 'Rainfall and river flow variability').

important predictor of flow in the Blue Nile and main Nile is the Southern Oscillation Index, expressed as the atmospheric pressure difference between Darwin and Tahiti.[13] Time series analysis shows a statistically significant correlation between low Nile flow, Indonesian drought, and years of summer monsoon failure in India and eastern China. The converse is equally true, with exceptionally wet years occurring synchronously in all of these regions. The lesson from this is that intervals of severe drought in the savanna countries of Africa are likely to coincide with intervals of reduced rainfall in a number of other regions of the globe, many of them densely populated. The same is true of exceptionally wet years and phases of widespread flooding (see Figures 2.1 and 2.2 and Table 2.1). What impact any putative global warming might have on the magnitude and frequency of future floods and drought is still a matter of intelligent guesswork.[14] Regardless of the ultimate explanation, there is growing empirical evidence that the incidence of extreme tropical floods seems to have increased during the 1980s and 1990s, so that what were once deemed to be 'one in a thousand year' flood events are now far more frequent.[15]

Legacy of past fluctuations in Nile headwaters: soils and salinity in central Sudan

The impact of past climatic changes in the Nile headwaters is evident in the pervasive depositional legacy in the lower Blue and White Nile valleys further downstream. At intervals during the past 25,000 years and more, the ancient channels of the Nile were constantly shifting course and changing in other more subtle ways. Between about 25,000 and 15,000 years ago, the Blue Nile was a highly seasonal river ferrying coarse sand and gravel from the highlands of Ethiopia across the alluvial plains of the Sudan and thence into southern Egypt. With the drying out of Lake Victoria in Uganda, the White Nile ceased to flow at this time, so that the main Nile in Egypt probably ceased to flow in the winter months, at least during very dry years. (Some winter flow from now dry Egyptian wadis is evident at this time, but is unlikely to have compensated for the loss of White Nile water.) Shortly after about 12,000 years ago, the Blue Nile began to carry and deposit fine silt and clay in the alluvial plains of the Sudan and Egypt, and perennial flow was restored to the White Nile and main Nile.[16]

A consequence of these changes in river regime is that the alluvial plains

[13] P. Whetton, D. Adamson, and M. A. J. Williams, 'Rainfall and river flow variability in Africa, Australia and East Asia linked to El Niño-Southern Oscillation events', in P.Bishop (ed.), *Lessons for Human Survival: Nature's record from the Quaternary*. (Canberra: Geological Society of Australia Symposium Proceedings, 1, 1990), pp. 71–82; D. Conway and M. Hulme, 'Recent Fluctuations in Precipitation and Runoff over the Nile Sub-basins and their Impact on Nile Discharge', *Climatic Change*, 25 (1993), pp. 127–51; F. A. K. Farquharson and J. V. Sutcliffe, 'Regional variations of African river flows', in *Water Resources Variability in Africa during the XXth Century*, pp. 161–9.

[14] Williams and Balling, *Interactions of Desertification and Climate*.

[15] V. R. Baker, J. M. Bowler, Y. Enzel, and N. Lancaster, 'Late quaternary palaeohydrology of arid and semi-arid regions', in K. J. Gregory, L. Starkel and V. R. Baker (eds), *Global Continental Palaeo-hydrology* (Chichester: Wiley, 1995), pp. 203–31.

[16] M. Williams, D. Dunkerley, P. De Deckker, P. Kershaw, and J. Chappell, *Quaternary Environments*, (London: Arnold, 1998).

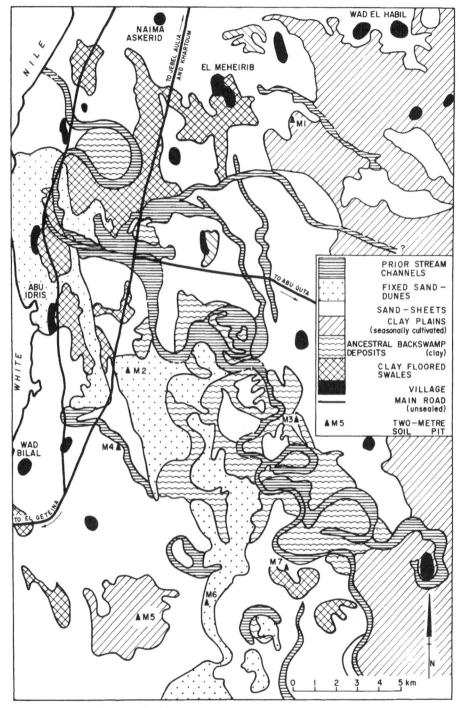

Figure 2.3 Landforms and soil-forming parent sediments in the north-western Gezira, Sudan

bounded by the lower Blue and White Nile rivers known in the Sudan as the Gezira (Arabic for island), cannot be considered as uniform clay plains with highly predictable soil properties.[17] In fact, these complex plains are covered with saline and non-saline clays, sand and gravel ridges, desert dunes, and local bedrock outcrops. [18]

Two-thirds of the export revenue of the Sudan comes from long-staple cotton grown in the Gezira, on less than 1 per cent of the total land area of the Sudan. Since cotton is intolerant of salt, it is important to be able to avoid regions where the subsoil is likely to be saline. The pattern of near-surface salinity in the soils bordering the lower White Nile does not reflect present-day climatic gradients of rainfall and evaporation, and so is hard to predict. Careful reconstruction of the alluvial history of the White Nile during the past 12,000 years suggests that the distribution of saline subsoils is a legacy of the depositional history of the river and can therefore be understood and predicted with some confidence. The zones of very high subsoil salinity coincide with buried shell beds. The fossil freshwater shells indicate times when a more extensive White Nile began to dry out. Desiccation resulted in the formation of saline carbonate deposits that were later buried beneath freshwater swamp clays.[19]

Deforestation and accelerated soil loss in the Ethiopian uplands

The discourse relating to deforestation and land degradation in Ethiopia is especially interesting because it shows how unproven assumptions can become the basis for well-meaning but sometimes ill-advised conservation and development policies.[20] The two most recent Ethiopian famines helped precipitate the end of Haile Selassie's long reign as well as that of Haile Mariam Mengistu. During and after both famines there were widely publicized assertions that overpopulation and indiscriminate destruction of the native forests and soils were the prime causes of the recurrent famines in Ethiopia. How valid are these arguments and upon what evidence are they based?

[17] Sir Alexander Gibb & Partners, *Estimation of Irrigable Areas in the Sudan 1951–3. Report to the Sudan Government* (London: Metcalfe & Cooper Ltd, 1954); Hunting Technical Services, *Report 6, Roseires Soil Survey Project. The White Nile, East Bank, Rabak to Khartoum: Soils and Engineering Reconnaissance* (Boreham Wood, UK: Hunting Technical Services, 1964).

[18] See Figure 2.3. Also M. A. J. Williams and D. A. Adamson, *The Origins of the Soils between the Blue and White Nile Rivers, Central Sudan, with some Agricultural and Climatological Implications.* Occasional Paper No. 6 (Khartoum: Economic and Social Research Council, National Council for Research, 1976); and W. A. Blockhuis, *Vertisols in the Central Clay Plains of the Sudan* (Wageningen, 1993).

[19] M. A. J. Williams, 'Soil Salinity in the West Central Gezira, Republic of the Sudan', *Soil Science,*105, 6 (1968), pp. 451–64; M. A. J. Williams and D. A. Adamson (eds), *A Land Between Two Niles. Quaternary Geology and Biology of the Central Sudan* (Rotterdam: Balkema, 1982); M. A. J. Williams, D. Adamson, B. Cock and R. McEvedy, 'Late Quaternary Environments in the White Nile Region, Sudan', *Global and Planetary Change,* 26, 1–3 (2000), pp. 305–16.

[20] A. Hoben, 'The cultural construction of environmental policy: Paradigms and politics in Ethiopia', in Leach and Mearns, *Lie of the Land*, pp. 186–208; E. Elias and I. Scoones, 'Perspectives on Soil Fertility Change: a Case Study from Southern Ethiopia', *Land Degradation & Development*, 10 (1999), pp. 195–206.

Consider first the assumed nexus between population, forest cover and accelerated soil erosion. I shall present one side of the argument first, as dispassionately as I can, noting that it contains more than a grain of truth, before I attempt to dissect out some of the all-pervasive inaccuracies, contradictions and non-sequiturs. The argument runs roughly as follows: Ethiopia has the potential to be one of the most successful agricultural nations in Africa, being richly endowed with deep, fertile volcanic soils, a range of microclimates and a great diversity of natural ecosystems. However, an abundant population (at least in the well-watered uplands), rapid rates of forest clearing this century, steep slopes and seasonally torrential rains have resulted in accelerated loss of soil from many upland regions. The dense network of deeply incised rivers is conducive to rapid evacuation of soil washed down from the deforested hillslopes, resulting in equally rapid silting up of reservoirs both large and small.

Hurni notes that even within the Simien Mountains National Park, the area of natural forest has declined from 56 per cent to 22 per cent of the Park area over the past 40 years.[21] A graphic description of the process of deforestation in the earliest years of the Park's existence is given by Clive Nicol, the first game warden appointed by the Imperial Ethiopian Government to manage the Park:

> Seen close, the destruction was incredible. The place looked like a First World War battleground. Everywhere the ground was strewn with hot ash, smoking debris, charred stumps, and partially burned tree trunks, lying about willy-nilly. Some of the felled trees had been sixty feet [c. 20 m] and more in height, and now they lay burned on the ground. I could hardly believe that a few men, with simple, blunt iron axes had felled so many huge trees. Cedar, olive, hagenia, Podocarpus, euphorbia, and many others. This year alone, Bogale and his sons and followers had cut and burned about three square kilometres of forest.[22]

He then went on, in what are unequivocally neo-Malthusian tones, to argue that these forest clearing practices were not sustainable.

> Disaster was surely inevitable for the Simien people. They had raped the land, the soil was impoverished and eroded away. Streams and rivers were dying and springs killed. Baboons were on the increase, together with rats and crop-eating birds. Already the people were barely surviving on what they raised. By destroying what remained, they could not last more than fifty years. Starvation, drought, and disease would cull off their numbers in the end. Surely it was better to have the people moved away now, while there was still time. A national park would benefit the country, and would be of lasting value.[23]

These assertions are supported by more recent work. The Swiss geographer and soil conservationist Hans Hurni has long maintained that, devoid of their protective forest cover, the mountains of Ethiopia are highly vulnerable to erosion. Traditional farming methods recognized that soil losses during cultivation were high and so allowed long years of fallow for the soils to recuperate. Mean annual rates of soil loss amounted to about 40 tonnes per hectare on mountain slopes, but attained rates of over 300 tonnes per hectare during cultivation years, or some 5–10 times more than in non-mountainous areas. The increasing demand for land has meant a reduction

[21] H. Hurni, 'Sustainable Management of Natural Resources in African and Asian Mountains', *Ambio*, 28, 5 (1999), pp. 382–9.

[22] C. Nicol, *From the Roof of Africa* (London: Hodder and Stoughton, 1971), p. 262.

[23] *Ibid.*, pp. 265–6.

in fallow to virtually zero and an expansion of the area under cultivation. Hurni cites an example from Gojjam Province where in one region the area cultivated rose from 40 per cent in 1957 to 77 per cent in 1995, while natural forest land decreased from 27 to 0.3 per cent.[24]

It thus seems that, despite the emotive language and apocalyptic tone of his writing, Nicol may have had a valid point. Removal of forest can alter the local hydrological balance, increasing runoff and soil erosion, and reducing infiltration and the perennial maintenance of springs and stream headwaters from subsurface throughflow and baseflow. The downstream effects were equally catastrophic, and were not always confined to Ethiopia.

Within ten years of its completion, the reservoir of the Roseires dam on the Sudanese Blue Nile near the border with Ethiopia had received such an influx of sediment from the Ethiopian highlands that its effective water depth had decreased by 10–15 metres.[25] Prior to completion of the dam in 1965, sediment load in the Blue Nile at Sennar amounted to 50–100 million tonnes a year. It declined to 40–70 million tonnes at Khartoum during 1967–9 as a result of sediment trapped in the Roseires dam reservoir.[26] By 1996, the capacity of the Roseires reservoir had been reduced by almost 60 per cent sedimentation and that at Khashm el Girba on the Atbara, by 40 per cent.[27] The Atbara rises in the Ethiopian highlands, close to the sources of the Blue Nile, where it is known as the Tekazze river.

It thus appears that, in the case of the Ethiopian highlands, deforestation in and around the Simien Mountains over the past 40–50 years has triggered a wave of accelerated soil loss from cultivated land and has led to siltation of reservoirs many hundreds of kilometres downstream. Is this a fair conclusion?

Detailed historical and agronomic studies in other regions of Ethiopia suggest that the reality may be more complex. McCann found that in the upland region around the former imperial capital Ankober, located high on the eastern escarpment overlooking the Afar desert to the east, the absence of trees goes back several centuries at least. He considered that this may reflect the naturally treeless nature of the vegetation in this area.[28] He also examined historical records for the area around Gera, in the forested southwestern uplands of the country. Recurrent intervals of clearing and forest regrowth were linked to sporadic military conquests and social upheavals during the past three centuries, with forest having now totally replaced land cleared and cultivated as recently as fifty years ago.[29] The concept of recent, progressive and linear deforestation in the Ethiopian highlands is therefore open to question.

A similar conclusion emerges in relation to land degradation. Elias and Scoones considered whether soil fertility was declining in the Wolayta region of southern

[24] Hurni, 'Sustainable Management of Natural Resources'.

[25] Minister for Agriculture, Sudan, personal comm., 1976.

[26] O. El Badri, 'Sediment Transport and Deposition in the Blue Nile at Khartoum, Flood Seasons 1967, 1968 and 1969', Unpublished M.Sc. thesis, Department of Geology, University of Khartoum, 1972; M. A. J. Williams and F. M. Williams, 'Evolution of the Nile Basin', in M. A. J. Williams and H. Faure (eds), *The Sahara and the Nile* (Rotterdam: Balkema, 1980), pp. 207–24.

[27] A. Swain, 'Ethiopia, the Sudan and Egypt: The Nile River Dispute', *Journal of Modern African Studies*, 35, 4 (1997), pp. 674–94.

[28] McCann, *Green Land*, pp. 84–92.

[29] *Ibid.*, pp. 93–104.

Ethiopia.[30] Using a combination of field-level evaluation of soil nutrient balances, archival data and personal interviews with farmers, they concluded that soil fertility decline was true of some localities, for certain farmers, over certain periods, but was far from universal. Scoones had reached a similar conclusion during his earlier fieldwork in Zimbabwe that led him to reject over-generalized assessments of environmental change and resource use.[31]

In the light of Hurni's observations, it is tempting to conclude that in order to achieve a more sustainable form of agriculture in the Ethiopian highlands, the local farmers need to be actively involved in soil and water conservation measures and a program of long-term re-afforestation of steeplands. However, programs of soil and water conservation and tree-planting are seldom if ever successful when imposed from above. They have certainly not worked well in Ethiopia.[32] The detailed historical, sociological, economic and agronomic studies by McCann, Hoben, Elias and Scoones, and others strongly indicate the need for good local research and for community involvement in planning their future.

Tectonically–induced erosion at K'one volcanic complex, Ethiopian Rift

Gully erosion is not uncommon in Ethiopia but there are almost no long-term monitoring programs in place to enable us to assess when and how rapidly different gully systems have developed. Unchecked gully erosion can lead to loss of arable and pastoral land, to increasing sediment loads in streams draining the affected areas and to more rapid runoff and reduced infiltration during heavy rains. For these very practical reasons it is important to know whether the gully erosion was triggered by human or non-human causes. Among the former causes we may list overgrazing, trampling by herds of cattle, sheep or goats, and removal of trees for firewood and charcoal. Non-human causes in Ethiopia may be climatic (for example, intense rain storms after prolonged drought), tectonic or volcanic.

The spectacular gullies incised into the clay-covered floor of one of the volcanic calderas at K'one in the Ethiopian Rift are a case in point (Photo 2.1). The sediments that mantle the floor of one of the volcanoes display a typical badland topography and are highly visible from the main road leading from Addis Ababa to Dire Dawa. Charcoal burners active in the vicinity of these spectacular gullies were initially blamed for causing the erosion. The September 1980 *SINET Newsletter of the Ethiopian Journal of Science* drew attention to the seasonal flood hazards in the Wolenchiti area and stressed the urgent need for applied research into runoff and erosion in this sector of the Ethiopian Rift. The ensuing discussion is based on some of my earlier work at K'one during February 1974 and January 1975 and is designed to clarify whether the gullies there were a result of human or non-human causes.[33]

K'one volcanic complex covers an area of 250 km[2] and consists of eight nested

[30] Elias and Scoones, 'Perspectives on Soil Fertility Change.'

[31] I. Scoones, 'The Dynamics of Soil Fertility Change: Historical Perspectives on Environmental Transformation from Zimbabwe', *Geographical Journal*, 163 (1997), pp. 161–9.

[32] Hoben, 'The cultural construction of environmental policy'.

[33] M. A. J. Williams, 'Recent Tectonically-Induced Gully Erosion at K'one, Metahara-Wolenchiti Area, Ethiopian Rift', *SINET: Ethiopian Journal of Science*, 4 (1) (1981), pp. 1–11.

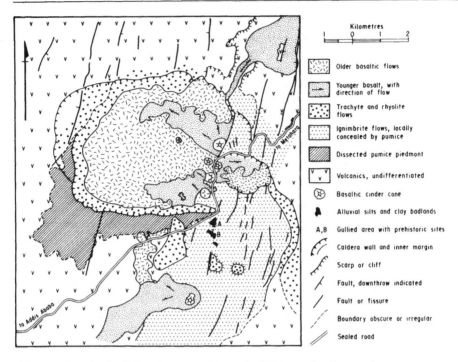

Figure 2.4a   Geological and geomorphic map of K'one volcanic complex

(Based upon field mapping and air photo interpretation in 1974-5, and upon J. W. Cole, 'Gariboldi volcanic complex, Ethiopia,' *Bulletin Volcanologique*, 33 (1969), pp. 566-78).

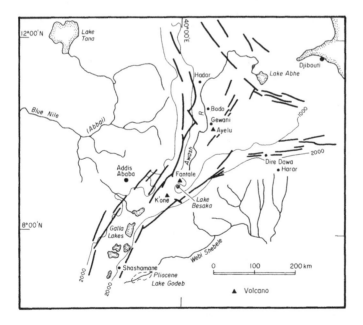

Figure 2.4b   Awash River Valley complex

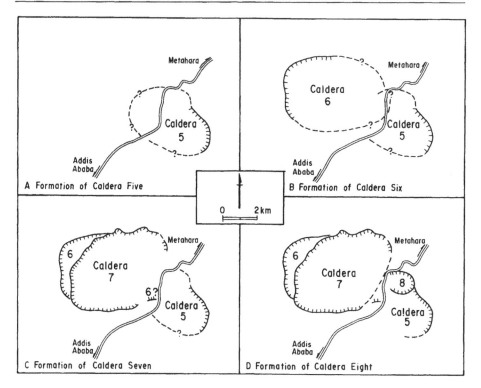

Figure 2.5 Hypothetical closing stages in the Quaternary evolution of K'one volcanic complex. (After M.A.J. Williams, 'Recent Tectonically-Induced Gully Erosion at K'one, Metahara-Wolenchiti Area, Ethiopian Rift'.)

volcanic calderas located along the Wonji Fault Belt within the Ethiopian Rift. Each caldera formed by the successive collapse of several large silicic volcanic cones. K'one lies 60 km northeast of Nazret and about mid-way between Wolenchiti and Metahara at 8°50'N, 39°43'E. The 600-800 mm annual rainfall falls mainly during June to August. Away from the steep and rocky crater rims the soils support locally dense *Acacia* woodland. The gullied clay-mantled floor of caldera 5 lies at an elevation of 1400 m (see Figures 2.4 and 2.5). It is instructive to note the location of the gullied areas A and B in relation to the regional geology (Fig. 2.4). Welded tuffs associated with the two most recent collapses (calderas 7 and 8, Fig. 2.5) line the floor of the earlier calderas. Basaltic eruptions followed the caldera collapse and formed cinder cones and scoriaceous flows. The centres of these eruptions run north-northeast, along the hinge line between calderas 7 and 8, and are probably associated with a line of fissure. The two youngest basalt flows cover the floor of calderas 7 and 8 and are very similar in appearance to the early nineteenth-century flows between Fantale volcano and Lake Besaka discussed in the next section. The point of this geological information is to indicate that volcanic and tectonic activity is far from over at K'one.

A welded tuff of probable Middle Pleistocene age (c. 125–750 ka BP) covers the floor of caldera 5 in which the gullies occur. Some 10–15 m of horizontally bedded

Photo 2.1 Badland topography in the floor of caldera 5 at K'one volcano, Ethiopia (Photograph by M. Williams, 1975).

Figure 2.6 Plane table map of K'one gully complex 'A' (see Fig. 2.4a) (Surveyed by the author and the late W.H Morton, January 1975; compass traverse of gully floors by P.M. Bishop and F.A. Street).

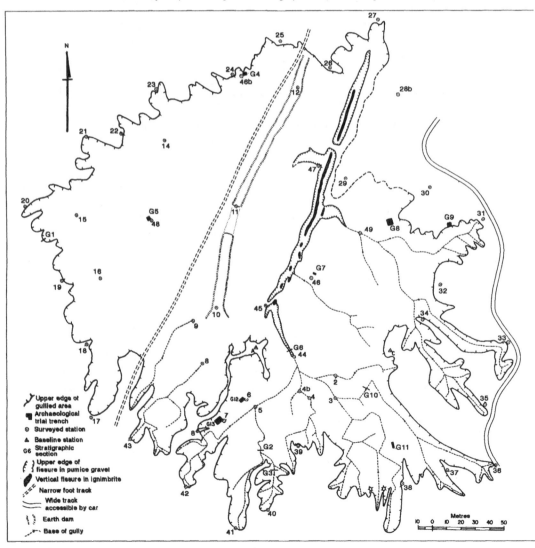

clays and tuffaceous silts and a basal layer of airfall pumice overlie the welded tuff. The clays and silts contain numerous Middle to Late Stone Age flaking floors. The welded tuff is cut by at least two narrow and fresh-looking tensional fissures, the deeper of which is over 36 m. Runoff into these fissures caused the gullying of the overlying sediments. When did this begin?

The gullies can be no older than the youngest horizontal sediments they entrench. No cut-and-fill structures or buried prehistoric gullies were observed. Every stratigraphic unit could be traced horizontally along its exposed length, and every gully was mapped by plane table survey and compass traverse (Fig. 2.6). The present episode of gullying is therefore unique relative to the depositional history of this area. Elsewhere in Ethiopia, Middle Stone Age sites date from > 100 ka to < 30 ka. We have two radiocarbon dates from the second youngest sedimentary unit. Ostrich eggshell from the unit is dated to 14, 670 +/- 200 years BP (I-8322). The date of 6, 810 +/- 120 years BP (SUA-462) for a hard pedogenetic carbonate nodule from this unit gives a minimum age for the onset of carbonate segregation within the loam. The onset of gullying, and the opening of the fissures, must likewise be younger than this date. This conclusion accords well with the known history of opening of tensional fissures in the Wonji Fault Belt in response to crustal extension and fault reactivation.[34] The present geodetically monitored rate of rift dilation is 5 mm/yr.[35] If we assume that the K'one fissures also widened at this rate, we obtain a maximum age of 300-500 years for the inception of the fissures. This estimate is in intuitive accord with the sharp edge of the fissure walls. It is further tempting to relate these tensional cracks to the volcanic activity further east dated to about 1820 AD.[36] There seems no reason to implicate the charcoal burners for what can only be described as tectonically-induced gully erosion at K'one. Furthermore, many of the gully sidewalls have attained a stable 30–35° angle of repose, consistent with a long history of gully extension and eventual stabilization. In such circumstances, soil conservation measures become irrelevant.

The Recent Fluctuations of Lake Besaka in the Ethiopian Rift

Lake Besaka occupies a small fault trough at the junction of the Afar Rift and the Ethiopian Rift, just west of Metahara (Fig. 2.4b). Fantale volcano overlooks the lake to the north and lavas from the 1820 AD eruption reached the northern shores of the lake (Fig. 2.7). The level of the lake has fluctuated during geologically recent times, attaining a maximum level of 10 m above the January 1974 datum at 11.2–11.4 ka.[37] Since this was a time of very high lake or river levels elsewhere in Ethiopia, Kenya, Uganda and the Nile basin, there seems no reason to doubt that this high level in

[34] I. L. Gibson, 'The Structure and Volcanic Geology of an Axial Portion of the Main Ethiopian Rift', *Tectonophysics*, 8 (1969), pp. 561–5.

[35] P. A. Mohr, *1973 Ethiopian-Rift Geodimeter Survey*. Smithsonian Astrophysical Observatory, Special Report, 385 (1974).

[36] W. C. Harris, *The Highlands of Ethiopia* (London: 3 vols., 1844), Vol. 3; D. R. Buxton, *Travels in Ethiopia* (London, 1949).

[37] M. A. J. Williams, P. M. Bishop and F. M. Dakin, 'Late Quaternary Lake Levels in Southern Afar and the Adjacent Ethiopian Rift', *Nature*, 267 (1977), pp. 690–3; M.A.J. Williams, F.M. Williams and P.M. Bishop, 'Late Quaternary History of Lake Besaka, Ethiopia', *Paleoecology of Africa*, 13 (1980), pp. 93–104.

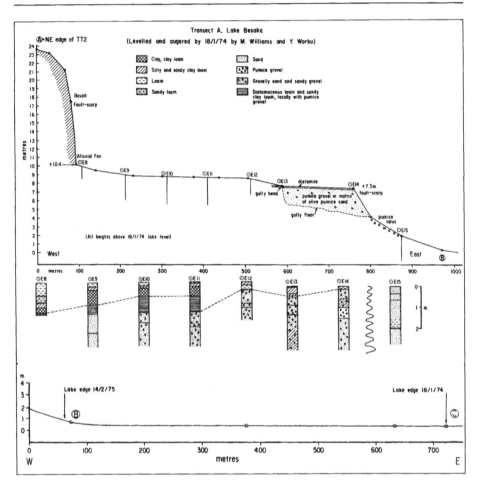

Figure 2.7   Levelled stratigraphic transect west of Lake Besaka
The transect runs due east from archaeological trial trench TT2. Note the change in lake-level between 18/1/74 and 14/2/75. (After Williams *et al.*, 'Late Quaternary History of Lake Besaka').

Lake Besaka was a result of more abundant summer rains and/or lower rates of evaporation.

If we accept a simple equation of high lake level equals humid phases and low lake level equals dry phases, as seems reasonable, then we might have expected to see a drop in the level of Lake Besaka during the height of the mid-1970s drought. Detailed levelling which I undertook in January 1974 and February 1975 revealed the opposite: in that time the lake rose 0.5m (Fig. 2.7). The lake continued to rise throughout 1976 and 1977, flooding both road and railway. The rise prompted rumours of illicit water diversion into the lake from nearby irrigation projects drawing water from the Awash River (Fig. 2.8). The Ethiopian Relief and Rehabilitation Commission expressed concern at the submergence of road and railway, as did other government agencies. In September 1980, Aberra Alemu of the Addis Ababa Geophysical Observatory reported in *SINET Newsletter of the*

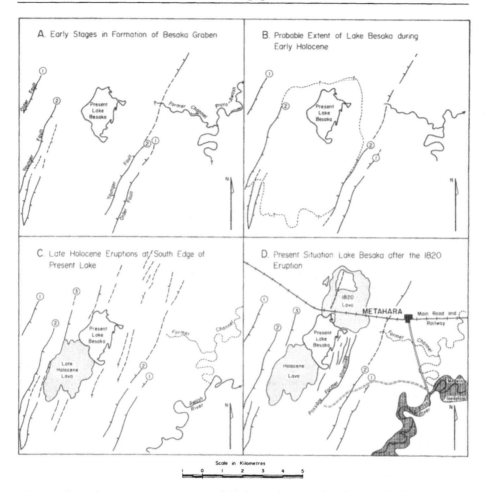

Figure 2.8 Schematic reconstruction of the late Quaternary evolution of Lake Besaka
Note the 1820 lava flow and the location of fault-scarp 3, from which there was increased hot-spring activity from 1974 onwards. (After Williams *et al.*, 'Late Quaternary History of Lake Besaka').

Ethiopian Journal of Science that Lake Besaka was one of two areas to which the Committee to study geological and geophysical hazards was now giving priority. To quote directly from the *Newsletter*:

> Metahara: the first problem in this area is the expansion of Lake Besaka. This lake has been reported to be expanding for the last few years. It has now expanded to the extent of over-flooding the highway and railway lines from Addis Ababa to Dire Dawa which run along the border of the lake. There is also a high probability that the town and the surrounding grazing area including the Abadir plantation will be overflooded by this lake in the near future. Other problems in this area caused by the lake such as the chemicals in the water affecting the growth of plants and the health of the inhabitants are of great significance. Therefore, there should be extensive studies on the Geochemical, Hydrogeological and Geological [hazards] of the area.

The change in water chemistry was real, although not caused by irrigation over-flow or Awash water diversion. Within two years of the initial rise in lake level, the

Photo 2.2 Hot springs near Fantale Volcano, Ethiopia
(Photograph by M. Williams, 1975).

lake had ceased to be brackish. The flamingoes that were once such a colourful feature of the lake migrated, presumably because the lake waters no longer sustained the brine shrimps and other organisms on which they fed. Fish and crocodiles now appeared.

As is often the case when rhetoric is used as a substitute for accurate observation, a few simple measurements by hydrogeologist Yemane Zecharias were all that was needed to identify the true cause of the puzzling rise in lake level.[38] Human mismanagement was not the culprit. The explanation was more prosaic: a resurgence of hot-spring activity (Photo 2.2) along the fault-scarp bordering the western lake margin, possibly a prelude to renewed eruptive activity from Fantale volcano to the north of the lake. Thus, within less than two years of the initial rise in lake level, Yemane Zecherias was able to demonstrate an increase in groundwater discharge from the thermal springs that issue from the foot of the lower of the two basalt fault-scarps which define the western margin of the graben. Overspill from the Awash-fed irrigated cotton project at Abadir, blamed as the culprit in 1975, could now be ruled out. Elsewhere in the middle and lower Awash basin the 1968–74 drought caused lakes to shrink (Lake Abhe) or to dry out (Lake Lyadu). Volcano-tectonic rather than climatic factors therefore seem likely to be the initiator of the increase in groundwater resurgence at Lake Besaka. It is possible that the greater hot-spring activity may be an early warning of future volcanic eruptions from Fantale, the last of which was early in the nineteenth century, and is evident in the very fresh, unvegetated lava at the northern margin of the lake.[39] Earlier observers

---

[38] Yemane Zecharias, pers. comm., 1976.
[39] Williams *et al.* 'Late Quaternary Lake Levels in Southern Afar'; *idem.*, 'Late Quaternary History of Lake Besaka'.

attributed this to an eruption in the 1820s.[40] On Figure 2.8 it is simply labelled the 1820 flow, although the precise year is not known.

We now turn from volcano-tectonic influences in the Ethiopian Rift to two very different case studies. One involves the erosion of fixed dunes in the Sahel zone of central Niger. The other concerns an unexpected response to flood events in the Jubba basin of western Somalia. Both are presented to illustrate my thesis that some subtlety is needed when deciphering the causes and possible future impacts of environmental changes in the savanna regions of Africa.

Cyclic Erosion of Fixed Dunes in Central Niger

Vegetated and stable dunes are a feature of the Sahelian borders of the Sahara and extend more or less unbroken from the Atlantic to the Nile.[41] The sandy soils of these dunes have trapped fine clay and silt particles derived from many millennia of dust storms and so are prized for growing millet and other crops. A number of the dunes are dissected and gullied. Is such dissection evidence of land degradation caused mainly by overgrazing and inappropriate methods of cultivation, or is it part of a suite of geomorphic processes characteristic of these particular savanna environments?

During the 1974–5 dry season, Dr Michael Talbot and I carried out an extensive survey on foot and by camel of the geomorphology of the Wadi Azouak basin. This account draws heavily on that work.[42] Our attention was attracted by the occurrence of a very recently formed series of sandy alluvial fans located on the lower slopes of fixed dunes between Tahoua and Abalak, roughly midway along the road from Niamey to Agades. Older fan sediments were exposed in the banks of the entrenched channels upstream of the most recent fan. Buried soils showed that previously active fans had become vegetated and stable.

Our interpretation of the most recent fans was that extremely intense rainfall early in the preceding wet season, confined to an area about 10-15 km radius, had triggered very high rates of runoff from the sparsely vegetated surface of the fixed dunes. This caused channel entrenchment above the fan apices and fan sedimentation below. The youngest buried soil predated a fireplace dated to 335+/-60 years BP (N-2129). The soil was regionally widespread and may have formed during a slightly wetter interval evident throughout the Sahel zone dated to about 150–350 years BP. The bank sections showed that episodic dune dissection, fan accumulation, soil formation and fan stabilization had been typical of the last few thousand years.

---

[40] Harris, *Highlands*; Buxton, *Travels*; P. Teilhard de Chardin and P. Lamare, 'Le canon de l'Aouache et le volcan Fantale', *Mémoire Société Géologique de France*, 14 (1930), pp. 13–20.

[41] A. T. Grove and A. Warren, 'Quaternary Landforms and Climate on the South Side of the Sahara', *Geographical Journal*, 75 (1968), pp. 438–60; M. R. Talbot, 'Environmental responses to climatic change in the West African Sahel over the past 20,000 years', in Williams and Faure, *The Sahara and the Nile*, pp. 37–62; J. E. Nichol, 'Geomorphological Evidence and Pleistocene Refugia in Africa', *Geographical Journal*, 165 (1999), pp. 79–89.

[42] M. R. Talbot and M. A. J. Williams, 'Erosion of Fixed Dunes in the Sahel, Central Niger', *Earth Surface Processes*, 3 (1978), pp. 107–13; *idem.*, 'Cyclic Alluvial Fan Sedimentation on the Flanks of Fixed Dunes, Janjari, Central Niger', *Catena*, 6 (1979), pp. 43–62.

We concluded that the 1974 phase of dissection was no greater in scale than previous phases and that there was no reason to believe that recovery would not once again be possible.[43] A similar inference is probably equally true for other parts of the Sahel, assuming that they have not been too severely overgrazed and depleted of their protective plant cover.

Potential Impact of River Regulation in the Jubba Valley, Western Somalia

My final example is somewhat more speculative than the six examples discussed earlier, but it does serve to illustrate the need for a very broadly based approach to environmental impact assessment of development projects.

In October 1988, I completed a 500 km foot traverse of the Jubba River right bank between Baardheere and Luq. The Jubba River drains the southeastern highlands of Ethiopia and flows from north to south through western Somalia. It is the only perennial major river in Somalia to reach the sea. As such, it was the focus of interest as a potential source of hydro-electric power for the capital Mogadishu and irrigation water for the adjacent savanna region. The proposed dam at Baardheere was expected to create a reservoir several hundred kilometres long and provide sufficient head of water for hydro-electric power generation. My role was to seek evidence of past environmental fluctuations in the valley as part of a wider survey of archaeological sites being conducted on behalf of the Somali National Academy of Sciences and Arts. The details of this work were submitted as a set of detailed geomorphic maps and accompanying report to the Minister responsible for the Baardheere Dam Project. Given the present unrest in Somalia, the project remains in limbo. My concern here is not with whether or not the dam is ever built, but with the limitations of orthodox civil engineering approaches to river regulation projects at the planning stage.

The headwaters of the Jubba River lie close to a number of streams that flow into the Oromo Lakes area of the Ethiopian Rift. Alayne Street mapped and dated a number of the strandlines surrounding these lakes.[44] The late Quaternary lake-level history of the Oromo Lakes is now thoroughly documented.[45] Since the rivers feeding into the Oromo Lakes rise in a similar part of the Ethiopian uplands as the Jubba headwaters, it is logical to expect a similar flood history. On this basis, we might expect to find evidence of high-level floods in the form of abandoned floodplains or river terraces flanking the Jubba River in Somalia. Oddly enough, I was unable to find any evidence of high terraces preserved along the Jubba. Being well aware that 'absence of evidence is not evidence of absence', I also searched every side valley I came to very thoroughly.

South of Luq, the river flows through a series of narrow limestone gorges dissected by tributary valleys that remain dry for most of the year. A series of low alluvial terraces of comparable age to the highest levels of the Ethiopian Oromo

---

[43] Ibid., 'Cyclic Alluvial Fan Sedimentation on the Flanks of Fixed Dunes'.

[44] F. A. Perrott, 'Late Quaternary Lakes in the Ziway-Shala Basin, Southern Ethiopia', unpublished Ph.D. thesis, University of Cambridge (University Microfilms International, Ann Arbor, MI). See especially Fig. 8.1, p.347 and Table 8.1.

[45] Gasse, 'Water resources variability' and idem., 'Hydrological Changes in the African Tropics'.

Lakes were present in a side valley abutting the main river upstream of Baardheere. The youngest of these contained an assemblage of sub-fossil freshwater mollusca shells, including *Biomphalaria pfeifferi* and *Bulinus truncatus*, two of the modern vectors of the schistosome parasite in its early developmental stages. The oldest of the alluvial terraces rose about 4 m above the modern floodplain. Heavy rainstorms on 19 and 21 October heralded the early onset of the drought-breaking summer rains. During these storms it was instructive to see how overflow from the main channel caused water to back up into the dry tributary valleys, leading to deposition of the fine red-brown clays and silts carried down in suspension from the Ethiopian highlands.

What conclusions can we draw from this brief overview of the October 1988 survey? First, the absence of high-level alluvial deposits from the past 30,000 years strongly suggests that the limestone gorges are honeycombed with caves and fissures and are highly permeable. In other words, the putative reservoir would be unlikely to fill to the level envisaged by the engineering consultants. This interpretation may be flawed, but the onus surely lies with the engineers to prove it wrong. Second, the level attained by the highest floods in the past is that likely to be reached by the reservoir. Hence, all existing arable land in the valley would be submerged, with little to show for this loss of land in the way of head for generating hydro-electric power. Third, and equally ominous, there is a high probability that schistosomiasis (bilharzia) would spread throughout the valley in the sluggish waters favoured as habitats by the snail vectors. It is hard to resist the conclusion that the thinking that went into the initial design of this project was somewhat limited. In particular, there was a remarkable lack of any ecological, geomorphic or historical awareness.

Conclusion

Specific conclusions have been drawn from each of the case studies discussed, together with some more general inferences. The two leading questions posed at the outset of this chapter have been answered with actual examples, so some recapitulation and some broader conclusions may now be useful.

Variations in the flow of big rivers like the Nile, Senegal and Niger are linked to global patterns of sea surface temperature in the Pacific, Atlantic and Indian Oceans, with ENSO events playing a significant but not overwhelming role in modulating floods and droughts in the Nile basin. Advance knowledge of anomalies in sea surface temperature in critical parts of the world's oceans can therefore be useful in predicting the future likelihood of above- or below-average flows in big rivers like the Nile.

Just as the discharge of the Nile has varied from year to year over the past century, so it has varied in earlier times in relation to global climatic fluctuations. The depositional legacy of these past changes is sometimes obvious, as in the case of former channels and former floodplains, and sometimes subtle, as with buried former strandlines and highly saline carbonate lenses. Without a clear appreciation of this depositional legacy, it is difficult to make much sense of the pattern of subsoil salinity in the central Sudan. Such knowledge is highly relevant to plans to extend the areas under irrigated cotton along the lower White Nile.

Irrigation in the Sudan depends upon a guaranteed supply of water from the Ethiopian highlands via the Abbai/Blue Nile and Tekazze/Atbara rivers. Reservoirs in the Sudan are silting up rapidly as a result of sediment brought in from the Ethiopian uplands. Deforestation and accelerated soil loss from cultivated land in the Ethiopian highlands have been held responsible for these rapid rates of siltation. Although there is some truth in this assertion, the reality is considerably more complex. Some parts of the Ethiopian uplands draining towards the Sudan have more forest today than they did fifty years ago, and soil fertility and soil depth are actually increasing in some places as a direct result of human management.

Equally, some sites in the Ethiopian Rift that display striking evidence of badland erosion have little or nothing to do with human impact. For example, detailed evidence from archaeology, radiocarbon dating and plane table surveys shows that the gullies in one locality are in fact relatively stable and were apparently initiated by an earthquake several thousand years ago.

The influence of tectonic and volcanic factors on recent environmental change is also evident in the recent history of a lake near the junction of the Ethiopian and Afar Rifts. During the height of the 1968–76 droughts, other lakes were drying out but this lake rose 50 cm between January 1974 and February 1975, and continued to rise thereafter, prompting fears of illicit water diversion into the lake from irrigation projects fed by the Awash River. In fact, the rise in lake level was a response to accelerated hot-spring activity, itself a possible precursor to future volcanic activity.

The role of other forms of non-human catastrophic events is also evident in the pattern of cyclic erosion of fixed dunes in the Sahel zone of Niger. Contrary to expectation, the cycles of gully erosion and alluvial fan deposition along the flanks of the fixed dunes are unrelated to human activities and are caused by severe and very localized cloudbursts which occur during otherwise prolonged regional droughts. The incidence of such events is extremely rare but the effects persist for many hundreds of years. Restoration of the bare dunes and gullies has occurred in the past as part of this rare but cyclic erosion and deposition of the dune flanks. Hence, res-toration of degraded dunes seems equally likely to occur in the future, setting natural limits to processes some have regarded as causing irreversible land degradation.

An ambitious plan to regulate the flow of Somalia's only river to reach the sea is examined in the light of field evidence collected by the author. Advocates for the scheme claim substantial potential benefits from hydro-electric power generation. A more cautious approach is advocated here. In particular, an equally plausible counter-view is that the reservoir would not fill to the envisaged level, would submerge what little arable land there is on the floodplain upstream of the proposed dam site, and would be instrumental in spreading schistosomiasis (bilharzia) throughout the valley.

The social, economic, political and biophysical contexts of each of the case studies discussed are of course very different, but a number of general themes are common to them all. Any coherent discussion of environmental change and land use in the savanna regions of Africa must try to integrate the physical, biological, social, economic and political aspects of local resource management practices over the past half-century, and sometimes a great deal longer. The examples chosen from Ethiopia, Niger, Somalia and the Sudan well illustrate the dictum that if we choose to ignore the legacy and lessons from the past, we remain doomed to repeat its mistakes.

PART THREE
Land Users & Landscapes

3

Fire in the Savanna
Environmental Change & Land Reform
in Northern Côte d'Ivoire

THOMAS J. BASSETT, ZUELI KOLI BI
& TIONA OUATTARA

Introduction

The Côte d'Ivoire savanna is widely believed to be experiencing severe degradation of its flora and fauna. Government officials, donor agencies, and NGOs view savanna vegetation as undergoing a progressive transformation from a wooded to grassy savanna, known as 'savannaization', and that desertification is taking place. The main agents of this assumed environmental degradation are believed to be farmers, hunters, and pastoralists who commonly use fire to burn vegetation for farming, hunting, and to improve pasture quality.[1] Indeed, burning is singled out as being particularly destructive to natural resources at multiple scales. In 1998 the High Commissioner of the Savanna Region declared bush fires (*les feux de brousse*) to be a major cause of environmental degradation at the global, national, and local levels. Speaking to members of three rural communities whose homes had been accidentally destroyed by fire set by hunters driving game, the Commissioner took the opportunity to inform his audience about these multi-scale dangers:

> The ozone layer which protects us from the sun's ultra-violet rays has been destroyed by us through carbon gas emissions generated by our diverse activities including bush fires. Today we are experiencing the harmful effects of climatic change in our severe Harmattan.[2]

He also decried fire's role in destroying the country's patrimony of tropical rain forests, savannas, wildlife, and farmland. He lamented that despite public campaigns to combat bush fires, destruction has continued unabated.

---

[1] T. Bassett and Z. Koli Bi, 'Fulbe pastoralism and environmental change in northern Côte d'Ivoire', in M. DeBruijn and H. van Dijk (eds), *Pastoralism under Pressure? Fulbe Societies Confronting Change in West Africa* (Amsterdam: Brill, 1999), pp. 139–59; *idem*, 'Environmental Discourses and the Ivorian Savanna', *Annals of the Association of American Geographers*, 90, 1 (2000), pp. 67–95.

[2] The Harmattan refers to the hot, dry, dusty winds that blow from the direction of the Sahara Desert in late December and January in Côte d'Ivoire. They are considered unhealthy because the dust is seen as a vector of diseases such as meningitis. A more severe Harmattan refers to a longer period of such conditions.

Côte d'Ivoire is rich because of its environment, its forest, its savanna, its rivers, its wildlife, its agriculture, and the work of its peasants. If we destroy all of this we do not have the right to exist. Fire is the enemy of peasants. Fire is the enemy of Côte d'Ivoire. Let us all say: No more fire![3]

This broad-scale attack on fire as an agent of environmental destruction appears to be unjustified on the grounds that there are few data to support certain claims, especially the view that fire is a leading cause of declining tree cover and biodiversity in the savanna region. Notwithstanding this major data gap, Côte d'Ivoire land-use planners and journalists have been greatly swayed by the savannaization idea.[4] Policy papers and the press commonly call for government intervention to stop the purported loss of tree cover and to restore biodiversity. Such measures include controlling bush fires, protecting national parks and reserves from poaching, stabilizing agriculture, planting trees, and forbidding logging north of the 8th parallel.

Land reform is seen as a linchpin of environmental conservation. National Environmental Action Plan (NEAP) documents at the national and regional levels repeatedly stress land titling as a condition for halting what is described as anarchic land clearing and burning. This destruction of tree cover is purportedly related to insecure land rights. The assumption is that land privatization will motivate individuals and communities to practise good stewardship.[5] The NEAP preparatory document for the Korhogo region explicitly links deforestation with bush fires and land tenure regimes.

We are witnessing more and more a reduction in the vegetative cover, due essentially to bush fires practised in an intensive and abusive manner, and which threaten to lead to the disappearance of certain varieties … (T)hese aggressions to the natural environment are essentially linked to
• the abusive exercise of customary rights;
• the extreme interpretation of two declarations of [the former] President Félix Houphouët-Boigny according to which on the one hand, '*The land belongs to the person who improves it*' and, on the other, '*that which has been planted by the hand of man must not be destroyed, no matter where*';
• and, especially, the absence of a rural land code.

The Ivorian government passed a new rural land law on 18 December 1998

---

[3] A. Timité, 'Le feu, l'ennemi No. 1', *Bulletin d'Information et de Sensibilisation Environmentale*, 2, 14 (1998).

[4] Current thinking among environmental policy-makers and educators is that the process of 'savannaization' (and ultimately desertification) entails considerable diminution in biodiversity of both plants and animals A good example of this view is found in the NEAP planning document for the Korhogo region: 'The vegetative cover is declining as a result of the practices of shifting cultivation, bush fires, and uncontrolled logging and overgrazing. As a general rule, bush fires annually pass through a large part of the region. The reduction in plant cover is accompanied by a replacement of the tree savanna by the grass savanna. In effect, associated with the disappearance of habitats is the threat of the disappearance of faunal species such as the panther, hartebeest, elephant, African buffalo, antelopes, etc. In sum, the loss of floristic biodiversity is accompanied by a loss of faunal biodiversity' (République de la Côte d'Ivoire, Ministère de l'Environnement et du Tourisme, Plan National d'Action pour l'Environnement, Cellule de Coordination, *Synthèse régionale: Région nord (Korhogo)* (Abidjan: PNAE-CI, 1994).

[5] T. Bassett, 'Introduction: the land question and agricultural transformation in Africa', in T. Bassett and D. Crummey (eds), *Land in African Agrarian Systems* (Madison: University of Wisconsin Press, 1993), pp. 3–31.

which was in part rationalized by such arguments. One of the remarkable aspects of this land legislation is how little public debate has taken place over its promulgation and application.[6] Yet, a major debate exists in the African rural development literature on whether indigenous land rights systems should be built upon or supplanted by land titling to promote natural resource conservation and economic growth. The modernization view promoted by the World Bank argues that land privatization is a motor of conservation and development because it frees land-holders to invest in land improvements without fearing that their investments will be appropriated by a third party.[7] Much emphasis is placed on the security that land titling gives to landholders and the efficiency of property markets in economic development.[8] Advocates of preserving indigenous land rights systems emphasize their flexibility and equity. They point out that conservation and development have historically taken place under customary tenure arrangements and that privatization does not automatically result in land improvements.[9] Jean-Pierre Chauveau illuminates the historical and domestic political and economic contexts of the 1998 Ivorian land law.[10] In this chapter we argue that the environmental degradation arguments used to justify this great land enclosure are highly questionable.

In the next section, we examine the relationship between bush fires and savanna vegetation in northern Côte d'Ivoire to assess whether the dominant notions of environmental degradation are accurate. This is followed by case studies of vegetation and wildlife trends in one savanna area from which we paint a very different picture of environmental change dynamics. In the final section we address the relevance of the new land law in the light of our findings. By pointing to the major directions and agents of change at the local level, and linking them to national and international policy discussions, we argue that recent land reform legislation is not the panacea that many believe will cure the environmental ills currently afflicting the Ivorian savanna.

Fire Regimes and Savanna Vegetation

Savannas owe their existence to a host of human and biophysical factors in which fire figures prominently. Tropical climates provide ideal conditions in which

[6] J. P. Chauveau, *The Land Question in Côte d'Ivoire: A lesson in history*, IIED Issue Paper No. 95 (London: IIED, August 2000), p. 24.

[7] K. Cleaver and G. Schreiber, *Reversing the Spiral: The Population, Agriculture, and Environment Nexus in sub-Saharan Africa* (Washington, DC: The World Bank, 1994); F. Falloux and L. Talbot, *Crisis and Opportunity: Environment and Development in Africa* (London: Earthscan, 1993).

[8] D. Atwood, 'Land Registration in Africa: The Impact on Agricultural Production', *World Development*, 18, 5 (1990), pp. 659–71; J. P. Platteau, 'The Evolutionary Theory of Land Rights as Applied to sub-Saharan Africa: A Critical Assessment', *Development and Change*, 27 (1996), pp. 29–86.

[9] Bassett, 'Introduction'; S. Migot-Adholla, P. Hazell, P. Blarel, and F. Place, 'Indigenous Land Rights Systems in sub-Saharan Africa', *World Bank Economic Review*, 5, 1 (1991), pp. 155–75. J. P. Platteau, 'Does Africa need land reform?' in C. Toulmin and J. Quan (eds), *Evolving Land Rights, Policy and Tenure Reform in Africa* (London: DFID/IIED/NRI, 2000), pp. 51–74.

[10] He argues that the diminished role of foreigners in Ivorian politics and the land demands made by a new generation of educated but unemployed youth, have been key factors behind the 1998 land law which allows only Ivorian citizens to own customary held land. Chauveau, *The Land Question*.

abundant biomass grows during the rainy season and then dries out and becomes highly flammable during the dry season. Violent dry-season fire checks the advance of woody plants in the landscape through its destructive power. Fire is so critical to the maintenance of grass cover that savannas are widely considered to be 'pyro-climatic' formations. Some savanna plant species not only tolerate fire (pyrophytes) but depend on it to prevent bush and tree species from spreading. A widely accepted idea is that if fire is suppressed, the forest will advance.[11] This does not mean that fire is the only variable determining the balance between woody and grass cover. Forest advance can also result from climatic change, grazing pressure, fire suppression, and tree protection and planting.[12] The spatial and temporal distribution of bush fires, land use, and soil conditions results in a patchwork landscape of different savanna types characterized by different tree-grass ratios.[13]

Perhaps because of its dramatic and destructive character, fire is widely viewed as the single most important factor in the creation and maintenance of savannas. Yet, despite its prominent role, the varied nature of fire and its uneven impact on vegetation are rarely discussed. This is particularly true of the policy literature in which burning is generally viewed as highly destructive of tree cover. Some savanna ecologists provide a more nuanced view of fire with their focus on differing fire intensities.[14] They point to the interplay of a variety of biophysical and human factors which produce more or less intense fires, or *fire regimes*. Fire intensity, in turn, greatly influences the ratio of woody to grassy species in African savannas.[15] The most important factors comprising a fire regime are:

1. *Fuel load*: The amount of combustible material varies with rainfall. As one moves from the Sudanian savanna to the Guinea savanna of Côte d'Ivoire, the quantity of biomass increases. This increase in fuel load can lead to more intense and thus more destructive fires in the Guinea savannas.[16]

2. *Fuel mixture*: The floristic composition of savanna plants influences a fire's intensity. Tall and dense grasses burn more intensely than low-lying herbaceous plants. Trees containing flammable resins also burn more hotly than non-resinous

[11] J. César, 'Les Feux de Brousse' in *Fiche Techniques d'Elevage Tropical*. 3. (Montpellier: IEMVT-CIRAD, 1990).

[12] J. Boutrais, 'Eleveurs, bétail, et environnement', in C. Blanc-Pamard and J. Boutrais (eds), *À la Croisée des Parcours* (Paris: Editions ORSTOM, 1994), pp. 303–20; J. Boutrais, *Hautes terres d'élevage au Caméroun* (Paris: ORSTOM, 2 vols., 1995); J. Fairhead and M. Leach, *Misreading the African Landscape: Society and Ecology in a Forest-Savanna Mosaic* (Cambridge: Cambridge University Press, 1996).

[13] Soil, water, and slope characteristics, as well as human influences such as livestock raising and farming, are also important determinants of this patchwork pattern.

[14] V. Bruzon, 'Les Savanes du Nord de la Côte d'Ivoire: Mésologie et Dynamique', Thèse de Doctorat, Université de Lille III, 1990; César, 'Les Feux de Brousse'; D. Gillon, 'The fire problem in tropical savannas', in F. Bourlière (ed.), *Ecosystems of the World 13, Tropical Savannas* (Amsterdam: Elsevier, 1983), pp. 617–41; J.-C. Menaut, 'The vegetation of African savannas', in Bourlière, *Ecosystems of the World 13*.

[15] B. W. van Wilgen, C.S. Everson and W. S. W. Trollope, 'Fire Management in Southern Africa: some examples of current objectives, practices, and problems', in J. G. Goldhammer (ed.), *Fire in the Tropical Biota: Ecosystem Processes and Global Challenges* (Berlin: Springer-Verlag. 1990), pp. 179–215.

[16] W. Erdelen, P. Müller, P. Nagel, R. Peveling, and J. Weyrich, *Implications écologiques de la lutte anti-tse-tse en Côte d'Ivoire nord et centre* (GTZ Project No. 87.2539.2. Final Report) (Saarbrucken: GTZ, 1994).

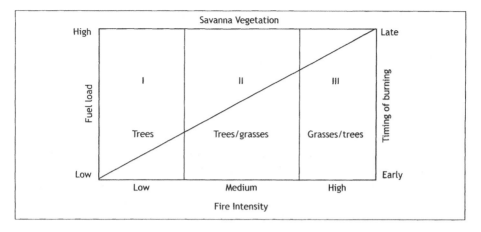

Figure 3.1 Fire regimes and tree/grass ratios

trees. Highly destructive crown fires are more likely to develop where grasses are high and dry and trees contain flammable resins.[17]

3. *Timing of fire*: This includes four components: seasonality, time of day, scheduling, and frequency. Late dry-season fires are more violent than early or mid-dry-season fires. The drier the biomass, the hotter the burn. Diurnal burns are also more violent than night-time fires. Convection winds characteristic of the hottest part of the day sweep fires up into the crowns of trees. In many localities, bush burning is regulated by a temporal-spatial schedule in which areas close to a village are burned first to create fire breaks; this early burn is followed by progressively later fires along the upper, mid- and lower slope of plateaux as soil moisture declines over the course of the dry season.[18] Infrequent fires (for example, once every 2-4 years) permit biomass to accumulate and result in the most violent fires.

4. *Land use*: Livestock raising and farming result in lower-intensity fires. Grazing pressure reduces the quantity of biomass that will burn.[19] Land clearing for farming dramatically reduces combustible material. Fields also function as fire breaks and prevent a fire from passing through an area, thus creating a patchwork of burned and unburned spaces.[20]

5. *Fire suppression*: Fire control experiments in savanna areas of Nigeria and Côte d'Ivoire show that, when burning is suppressed, fire-sensitive trees and bushes successfully compete with grasses and eventually eliminate them. Since grasses are the most important fuel in controlling trees, the savanna is eventually replaced by forest under these conditions.[21]

---

[17] César , 'Les Feux de Brousse'.

[18] Bruzon, 'Les Savanes'; O. Hoffman, *Pratiques pastorales et dynamique du couvert végétale en pays Lobi (Nord est de la Côte d'Ivoire)*, Collection Travaux et Documents 189 (Paris: ORSTOM, 1985).

[19] Boutrais , 'Eleveurs'.

[20] J-C. Filleron, 'Essai de géographie systématique: les paysages du Nord-Ouest de la Côte d'Ivoire', Thèse doctorat d'État de l'Université Toulouse-Le Mirail.

[21] E. Adjanohoun, *Végétation des savanes et des rochers découvertes en Côte d'Ivoire Centrale*, Mémoire ORSTOM 7 (Paris: ORSTOM, 1964); Gillon, 'The fire problem'.

6. *Rainfall*: There is more biomass to burn if rainfall during the previous rainy season was abundant. Conversely, below-average rainfall will result in smaller fuel loads the following year.

7. *Spatial pattern*: Savanna vegetation takes on a mosaic pattern in the landscape as a result of burning regimes influenced by a temporal-spatial calendar (see factor 3), natural and human-made firebreaks (for example, streams, roads), and grazing pressure. Depending on the history of past burns and the spatial patterning of savanna types (for example, grass vs. wooded savanna), fires will burn more or less intensely. This patchwork pattern also offers a diverse set of habitats for wildlife.[22]

The concept of fire regimes encourages us to consider bush fires in a more subtle and contingent manner (Figure 3.1). In contrast to the popular view in which bush fires are described in monolithic terms as highly destructive agents of environmental change, we suggest a more cautious approach that considers the *actual* intensity and impact of fire on savanna vegetation. Rather than viewing fire as 'public enemy number 1', we asked our informants if the savanna landscape had changed since they were young, and if so, how and why. We were presented with a remarkably different picture of vegetation and bush fire dynamics than the one found in environmental planning papers and NGO publications.

Land-use Patterns and Vegetation Change in the Katiali Region

Interviews with farmers and herders of Katiali in 1997–8 revealed a considerably more sophisticated understanding of fire and landscape evolution than that expressed by government ministers and environmental educators.[23] More than two-thirds of the 38 household heads interviewed stated that the landscape had become *more wooded* in their lifetime. Informants commonly said that twenty to thirty years earlier they used to be able to see for great distances in the bush because of the low density of trees. Today, they find trees are everywhere hemming in their view of the countryside. The reasons most often given for this landscape transformation are changing fire regimes and cattle grazing.

Livestock raising has significantly increased in northern Côte d'Ivoire since the early 1970s when increasing numbers of Fulbe pastoralists entered the country from neighboring Mali and Burkina Faso during the major drought affecting the sudano-sahelian region.[24] The data for the Katiali area show the basic trend in Fulbe cattle population growth in the 1980s and 1990s. Fulbe herders decided to stay in the country in view of the various services offered to them by the national livestock development agency (SODEPRA).[25] The 1970s and 1980s also witnessed consider-

---

[22] A. J. Belsky, 'Spatial and temporal landscape patterns in arid and semi-arid African savannas', in L. Hansson, L. Fahrig and G. Merriam (eds), *Mosaic Landscapes and Ecological Processes* (London: Chapman and Hall, 1995), pp. 32–56.

[23] For a fuller presentation of the interview results and methodology, see Bassett and Koli Bi, 'Environmental Discourses'.

[24] T. Bassett, 'Fulani Herd Movements', *The Geographical Review*, 76, 3 (1986), pp. 233–48.

[25] For details on the history and activities of SODEPRA, see T. Bassett, 'Land use conflicts in pastoral development in Northern Côte d'Ivoire', in Bassett and Crummey (eds), *Land in African Agrarian Systems*, pp. 131–56.

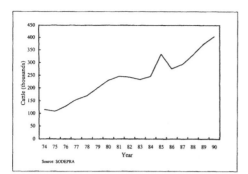

Source: SODEPRA

Figure 3.2 Fulbe cattle population growth, 1973–90

able growth in the number of oxen and ox plows in northern Côte d'Ivoire. Peasant farmers purchased oxen with credit on future cotton earnings through the parastatal cotton company, *La Compagnie Ivoirien du Développement des Fibres Textiles* (CIDT).[26] Cotton earnings also allowed some farmers to invest in cattle which are typically placed in a collective herd and guarded by a salaried Fulbe herder. Figure 3.2 shows the general trend in Fulbe livestock numbers over the period 1973–93.

The most immediate impact of greater numbers of cattle grazing in the region has been a reduction in plant biomass. In particular, grazing pressure has been greatest on perennial savanna grasses such as *Andropogon gayanus*, *Andropogon schirensis*, and *Schizachyrium sanguineum*. One measure of the decline in these species is the difficulty farmers experience in finding these grasses to thatch the roofs of their houses. Another indication of reduced grass cover is the lower intensity of fires reported by informants. A consensus view is that bush fires are not as intense as they used to be and that this changing fire regime is largely due to the reduced plant cover associated with cattle grazing. Respondents also noted that today's fires are set earlier in the dry season than they were in the past, and linked this practice to Fulbe pastoralists. Indeed, it is common practice among herders to burn lignified and unpalatable grasses at the beginning of the dry season to encourage a flush of highly nutritious pasture.[27] One consequence of early burns is that fires are less intense and a lower percentage (as little as 25 per cent) of grass cover is burned.[28]

We hypothesize that the combination of grazing pressure and early fires has resulted in a less intense fire regime that has created conditions favorable to bush and tree encroachment. Reduced grass height and combustible material means that woody plant seedlings are better able to establish themselves. Not only are seedlings not destroyed by fire but the reduced grass height also allows photosynthesis to take place more efficiently. César notes further that perennial grasses are weakened when they are not allowed to build up their underground reserves as cattle graze regrowth after early burns.[29]

The expansion in cultivated area is another land-use factor configuring this new fire regime. Farming systems surveys conducted in Katiali over the 1980s and 1990s show that, during this period, the cultivated area has more than doubled largely as a result of the use of ox plows. On the one hand, fields function as fire breaks and thus protect tree seedlings growing downwind. Trees are also more likely to

---

[26] The Ivorian Textile Fibers Development Company.

[27] Gillon reports that the nutritive quality of young grass leaves is high in protein, calcium, phosphorous. Gillon, 'The fire problem', pp. 621–2.

[28] *Ibid.*, p. 619.

[29] César, 'Les Feux de Brousse', p. 7.

establish themselves more successfully in fallow fields where impoverished soils impede the development of sufficient biomass for violent bush fires.[30]

To see if local perceptions of environmental change were supported (or not) by scientific data, we analyzed aerial photographs of the Katiali area for different periods. Figure 3.3 shows the results of our interpretation of aerial photos at the scale of 1:50,000 for the years 1956, 1979 and 1989. The trends are very clear for the categories of open bush, woodland,

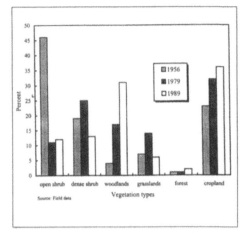

Figure 3.3 Land cover change, Katiali, Côte d'Ivoire

and cropland. Corroborating our informants' experience, the area classified as open bush declined from 46 per cent of the land cover in 1956 to between 11 and 12 per cent in 1979 and 1989. The area in woodland, which includes both wooded savannas and open forests, expanded from 4 per cent in 1956 to 17 per cent in 1979 and then to 31 per cent of the land cover.[31] During the three decades from 1956 to 1989, the area in cropland, including fallow, rose from almost a quarter (23 per cent) of the area to more than a third (36 per cent).

In summary, based on interviews with land users and aerial photo analysis, we have strong evidence at the local level that the landscape of the Katiali area has become more wooded since the mid-1950s. Similar results were obtained using the same methodology for a more densely populated community south of Korhogo where we conducted a land-use and land-cover study.[32] These findings contrast with the dominant view that the savanna landscape has become less wooded and that savanna grasslands are spreading. If the savanna is not evolving in the direction that is commonly assumed, then what does this say for wildlife dynamics? Is wildlife declining as a result of habitat changes, or are their numbers influenced by a different set of pressures?

Hunting Pressure and Wildlife Change

Information on trends in wildlife populations was collected using two principal methods: interviews with senior hunters, and commercial game market studies for the years 1981–2 and 1997. The latter provide a proxy quantitative measure of the

---

[30] I owe this observation to Jean Charles Filleron, pers. comm., 22 June 1998, Abidjan.

[31] The difficulty in distinguishing these two savanna types at the scale of 1:50,000 required that we merge these two categories into one which we have labeled 'woodland'. The French terms for these two savanna types are *savane boisée* and *forêt claire*.

[32] This comparative study was conducted in Tagbanga in the sub-prefecture of Tioroniaradougou in the Department of Korhogo: Bassett and Koli Bi, 'Environmental Discourses'.

relative abundance of wildlife over a 16 year period, while the main informant interviews provide a longer and more qualitative view of wildlife trends.

The hunting market study was carried out in the Katiali region in the community of Kakoli – a pseudonym for the location where regional hunters sold bush meat to Korhogo-based merchants.[33] Bush meat merchants have regularly visited Kakoli since the mid-1970s. They buy game at higher prices (25–100 per cent higher) than those obtained locally. Prior to the mid-1970s, game was sold to individuals or distributed to kin within the community.

The data for 1981/2 were collected by Bassett from a hunter who was present at the weekly game market. The 1997 data were recorded twice weekly by a research assistant from the bush meat buyer. The number of hunters selling game did not change significantly for these dates.[34] The type of data collected for both years include: date of sale, game species marketed, and the selling price. Interviews with the top hunters of Kakoli focused on the relative abundance of game, hunting techniques, and relationships between habitat change, hunting pressure, and species abundance. A sample of 38 households was also surveyed on meat consumption and for their perceptions of wildlife population trends.

The striking contrast in monthly game sales for the two time periods is evident in Figure 3.4. Two patterns stand out. First, the total number of animals sold dropped by 41 per cent between 1981 and 1997. Second, the seasonal peak in game sales recorded for 1981/82 flattened out by 1997. Just six animals accounted for 92 per cent of sales in 1981/82 and 81 per cent in 1997 (Figure 3.5). However, the top three game species sold in 1981 (Grivet monkey (*Cercopithecus aethiops* var. *sabaeus*), Cane rat (*Thryonomys swinderianus*), and Guinea fowl (*Numida melagris*)) had virtually disappeared by 1997. Their market position was taken by the Patas monkey (*Erythrocebus patas*), Bushbuck (*Tragelaphus scriptus* (Pallas)), and Crested porcupine (*Hysterix cristata*). However, the numbers of the last two species declined in absolute terms by 43 and 21 per cent respectively from 1981/2 levels.

Figure 3.6 shows the total number of animals sold for the two years by species. In addition to the major shifts signaled above, this figure also shows an increase in the number of small game animals (Gambian rat, genet, hare) in the 1997 market. When asked why they did not market these animals in 1981/2, hunters replied that larger game was more plentiful then. An indication that hunting pressure has had a major impact on wildlife populations is the fact that hunters spend more time hunting today but bring home fewer and smaller game. Hunters expressed more than a little frustration about this declining return to effort. When asked why there were fewer grivet monkeys, guinea fowl, cane rats, and hares marketed in 1997 in comparison with 1981/2, the seven top hunters of Kakoli gave two general reasons: 'We have killed everything'; and 'There are more fields and cattle in the bush, which has caused the animals to move away'.[35] More than a few hunters attributed the decline in guinea fowl to poisoning by farmers who place poison-laced rice and maize seeds at the borders of fields to reduce crop damage by these grain-eating birds.

[33] A fictitious name is used to protect informants from prosecution by Ivorian authorities. Hunting was made illegal in 1974 although farmers can shoot and trap animals if they are causing crop damage.

[34] A total of 44 hunters sold game to the merchant in 1981/2. The number in 1997 was 47.

[35] Interviews with P. S. and B. T., Kakoli, 28 June 1998.

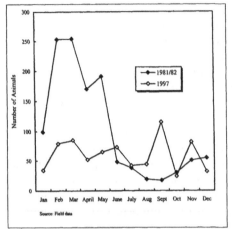

Figure 3.4 Monthly sales of commercial game, Kakoli, 1981/2 and 1997

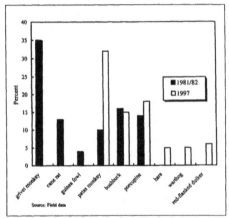

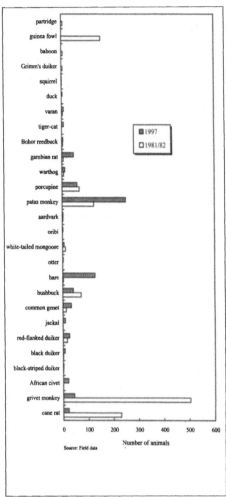

Figure 3.5 Top 6 animals marketed, Kakoli, 1981/2 and 1997 (% of total sales)

Figure 3.6 Commercial game sales by species, Kakoli, 1981/2 and 1997

Hunters spoke frankly about what they viewed as a 'slaughter' of game. Despite the decline in wild animals, hunters 'will shoot anything that moves in the bush.'[36] Merchants will buy virtually everything that is brought to market and hunters will continue provisioning these markets as long as there are animals in the bush and the price is attractive. Hunters also use better rifles today than they did only 15 years earlier. In 1981/2 muzzle loaders were commonly used for hunting; today, the most common rifle is the 12-gauge shotgun.

Hunters were specifically asked if the more wooded landscape had affected animal behavior, especially relative abundance. The consensus view was that the expansion of cattle and cropland was more important than increasing tree cover in

[36] Ibid.

Photo 3.1 Kakoli hunter holding a grivet monkey
(Photograph by T. Bassett, 1982).

Photo 3.2 Kakoli hunter (*donzo*) in a savanna woodland
(Photograph by T. Bassett, 1995).

the decline in some animals. One hunter noted 'there are fields that are cleared by rivers where the grivet monkey used to hide in the [gallery] forest'.[37] Cattle are widely viewed as major agents of habitat change. 'There used to be areas that were so closed in [by vegetation] you couldn't get inside them. That's where the animals hid. Cattle have opened up all of these areas and the animals have fled.'[38] Cane rats, in particular, have seen their grassy upland nesting habitat modified by cattle; they are now hunted in lowland areas where they hide in burrows. Many hunters used to use fire to drive game into a cul-de-sac where they were trapped and killed. This technique is rarely used today because 'there is so little grass left [to burn] because of cattle grazing'.[39]

A sample of 38 households was surveyed about the relative abundance of game and the reasons for its decline. When asked if there was more or less game in the community today in contrast to 16 years previously, everyone agreed there was less. The principal causes of this decline were believed to be: too many hunters killing animals (38 per cent of the responses); the departure of animals from the area (25 per cent); and habitat changes linked to cattle grazing (21 per cent). Informants also noted that the scarcity of bush meat in the village was linked to hunters' preference to sell game wholesale to Korhogo-based merchants rather than to consumers in the village. A few hunters acknowledged that they had to 'hide' meat from their family in order to sell it to merchants.[40]

This recent decline in diversity and total numbers of wild animals is part of an historical pattern of large game loss. Elderly informants talked about the larger mammals they hunted in the 1950s that were no longer seen in the late 1990s. These included the Roan Antelope (*Hippotragus equinus* (Desmarest)), African Buffalo (*Syncerus caffer* (Sparrman)), Bubal Hartebeest (*Alcelaphus buselaphus major*), Kob (*Kobus (Adenota) kob*), Leopard (*Panthera pardus*), Brown Hyaena (*Hyaena brunea* (Thunberg)), and African Elephant (*Loxodonta africana*). The last time a Roan Antelope was sighted (and killed) was in 1992. The last Bubal Hartebeest was shot in 1991.

In summary, the relative scarcity of game in the Katiali-Kakoli area is linked to a combination of sustained hunting pressure and habitat change. Hunters provision weekly game markets in Kakoli to increase their incomes and status. The severity of hunting pressure is evident in both quantitative and qualitative terms. The total number of animals sold dropped precipitously and some species virtually disappeared between 1981 and 1997. The most important habitat changes result from land clearing for farming and reduced grass cover linked to cattle grazing. Development of both agriculture and livestock has received considerable support from the Ivorian government since the mid-1970s. The irony of current government statements decrying environmental degradation is twofold: first, the popular image of environmental change is wrong; and second, the dynamics of environmental transformation are directly linked to government rural development policies. The question that needs to be asked is what kind of environmental policies should be encouraged in light of this new information?

[37] Interview with K. Z., Kakoli, 29 June 1998.
[38] Interview with A.T., Kakoli, 2 July 1998.
[39] Interview with P.S., Kakoli, 28 June 1998.
[40] Interview with S. T., Kakoli, 3 July 1998.

Environmental Change and Policy

Environmental planning in Côte d'Ivoire began in earnest in the early 1990s on the eve of the United Nations-sponsored conference on environment and development (the 'Earth Summit') held in Rio de Janeiro in June 1992. The Ministry of the Environment and Tourism organized a preparatory national conference in May in concert with the World Bank which was urging, if not requiring, its aid recipients to create national environmental action plans (NEAPs).[41] A second national conference took place in November 1994 at which the NEAP co-ordinating committee presented a White Paper that summarized the results of a series of regional workshops at which local perceptions of environmental problems were solicited.

The committee's aim to include representatives of government and civilian organizations in these meetings resulted in their being dominated by civil servants, with an under-representation of non-governmental organizations and ordinary farmers and herders. This limited NGO participation was largely the legacy of an authoritarian state and single political party which did not encourage popular participation in planning and policy-making.[42] In light of this tradition, combined with the reticence of most (illiterate) farmers and herders at these regional meetings, the voices and views of rural land users were not adequately expressed. Consequently, the environmental problems identified for the savanna region, such as reduced tree cover and the expansion of grassy savannas (viz. 'savannaization'), are more a reflection of received ideas rather than a consensus view of rural land users. By utilizing an alternative methodology (household surveys, historical aerial photo interpretation, review of the scientific literature), we arrived at a different understanding of the nature and direction of environmental change. This mis-identification of environmental problems raises serious questions about the efficacy of policies designed to address what strike us as erroneous views. At worst, these policies can exacerbate existing problems, create new ones, and entail regulations that unnecessarily restrict and penalize sound land-use practices such as burning.[43] Our main concern in the final section is to reflect upon appropriate interventions that policy-makers might consider in light of the dynamic of bush and tree encroachment and wildlife decline in the Korhogo region.

Whose Management Goals and Towards what Land Cover?

Any environmental policy and management strategy should reflect at least two understandings: first, a recognition that an environmental problem exists, and, second, a common goal towards which a management plan should work. The first understanding, and the one that has occupied this paper up to now, raises a variety of philosophical, methodological, and political questions that other authors have addressed in more general terms.[44] How we know that environmental problems

---

[41] Since the late 1980s, the World Bank has required countries receiving low-interest loans through its International Development Association to prepare a plan as a condition of funding.

[42] Multi-party politics and limited freedom of the press were tolerated only after 1990.

[43] Fairhead and Leach, *Misreading*, pp. 237–60.

[44] P. Blaikie, 'Changing Environments or Changing Views? A Political Ecology of Developing

Photo 3.3 Mural outside the Community-Based Natural Resource and Wildlife Management Project (GEPRENAF) in Ferkéssédougou, 'Nature Yesterday, Nature Today' (Photograph by T. Bassett, 2000).

exist is as much a social issue as a scientific one. In fact, it is not possible to separate the two since there are always social (political, economic, cultural, etc.) motivations as well as epistemological issues behind the identification of natural resource management concerns.

The second element, concerning management goals, assumes some consensus (and invariably conflict) on how a resource should be managed. Should the Ivorian savanna be managed for more or less tree cover? More trees will probably be in the interests of foresters, loggers, charcoal makers, and conservationists managing land against soil erosion. Farmers who practise slash-and-burn cultivation may also view trees more favorably, since their litter and ashes will enrich the soil for cultivation. Livestock producers, wildlife managers, and tourism operators may prefer more open grassy spaces in which pasture quality and game viewing are more valued. In the end, the decision to manage natural resources in one way and not another reflects a political process in which the interests of some groups eventually dominate.

What are the natural resource management priorities for northern Côte d'Ivoire? It depends on with whom you speak. Foresters will talk eloquently about the advantages of planting fast growing trees like eucalyptus to protect soils against erosion and to ensure fuelwood supplies. The defenders of wildlife seek more resources to guard national parks against poachers. They also promote the regulation of hunting by issuing different types of hunting licences and game preparation permits for restaurants. Peasant farmers will focus discussion on weeds and different ways of controlling them in their fields. They will certainly bring up the persistent problem of crop damage caused by Fulbe cattle herds and the vexing issue of compensation. Range managers will emphasize the need to practise some form of rotational grazing and pasture burning program to increase the percentage of perennial grass species preferred by cattle. Herders will talk about conflicts with peasants

44 (cont.) Countries', *Geography*, 80, 348 (1995), pp. 203–14; P. J. Taylor and F. W. Buttel, 'How Do We Know We Have Global Environmental Problems? Science and the Globalization of Environmental Discourse', *Geoforum*, 23, 3 (1992), pp. 405–16.

Photo 3.4 Cattle in heavily grazed area invaded by *Annona senegalensis*
(Photograph by T. Bassett, 1982).

Photo 3.5 Burnt fallow showing grass regrowth and *Isolberlina doka* seedlings
unscathed by fire
(Photograph by T. Bassett, 1997).

over access to grazing land and the importance of mobility to their herd's health and productivity. Although they are aware of rangeland degradation in the form of tree and bush encroachment, herders seemed more concerned about access to dry-season rangelands 75–150 kilometers to the south than in the decline in perennial grass species locally. They will also talk at length about tensions with farmers over crop damage but deny any role in this conflict. Environmental policy analysts involved in the preparation of the Côte d'Ivoire NEAP emphasize the importance of tree cover in reversing the perceived land degradation processes of savannaization and desertification. Fire suppression, tree-planting, and more vigorous law enforcement are viewed as necessary measures to 'save trees'. The NEAP's fixation on tree cover is a legacy of the colonial period and the dominance of the savannaization narrative in contemporary government discourses on the environment.

In summary, there are as many environmental management goals as there are land users and land-use planners. Whose management goal will receive priority will be determined by the relative power of each interest group in advancing its agenda. The lack of good data on environmental change has hindered consensus-building around priorities and appropriate policies. This has left the door open to such powerful interests as the World Bank and Caisse Française de Développement to promote their neo-liberal economic agendas within the framework of national environmental planning. The prominence given to land privatization as conditions of 'sustainable' resource management is a case in point.

Environmental Degradation, Land-use Planning, and Land Reform

On 28 December 1998, the National Assembly of Côte d'Ivoire voted to change the nation's land laws affecting the rural areas. The new law replaced the 1963 land code which was never applied, in part because it failed to recognize 'customary land rights'.[45] The 1998 law recognizes customary rights in that land can be registered by groups 'exercising collective rights to communal lands'.[46] The rationale for a new land law had been repeatedly made in government planning and project documents since the late 1980s. For example, in the White Paper and final report of the country's environmental action plan, land titling is presented as a strategy for achieving 'sustainable agricultural development' and environmental protection.[47] Proponents of land reform argue that the co-existence of customary and modern land rights systems has only resulted in land-use conflicts leading to poor agricultural performance, environmental degradation and deaths. A recurring theme is that land titling will bring peace, agricultural growth and environmental conservation.[48] This argument

---

[45] J. Oble, 'Je soutiens que seuls les Ivoiriens soient propriétaires terriens', Le Jour, No. 997, 28 May 1998. To obtain an official land title, a group must formally prove that its claims are not contested, pay the costs of a survey, and finally register the land holding with the administrative authorities.

[46] République de la Côte d'Ivoire, Ministère de l'Agriculture, Recueil des Textes Relatifs au Domaine Foncier Rural (Abidjan, 1999).

[47] République de la Côte d'Ivoire, Ministère de l'Environnement et du Tourisme, Plan National d'Action Pour l'Environnement, Cellule de Coordination, Livre blanc de l'environnement de la Côte d'Ivoire (Abidjan: PNAE-CI, 1994), pp. 120-21; idem., Plan d'action environnemental de la Côte d'Ivoire: 1996-2010 (Abidjan: PNAE-CI, 1995), pp. 16, 18.

[48] T. Bassett, 'Mapping the terrain of tenure reform: the rural land holdings project of Côte d'Ivoire',

is also made by the World Bank in its general policy papers and country studies.[49]

The ground work for rural land titling began in 1989 when the Plan Foncier Rural (Rural Landholdings Project) began surveying and mapping village boundaries and agricultural lands in five pilot zones. Funded by the World Bank and the French Fonds d'Aide et de Coopération (FAC) and the Caisse Centrale de Coopération Economique (CCCE), the Plan Foncier used large-scale aerial photographs, geographical positioning systems technology, and household surveys to associate specific land parcels with land users and land 'owners'. The origin and nature of the land rights held by individual farmers was noted on survey forms. Project documents consisted of a land use/holding map of each village territory within the pilot zones and a land register containing the names of individual land users and land holders.[50] At the conclusion of the three-year pilot phase in 1992, the project was extended to other areas of the country.

From its inception, donors perceived the Plan Foncier as providing basic information to the government, developers, and local groups for the purposes of economic development. It was also seen as contributing to the decentralization of rural development by providing communities with tools for land-use planning and natural resource management. In this way, the Plan Foncier was and continues to be viewed as a foundation for village land management (*gestion des terroirs villageois*).[51] It will also play a role in the implementation of the 1998 rural land code. Its land registers will facilitate the formal land registration process and will be updated as land titles are issued and land ownership changes hands.[52]

What remains unclear is how land titling will address the two environmental degradation issues addressed in this study: bush and tree encroachment and wildlife decline. Too often, land reform is considered the panacea for a host of rural development and environmental problems that require quite different solutions.[53] We contend that this is true in the case of northern Côte d'Ivoire where environmental planners have promoted land reform to fix what we consider to be an imaginary environmental problem – savannaization. Put simply, the idea that land titling will fix the problem of savannaization has no logical basis. Similarly, the notion that land titling will reduce land-use conflicts between peasant farmers and Fulbe herders is based on a misunderstanding of the basis of this conflict – uncompensated crop damage. We argue that requiring Fulbe herders to compensate farmers

---

[48] (cont.) in J. Stone (ed.), *Maps and Africa* (Aberdeen: Aberdeen University African Studies Group, 1994), pp. 128–46.

[49] Falloux and Talbot, *Crisis*; World Bank, *Côte d'Ivoire: vers un développement durable*, Rapport No 13821-IVC (Abidjan: World Bank, 1994).

[50] For a more detailed and critical analysis of this project, see Bassett, 'Mapping'; J. P. Chauveau, P-M. Bosc and M. Pescay, 'Le plan foncier rural en Côte d'Ivoire', in P. Lavigne Delville (ed.), *Quelles politiques foncières pour l'Afrique rurale? Réconcilier pratiques, légitimité et légalité* (Paris: Karthala/ Coopération Française, 1998), pp. 553–82.

[51] This link became all the more apparent in June of 1997 when the Plan Foncier was integrated into the national land management and rural infrastructure project (*le projet national de gestion des terroirs et d'équipement rural* or PNGTER).

[52] Interview, Plan Foncier, Abidjan, 23 June 1998.

[53] J. Bruce, 'Do indigenous tenure systems constrain agricultural development?' in Bassett and Crummey, *Land in African Agrarian Systems*, pp. 35–56; H. W. O. Okoth-Ogendo, 'Some Issues of Theory in the Study of Tenure Relations in African Agriculture', *Africa*, 59, 1 (1989), pp. 6–17.

for damaged crops will be more effective in resolving this conflict than land titling.[54] Land titling could very likely increase land-use conflicts, because its logical result will be to restrict the mobility of Fulbe herds in 'privatized' areas.

Fulbe herds have relatively free access to rangelands under the existing conditions of non-registered land holdings. This flexibility is important to herd health and productivity, given the spatially and temporally shifting pattern of disease hazards and pasture quality.[55] As land titling spreads, herd mobility and thus health will be threatened. Inevitably, new, more sedentary forms of livestock raising will develop, based on arrangements between individual herders, farmers, and communities. New range management practices such as rotational grazing and late burning may evolve under these conditions. This restructuring of livestock raising along new socio-spatial arrangements is the objective of a PNAGER project focused on the north.[56] It also conforms to the World Bank's neo-liberal economic development policies in which land titling is considered the cornerstone to agricultural intensification and environmental conservation. The limited know-ledge of the dynamics of environmental change shown by land-use planners and reformers alike, does not inspire confidence that the new land laws will bring about the expected social, economic, and environmental benefits. What is certain is that the rural land law will create new political-administrative entities, rules, and regulations within rural communities that will modify existing institutions and practices of land allocation. These institutional innovations, linked to new governance structures at the local, regional and national scales, present new opportunities for individuals and groups to gain (and lose) access to and control over productive resources. The general question is how resource access, use and management patterns will change as a result of these interactions, as different actors interpret and modify these new rules and regulations.

Conclusion

The wave of land registration and titling programs currently passing through sub-Saharan Africa is commonly rationalized by its expected environmental and economic development benefits. What is striking about contemporary environ-mental discourses is the lack of connection between assumed processes and agents of environmental change and reality. There is a politics to this gap in environmental knowledge. Various actors (the state, urban elites, aid donors) have a stake in 'finding' environmental degradation and blaming rural land users as its unruly

---

[54] T. Bassett, 'The Political Ecology of Peasant-Herder Conflicts in Northern Ivory Coast', *Annals of the Association of American Geographers*, 78, 3 (1988), pp. 453–72; T. Bassett, 'Hired Herders and Herd Management in Fulani Pastoralism (Northern Côte d'Ivoire)', *Cahiers d'Études Africaines*, Nos. 133-135 (1994), pp. 147–73.

[55] Bassett, 'Fulani Herd Movements'.

[56] The *Programme National de Gestion de l'Espace Rural* (PNAGER) – *Operation Nord*, funded by the Caisse Française de Développement (CFD), aims to assist rural communities in integrating farming and livestock raising in a complementary manner through its Agro-Pastoral Units project. See C. Barrier and A. Hugon, *Programme Nationale de Gestion de l'Espace rural (PNAGER) – Operation Nord – Côte d'Ivoire: Rapport d'Evaluation* (Paris: Caisse Française de Développment, 1994).

agents. In this way, environmental crises provide the necessary pretext for outside intervention. In the last two decades of the twentieth century, this intervention has most often taken the form of structural adjustment programs in which highly indebted African states have been required to change their policies and restructure their economies according to neo-liberal economic principles. Whether or not environmental change is linked to land rights systems is not the point. The neo-liberal agenda takes as a basic tenet that all developing economies must embrace the principle of private property. Thus, the tragedy of the commons can be cited as the basis of wildlife decline even though that global environmental narrative fails to capture the complex political ecology of game depletion. The objective of state planners is to better manipulate people and appropriate resources for the purposes of taxation, accumulation, and other ends. This largely fiscal goal requires the reshaping of rural economies and spaces to form grids of control and extraction. These new spaces of conservation and development are structured according to liberal market institutions and practices. Local environmental knowledge and change dynamics are largely irrelevant to these global environmental managers. One of the objectives of this chapter is to point to the discrepancies between totalizing discourses and local understandings and interpretations of environmental change. By doing so, we seek to demonstrate that there is a politics to representing the African environment that reveals more about the interests of a myriad of stakeholders than it does about the actual dynamics and direction of environmental change.

Appendix 3.1 Common and Scientific Names of Games Species in the Kakoli region

Aardvark (*Orycteropus afer*)
African civet (*Viverra civetta*)
African hare (*Lepus whytei*)
Black duiker (*Cephalophus niger*)
Black-striped duiker (*Cephalophus dorsalis*)
Bohor reedbuck (*Redunca redunca*)
Bushbuck (*Tragelaphus scriptus*)
Cane rat (*Thryonomys swinderianus*)
Genet (*Genetta genetta*)
Gambian rat (*Cricetopmys gambianus*)
Grivet monkey (*Cercopithecus aethiops*)
Guineafowl (*Numida melagris*)
Jackal (*Canis sp.*)
Oribi (*Ourebia ourebi*)
Patas monkey (*Erythrocebus patas*)
Red-flanked duiker (*Cephalophus rufilatus*)
Serval (*Felis serval*)
Porcupine (*Hystrix cristata*)
Warthog (*Phacochoerus aethiopicus*)

4

Investing in Soil Quality
Farmer Responses to Land Scarcity
in Southwestern Burkina Faso

LESLIE C. GRAY

Introduction

With increasing commercialization, population growth and land scarcity, the nature of land relations is changing in southwestern Burkina Faso. Local farmers, facing massive migration into the region and the disappearance of their once abundant land resource, no longer willingly leave fields in fallow or give fallow fields to strangers. Migrant farmers, both newly arrived and long settled, have greater difficulty gaining access to land, and when they do, find it harder to keep. Demographic and socio-economic transformations have resulted in land scarcity and in uncertainty about the rules and land rights of different groups.

Government and non-governmental programs work under the assumption that this uncertainty about land rights exacerbates the problem of land degradation.[1] This perception has resulted in policies that attempt to alter local land rights. New statutory reforms allow farmers to register land.[2] Local village projects implemented under the *Gestion des Terroirs* banner seek to redefine land holding relations by means of zoning and land reallocation. These policies have their roots in the belief that indigenous tenure systems exacerbate the problem of land degradation and constrain agricultural development because they do not encourage farmers to invest in soil quality.[3]

[1] PNGTV (*Programme National de Gestion des Terroirs Villageois*), Cellule de Coordination, 'Rapport de Synthèse et d'Analyse des Expériences Pilotes de Gestion des Terroirs Villageois', 1989; C. Toulmin, *Gestion de Terroir Principles, First Lessons and Implications for Action*, UNSO discussion paper (London: Drylands Programme, International Institute for Environment and Development, 1993).

[2] The RAF (*Réorganisation agraire et foncière*), an initiative put forward by the revolutionary regime of Thomas Sankara in 1984, decreed that all land is 'owned' by the state. The RAF of 1991 allowed farmers to register farm land as private property. This has not happened. While statutory reforms have in practice had little effect on how land is allocated, they have nonetheless given the appearance that the government might act and created concern at the local level.

[3] G. Feder and R. Noronha, 'Land Rights and Agricultural Development in Sub-Saharan Africa', *World Bank Research Observer*, 2, 2 (1987), pp. 143–9.

Figure 4.1 Map of study area in Burkina Faso

While much empirical research now questions this relationship between land tenure and agricultural investment, the narrative that the individualization of land rights leads to more investment in land quality and improved land management practices permeates local and national policy.[4]

The narrative that uncertainty about land rights has led to widespread land degradation appears to be misplaced in southwestern Burkina Faso.[5] Indeed, this chapter argues quite the opposite: many farmers are responding to both land scarcity and what they perceive as uncertain rights by investing in agricultural practices that promote soil regeneration. They apply manure to their fields, construct anti-erosion structures, and implement agroforestry techniques in an attempt to strengthen rights to land. By investing in soil quality, farmers are able to farm land for longer periods, thus increasing and creating land rights in an arena of

---

[4] D. Atwood, 'Land Registration in Africa: The Impact on Agricultural Production,' *World Development*, 18, 5 (1990), pp. 659–71; T. Bassett, 'Introduction: The land question and agricultural transformation in sub-Saharan Africa,' in T. Bassett and D. Crummey (eds), *Land in African Agrarian Systems* (Wisconsin: University of Wisconsin Press, 1993); and S. Migot-Adholla, P. Hazell, B. Blarel, and F. Place, 'Indigenous Land Rights Systems in Sub-Saharan Africa: A Constraint on Productivity?', *The World Bank Economic Review*, 5, 1 (1991), pp. 155–75. For a discussion of development narratives, see E. Roe, '"Development Narratives" or Making the Best of Blueprint Development,' *World Development*, 19, 4 (1991), pp. 287–300.

[5] A.-S. Braselle, F. Gaspart, and J.-P. Platteau, *Land Tenure Security and Investment Incentives: Further Puzzling Evidence from Burkina Faso*, Faculté des Sciences économiques, sociales et gestion, 1998; R. Ouedraogo, J.-P. Sawadogo, V. Stam and T. Thiombiano, 'Tenure, Agricultural Practices and Land Productivity in Burkina Faso: Some Recent Empirical Results', *Land Use Policy*, 13, 3 (1996), pp. 229–32.

growing land scarcity and uncertainty. This chapter, therefore, challenges the oft-propounded idea that investment in soil quality depends on secure land rights. Instead, land rights and investment develop simultaneously; farmers invest in soil quality as part of active strategies to secure control of land.

Investment in soil quality, however, is not a strategy that is evenly distributed. Farmer decision-making depends not only on issues such as access to resources, in particular wealth, land and labor, but also on social standing within communities. These influence the types of choices farmers make as well as the amount of different resources they can allocate to different strategies. For example, wealthier farmers who invest in livestock are more likely to apply manure to fields. Poorer farmers may use less manure but they also use less animal traction which can degrade soil bio-physical status and result in lower tree densities on fields. These types of choices farmers make have different implications for soil quality.

We also see inter-village and inter-ethnic differences in investment choices. The village where land is in most scarce supply is also the one where farmers are making the largest investments at the same time as they are asserting more and more individual control over land. They use inputs, leave trees on their fields and build conservation structures at higher rates than farmers in other villages, thus intensifying their production systems. Second, farmers with the least secure access to land – particularly Mossi migrants who are both at risk of being asked to leave land they are cultivating and have difficulty maintaining control over fallow land – make the most investments in land, creating and increasing their rights over land in the process. Investment is used as a political strategy by migrant and local farmers to create rights in land to which they have few formal rights.

This chapter examines land scarcity, changing land rights and the land investment strategies of farmers in three villages in the province of Tui, one of the primary cotton-growing regions in southwestern Burkina Faso. All three villages – Sara, Dimikuy and Dohoun – are inhabited by both Bwa and Mossi farmers. The Bwa are considered the ethnic group indigenous to the area,[6] although in sheer population numbers Mossi migrants now dominate local Bwa in much of the area. In the three villages, 106 Bwa and Mossi farmers were interviewed about agricultural practices, access to land and perceptions about changes in land use. The sample was stratified by ethnicity; half of the households selected were from Mossi migrant communities and half from the local Bwa community.

The next section will introduce the study area, describing how local environmental conditions as well as external changes have created the conditions of land scarcity. The third section links changing local land rights to soil quality investment strategies. The final section draws some conclusions.

Population, Agriculture and Landscape Changes in Sara, Dimikuy and Dohoun

Southwestern Burkina Faso is an area of fast-paced demographic and technological change. In the province of Tui, population has doubled in the span of a decade,

---

[6] Little is known about the dates of establishment of Bwa settlements in this region; estimates range from the tenth to the fifteenth century: see J. Capron, *Communautés Villageoises Bwa: Mali, Haute Volta* (Paris: Institut d'Ethnologie, 1973).

mostly due to the migration of Mossi farmers from the more drought-affected and heavily populated northern regions of the country. At the same time, cotton and maize have emerged as the most important agricultural commodities and have been accompanied by the adoption of the plow. These changes are the basis of an economic boom that has made this region among the wealthiest in Burkina Faso.

An important part of the agricultural success story of southwestern Burkina Faso, generally thought of as the breadbasket of the country, is its favorable environment. The region is characterized both by high rainfall and relatively fertile soils. It lies in what is considered to be a transition zone between the northern fringes of the Guinea and the southern fringes of the Sudan Savanna zones, with a yearly average rainfall of 900mm distributed over a rainy season beginning in May and ending in October.[7]

The high level of fertility of the local soils is related to the local geology. Two rock formations known as the *Birrimien* complex run from northeast to southwest in the study area. The first formation, the *Schistes Birrimiens*, is made up of sedimentary rocks that have been metamorphosed over time. The bulk of soils are characterized as *sols ferrugineux* in the French taxonomic system. These highly weathered soils of moderate to low fertility and low base status correspond to Alfisols in the US Department of Agriculture soil taxonomic system. The second formation, *Roches Andestiques,* is volcanic in origin. Soils that develop from this rock material, *sols bruns eutrophes* in the French taxonomic system, are richer in basic cations and therefore more fertile. The soils roughly correspond to Vertisols in the USDA taxonomic system. Most of the soils found in the three-village study area are *sols ferrugineux*, with some *sols bruns eutrophes* near the village of Dohoun.[8]

The dominant tree species include *Butyrospermum parkii* (karité), *Parkia biglobosa* (néré), *Pterocarpus erinaceua*, *Terminalia avicennoides*, *Deterium microcarpum* and *Ximenia americana*. The primary species of grass found are *Andropogon gayanus, Andropogon pseudapricus, Pennisetum purpureum* and *Pennisetum pedicellatum*.[9]

The local agricultural landscape is striking. Cultivated fields are dotted with karité and néré trees. Many villages are ringed by fields with *Acacia albida* trees. In some villages, terraces or terrace remnants remind us of an earlier time when agriculture was practised in a very intensive manner despite low population densities. Tersiguel and Capron both suggest that these intensive near fields are artifacts of a time when villagers cultivated close to their homes to protect the inhabitants against the incursions of other groups.[10]

The local Bwa have garnered the reputation in the Francophone literature of Burkina Faso as very good farmers. Savonnet helped create this perception, by

---

[7] J. M. Kowal and A. H. Kassam, *Agricultural Ecology of Savanna: A Study of West Africa* (Oxford: Clarendon Press, 1978).

[8] Philippe Morant, 'Caractérisation de la Fragilité Ecologique et des Potentialités Agronomiques de la Région de Hounde au Burkina Faso: Utilisation de Différentes Technique de Diagnostic', Thesis, L'Institut National Polytechnique de Lorraine, 1991.

[9] *Ibid.*

[10] Capron, *Communautés Villageoises Bwa*; and P. Tersiguel, 'Boho-Kari, Village Bwa: Les Effets de la Mécanisation dans l'aire Cotonnière du Burkina Faso', Thèse de Doctorat de Géographie, Université de Paris X, 1992.

describing their agricultural system as 'perfection',[11] because of their complex near field/far field system that combined an array of intensive and extensive farming techniques. Farms were laid out in concentric circles, with each successive aureole planted less intensively. Fields closest to villages were generally scattered with *Acacia albida* trees under whose canopies crops such as millet, peppers, and tobacco were planted. Often these near fields were accompanied by constructed terraces and stone lines on sloped lands. The fertility of fields closest to the village was enhanced with inputs of manure and other forms of composted organic material. Moving out from the village, one found a progressively more extensive agricultural system. In the fields farthest from the villages farmers practised shifting cultivation, planting millet and sorghum on land until fertility and weeds became a problem and then returning it to fallow. The near field/far field system that Savonnet describes, however, is not unique to the Bwa. Similar production systems are found among other groups in western Burkina Faso. Prudencio details a similar system of intensive and extensive production in Mossi areas.[12]

Today, we find in the cotton-growing region that some Bwa farmers continue to farm in this manner. Some villages have maintained the near/far field system, with terraces, stone lines, and intensively manured fields near villages and extensively farmed fields farther away. The vast number of villages, however, have abandoned their intensive near-village fertility management practices, ceasing the near field/far field systems and expanding their territory under cultivation, concentrating on maize and cotton cultivation.[13]

Both local Bwa farmers and Mossi migrants have enthusiastically adopted cotton production and animal traction. In the early 1970s few farmers were using animal traction, but this trend changed in the early 1980s. Now, 60.4 per cent of the farmers in the study sample have two or more oxen and 65 per cent have plows. This expansion of cotton and maize production has affected the agricultural system in several ways. First, many farmers are concentrating the bulk of their effort on fields farther from villages, because they are larger and provide sufficient space for ox-plowing. Second, as farmers gain wealth through cotton and maize production, they are moving out of the closely settled inner areas of villages, preferring to build new houses in the more spacious outer limits. Both of these trends have contributed to the abandonment of intensively farmed near fields.

Population growth has been instrumental in changing landscapes in the region. Internal migration of Mossi migrants has been the primary cause of population growth in southwestern Burkina Faso. The droughts of the 1970s and 1980s saw the beginning of a massive population shift within the country, as many Mossi migrated to the less drought-prone and more land-abundant areas in the south and southwest.[14] Table 4.1 illustrates these population changes for the villages of Sara,

[11] G. Savonnet, 'Un Système de Culture Perfectionné, Pratiqué par les Bwabas: Bobo-Oulé de la Région de Houndé (Haute-Volta)', *Bulletin de l'IFAN,* XXI, sér. B, 3-4 (1959), pp. 425–58.

[12] C. Prudencio, 'Ring Management of Soils and Crops in the West African Semi-arid Tropics: The Case of the Mossi Farming Season in Burkina Faso', *Agriculture, Ecosystems and Environment,* 47 (1993), pp. 237–64.

[13] Morant, 'Caractérisation de la Fragilité Ecologique et des Potentialités Agronomiques'; Tersiguel, 'Boho-Kari, Village Bwa.'

[14] In the village of Bare described by Şaul (1993), many migrant farmers are wealthier outsiders, often

Table 4.1   Population change in Sara, Dimikuy and Dohoun

Villages	1995	1985	1975	Annual growth rate 1975–85
Dimikuy	916	812	358	8.1%
Sara/Basse	n/a	5608	2447	8.3%
Dohoun	n/a	2342	1220	6.5%

Dimikuy and Dohoun where population more than doubled from 1975 to 1985. The very high population growth rates, furthermore, could only be possible through in-migration.

Inhabitants of all three villages sense that migration has slowed in recent years. Bwa villagers in Dimikuy, who control formal political power and the land, state that they no longer allow Mossi migrants to settle. Indeed, population figures for Dimikuy confirm this reduction in settlement. Likewise, in Sara, there is no more land available for new settlers. In Dohoun, however, migrants are still allowed to settle, though they are frequently given land of poor quality. Another way of examining these differences is to look at the average length of residence of migrants. In Sara the average length of migrant residence is 26.3 years, while migrants in Dimikuy and Dohoun have an average length of residence of 15.8 and 11.7 years respectively. These numbers confirm perceptions that migrant communities in Sara are longer established than those in either Dimikuy or Dohoun.

These technological and demographic transformations have resulted in an extensification of the land under production and now, under conditions of land scarcity, intensification. Aerial photographs illustrate the extent to which landscapes have been transformed in two of the three study villages.[15] Photographs obtained for Sara and Dimikuy, taken at different scales, 1:50,000 in 1981 and 1:20,000 in 1993, indicate many changes in surface area cultivated from 1981 to 1993. Table 4.2 shows the changes in territory under different cultivated and uncultivated states from 1981 to 1993, describing the amount of land area (in hectares) under each type of land-use system. In Dimikuy, where population more than doubled from 1975 to 1985, the area under cultivation has increased from 414 to 901 hectares in 12 years, mirroring the magnitude of population change. Furthermore, the amount of land under bush savanna has decreased by 50 per cent. The amount of land classified as degraded, which essentially indicates a color change and bare soil, has increased, however, more than fivefold, from 18 to 122 hectares.

Sara saw very little change in fields under cultivation from 1981 to 1993. This

---

[14] (cont) retired or still active civil servants, who live in Bobo-Dioulasso and maintain a farm in the country. This situation did not exist in the study area which was relatively far from a main city center (the closest large city was Bobo-Dioulasso which was 70km. from Sara and almost 100 from Dohoun); most migrant farmers were from the drought-affected region in the north of the country: M. Şaul, 'Land custom in Bare: agnatic corporation and rural capitalism in Western Burkina', in Bassett and Crummey, *Land in African Agrarian Systems*, pp. 75–100.

[15] L.C. Gray, 'Is Land Being Degraded: A Multi-Scale Perspective on Landscape Change in Southwestern Burkina Faso', *Land Degradation and Development*, 10, 4 (1999), pp. 327–34.

Table 4.2 Hectares under land type, 1981 and 1993

	Dimikuy		Sara	
	1981	1993	1981	1993
Fields	414	901.86	571.5	594
Fallow	109	105.72	176	146.4
Tree Savanna	160	201.34	1215.5	1182.6
Bush Savanna	1207	599.48	330	424
Wooded Savanna	0	0	163	9.6
Degraded Soil	18	122.2	60	137.2
Bare Soil	0	1.2	0	46.4
Total	1908	1931.8	2516	2543

reflects the fact that the land area in Sara is for the most part saturated. Much of the area under tree savanna is uncultivable, located on hills with hardpans and stone surfaces. There has been a decrease in area under wooded savanna of 94 per cent during the period. This might reflect the fact that a whole new area, which had not been cultivated since the 1940s, has recently been deforested.

One of the results of increases in land area under cultivation and lessening of available land is shortened fallow periods. In interviews, many older farmers indicated that only 30 years ago they rarely cultivated a field for more than 5–7 years. After cultivation, they would leave a field fallow for periods ranging from 15 to 30 years before coming back to it. This natural regeneration through fallowing is no longer possible. In all three villages, there are areas of almost permanent cultivation. Around the village of Sara, most fields have been cultivated since the 1970s, some even since the 1950s. While farmers do leave fields in fallow, they are rarely able to leave them for more than 10–15 years. This situation is reflected in Table 4.3 which illustrates the average number of years fields have been cultivated for the three villages. Fields in Sara have, on average, been cultivated for longer periods than in Dimikuy or Dohoun.

Table 4.3 Average age of field in cultivation by village and ethnic group (years)

	Sara	Dimikuy	Dohoun
Bwa	14.8	12.0	11.3
Migrant	16.7	9.0	11.4

Changing Land Rights and Soil Quality in Southwestern Burkina Faso

Reduction in available farm land has created greater uncertainty regarding land rights for both Bwa and Mossi farmers. This section will describe how local land rights are conceived, the ways in which land scarcity is leading to changes in rights, and the strategies that farmers are using to secure land rights.

Land Rights of Bwa and Mossi Farmers. Most Bwa farmers gain access to land through membership in corporate groups,[16] although many Bwa borrow land as well. The amount of land that each corporate group controls has been determined to a great extent by both the historical sequence in which groups settled and the political influence they have achieved.[17] The families that settled a village early gained control over the most land. However, in many villages late-arriving families also control large areas of land. Villages, whose political strength was derived mostly from the number of inhabitants, often invited outside groups or villages in, by offering them land and protection.

The framework described by Boutillier for Burkina Faso accurately describes the land rights of Bwa farmers.[18] These rights generally include rights to permanent usage of land belonging to the larger group and of land that has never been claimed by anyone. Bwa also widely borrow and lend land to members of other descent groups[19] as well as to strangers. The rights of a Bwa farmer to borrowed land essentially become permanent over time, depending on the length of cultivation. It would be very difficult for a lender to try to reclaim borrowed land that has been passed on through generations.

We see this diversity of land rights in Sara, Dimikuy and Dohoun. Bwa farmers indicated that they obtained 43 per cent of the fields surveyed from their descent group. A substantial fraction of Bwa farmers borrow fields from outside their group; farmers in the sample indicated that 46 per cent of their fields came from land belonging to other descent groups. The incidence of 'borrowing' land between members of the same groups is low; only 4 per cent of fields were borrowed in this way. As mentioned above, borrowed land can often have considerable security of tenure associated with it. For example, in Sara, about 15 per cent of fields cultivated by Bwa farmers are on non-family land that has been passed on through generations. The borrower thus develops permanent control over land over time, and leaves it to his children.

[16] The Bwa are often categorized among the so-called 'stateless' societies of West Africa. Savonnet-Guyot describes Bwa social organization as *l'espace politique villageois*, where the main political entity is the village, with what Capron calls the 'house' at the center: Claudette Savonnet-Guyot, *État et Sociétés du Burkina: Essai sur le politique Africain* (Paris: Éditions Karthala, 1986); Capron, *Communautés Villageoises Bwa*. The house is a large unit organized along descent and residence lines, based on patrilineal filiation and on settlement. Men generally enter through birth and women through marriage. The house has historically been the locus of production, around which labor and other relations are organized. The size of a house is variable; when one becomes too large, it generally splits off into another autonomous unit. While the logic or organization of houses is along descent lines, they are far more flexible than kinship structures. People can join and leave houses; their formation and dissolution are based on negotiation and circumstance.

[17] Peter Matlon, 'Indigenous land use systems and investments in soil fertility in Burkina Faso', in John Bruce and Shem Migot-Adholla (eds), *Searching for Land Tenure Security in Africa* (Dubuque, IA: Kendall Hunt, 1994).

[18] J. L. Boutillier, 'Les Structures Foncières en Haute-Volta', *Études Voltaïques*, Mémoire 5 (Ouagadougou: Centre IFAN-ORSTOM, 1964).

[19] The term 'descent group' is used here to refer to the 'house' structure, based on both residence and descent, that Capron describes: Capron, *Communautés Villageoises Bwa*.

For Mossi migrants, rights to land are very different. Mossi migrant farmers, by definition, have no land of their own. They are strangers; all the land they obtain is borrowed. A Mossi can generally obtain land from his Mossi host or directly from a Bwa. Many of the Mossi borrow land from their local Mossi sponsor (or *tuteur* in French) who has 'presented' them to Bwa farmers and allocated them land.

Of the 81 fields of farmers in the sample, 57 per cent were borrowed directly from Bwa, while 40 per cent were borrowed from the farmers' Mossi hosts. On only 2 per cent of fields did farmers indicate that they had inherited the fields. The exception, however, is in Sara, where farmers claim that about 5 per cent of their fields have been inherited. This is surprising, because Mossi strangers should theoretically have no permanent claim to land. Generally, though, it was only farmers who had resided in the village for several generations and had in several cases intermarried with Bwa who claimed that they had inherited their land.

Within the framework of borrowing, farmers can have more or less secure land-tenure status. One determinant of this is how the field was obtained. Land that is obtained directly from a Bwa farmer is generally more secure than land that a Mossi obtains from another Mossi. In several villages this third-party borrowing between Mossi is considered to be inappropriate; one Bwa farmer told me that if a Mossi lent land to another Mossi, he would expel them from the land.

Rights of Mossi migrants to borrowed land strengthen over time. Basically the longer a farmer is able to stay on a field, the harder it is for someone to take it away from him when he tries to leave it fallow. Mossi farmers in all three study villages indicated that, if a field were farmed for a long period of time, then it would be difficult, if not impossible, for someone to take it away from them. These permanent use rights are increasingly recognized by local government officials; in cases where Bwa farmers have attempted to evict Mossi farmers from fields they have occupied over time, the government has generally intervened in favor of the migrants.

Certain types of agricultural investments lead to more permanent control over land. One way of securing almost permanent access or ownership over land is by planting trees. Laurent and Mathieu describe how 'the investment of work for long-term future benefits is seen as the symbol and the sufficient condition to give long-term rights to the area to be planted'.[20] For that reason, migrants are expressly prohibited from planting trees. While there is a strict norm against Mossi migrants planting trees, in several instances when farmers are long-term residents of a village they have planted trees and no one has challenged their right to do so. On the other hand, when newer migrants attempt this same behavior they are threatened with eviction. A recent large land conflict in Sara resulted in Mossi migrants being evicted from fields in an area where newer migrants were planting trees in an attempt to exert more permanent control over land.

The implication of land scarcity for soil quality. Farmers overwhelmingly perceive land scarcity as a problem. There are fewer forested areas now than in the past, owing to

---

[20] P. J. Laurent and P. Mathieu, 'Authority and conflict in the management of natural resources: a story about trees and migrants in Southern Burkina Faso', in: *Proceedings of the Seminar, 'Improved Natural Resources Management. The role of the state versus that of the local community'* (Roskilde University, 1993), p. 6.

increases in agricultural fields. In interviews, Bwa farmers tended to blame this situation on Mossi migrants. One Bwa farmer in Dimikuy asserted that 'many trees have been cut down since we have been invaded by the Mossi. Before it was easy to find wood for housing but now the Mossi have cut it all. Before when you cultivated a field you would leave it when it was tired, but this is no longer possible.'

Mossi farmers contended that forested areas have declined because of more people but also because there has been less rain. The ultimate effect of this situation is that farmers cultivate their fields for longer periods. Tamini Antoine, a Bwa farmer from Sara, explained how:

> there are a lot of people so we can't leave land in fallow. In the past, a person would leave a field fallow when the yields declined or if there were too many weeds. Before there were few people and lots of places. You could farm for 3-5 years and you would leave. Now you stay until the field is totally tired and then you move.

With land scarcity, fallowing fields, the primary mechanism for soil regeneration, has become increasingly difficult. Of immediate concern to most farmers is that they retain control over land if they decide to leave it fallow. The ability to do this depends on many things. Within Bwa and Mossi communities, how and from whom a field was obtained is crucial in determining the extent to which an individual can leave a field fallow and bring it back into cultivation again. Land that is cultivated by a farmer and belongs to his descent group is generally more easily left fallow and cultivated again than land that is borrowed.

Now, however, this situation is beginning to change. More and more, Bwa farmers complain that it is harder to leave their fields fallow. Because land is a gift from the ancestors, refusing land to people, especially close relatives, is viewed as inauspicious. In the past, many Bwa held this attitude towards Mossi farmers as well, although this has now changed. It still prevails, however, towards family and fellow villagers. It is almost impossible for Bwa to refuse fallow land to a relative. One Bwa farmer in Dohoun, Kahoun Besse, stated that he could leave a field fallow for about ten years, refusing requests from kin by declaring his intention to retake the field at some point in the future. After about ten years, the pressure from relatives would be too great.

Land scarcity has implications for land rights, especially for Bwa farmers who are exerting more and more individual control over land. In Sara, land scarcity has led to a situation where a large number of fields have become permanent or semi-permanent. Land is still controlled by descent groups, but since fields are rarely left fallow, they do not go back to the corporate pool. Bwa farmers are afraid to leave them in fallow, because then members of their families who have rights to the land will come and ask them for it. Instead, they will frequently not leave an entire field in fallow, but instead rotate plots on their fields. While Bwa in Sara still assert that it is impossible to 'own' land, that it belongs to the corporate group in trust for their descendants, fields are being passed on from father to son. Scarcity of land has led to a trend towards keeping land within families and has increased individual control over land. Many Bwa farmers in Sara indicated that they would not leave land in fallow at all, because relatives would ask them for its use. One farmer explained to me how he had a field that had been fallow for two years. Five people so far had asked to borrow the field – three Mossi and two Bwa. He said that it had been easy to refuse the Mossi but not the Bwa. He will probably cultivate the field again, earlier than he would like, to prevent having to hand it over to a relative or friend.

In all three villages, Mossi migrants fear putting land in fallow because someone might try to take it away from them. Practically all Mossi farmers in Dohoun declared that they would never leave their fields entirely in fallow, for fear that, if they did, their Bwa hosts would confiscate their land holdings. In the past, however, this was not the case. A migrant farmer could generally choose the parcel of land he wanted to cultivate and leave it in fallow, safe in the knowledge that no one would take it away from him. Korbeogo Raogo, a Mossi farmer from Sara, explained to me how, when the first migrants came, land was abundant and given freely. He indicated that in the past they could cultivate anywhere they wanted and clear as much land as they wanted. Now he related how the situation appears to have changed:

> Before, a Bwaba gave you a forest to cut and you could cultivate it all your life. If another migrant came, you gave him a piece of your land and later he would go ask the Bwa for land. But that doesn't happen much any more because there isn't much land left. Even the Bwa themselves don't have enough land.

The strategy of not allowing migrants to put fields into fallow is now openly used to ensure that they will not settle definitively in the area. One farmer acknowledged that 'if a Mossi's land isn't fertile, they can't give it to others and they can't leave it in fallow so they have to leave. This is how we make them leave.' Another Bwa farmer, Kahoun Zibita, confirmed this analysis remarking:

> A Mossi can't leave a field in fallow. Usually we give Mossi land that is of poor quality to prevent them from staying for a long time. If we give them good land, then they would stay for a long time. If we give them bad land, they don't have the means to buy fertilizer and so they'll leave.

Mossi farmers react to these restrictions against leaving their fields in fallow in several ways. Most commonly, a Mossi farmer will never leave his entire field in fallow. Like many Bwa, they will rotate plots on their larger fields, leaving a plot in fallow while continuing to cultivate other parts of the larger field. For example, it is common for a farmer to have a field of ten hectares, but to cultivate only three or four hectares, leaving the rest of the field in fallow.

The general response to land pressure and a reduced ability to fallow fields by both Mossi and Bwa farmers is intensification. Many Mossi farmers' fallows are improved fallows. Several farmers in the study applied manure to fallow fields a year or two before they cultivated them, enhancing the fertility of the soil in general as well as increasing the biomass of the shrubs and grasses present in the fallow field. The manure that is applied to both fallow and cultivated fields is an intensive strategy to replace fallow. The next sub-section will describe some of the strategies farmers use in the face of increasing land scarcity and uncertainty about land rights.

Investment in soil quality. We saw above how granting land of poor soil quality to Mossi migrants is an active strategy used by many Bwa to deny migrants control over land. In a situation where local custom prevents Bwa farmers from denying unused land to Mossi migrants, soil quality has become a key political strategy for denying permanent access to land.

Photo 4.1 Women harvesting cotton

(Photographs 4.1–4.5 by Leslie C. Gray).

Photo 4.2 Terraced field with young *Acacia albida* trees.

Photo 4.3 Preparing a field with an ox–drawn plow.

Photo 4.4 Market women selling agricultural products.

Photo 4.5 Maize interspersed with karité trees.

This chapter will argue conversely that in a situation where farmers cannot claim individual permanent control over land, they use investments in soil quality to create rights in land. This is a strategy used by both Mossi and Bwa farmers under conditions of increasing uncertainty and competition over land. Land shortages and the inability to leave cultivated fields fallow for long periods of time have led farmers to intensify their agricultural production. By improving soil quality, they increase the length of time they can farm a field. The longer a farmer cultivates a field, the harder it is for him to be asked to leave it, and the easier it becomes for him to put it down to fallow and then reclaim it. Migrants in Sara who had cultivated parcels of land for long periods of time openly stated that no one would try to take the land away from them if they left it fallow. It is widely recognized that claims to land are strengthened through physical occupation.

Not all farmers, however, are able to make the types of investment necessary to cultivate land indefinitely. There are differences among the three villages, between members of different ethnic groups, and within ethnic groups, depending on access to resources and social standing within local communities.

The soil quality investment practices examined are manure use, fertilizer use, anti-erosion structures and the practice of leaving trees in agricultural fields. Most of these practices are recognized by both ethnic groups as beneficial to soil quality. None of these practices are limited to either Mossi or Bwa systems. Despite the widespread notion that Bwa farmers have better agricultural practices, the knowledge of farming practices that Mossi migrants bring with them is similar to that of local Bwa farmers. Prudencio describes soil and crop management systems in the semi-arid home regions of many of the Mossi migrants that are surprisingly similar to the near/far field system of local Bwa farmers. [21]

Table 4.4 Correlations among socio-economic variables and manure use, fertilizer use, and tree density

	Man/ha	Fert/ha	Trees/ha	Animal value	# Ag. implements	Educa-tion	% Area plowed
Man/ha.	1.00						
Fert/ha.	.13	1.00					
Trees/ha.	.02	.09	1.00				
Animal value	.26***	.10	−.09	1.00			
# Ag. implem	.22**	.18**	−.09	.68***	1.00		
Education	.17*	.27***	.12	.18**	.34***	1.00	
% Area plowed	.09	.30***	−.16*	.35***	.36***	.36***	1.00

Note: *significant at LE .05, **significant at LE .01, ***significant at LE .001

Table 4.4 presents the results of a correlation matrix examining the relationship between manure and fertilizer use, tree density and socio-economic status. Animal value and number of agricultural implements (carts, plows, and other animal implements) represent accumulated wealth. Manure use per hectare in general is highly related to ownership of animals, number of agricultural implements and education. Fertilizer use per hectare is related to education and to area plowed. The number of trees per hectare is negatively related to area plowed. The more intensively plowed a field, the fewer the trees. In summary, farmer wealth is highly correlated with manure and fertilizer use, while tree densities are negatively correlated with plowing.

Table 4.5 Percentage of fields where manure was applied

	Sara	Dimikuy	Dohoun	Total
Migrant	.34 (n=41)	.42 (n=31)	.29 (n=35)	.35 (n=107)***
Bwa	.20 (n=40)	.12 (n=24)	0 (n=18)	.12 (n=82)
Total	.27 (n=81)	.27 (n=55)	.19 (n=53)	

Note: *** Significant at p<.001

Table 4.5 breaks down manure use by village and ethnicity, revealing that approximately 27 per cent of fields in both Sara and Dimikuy are manured, while only 19 per cent of fields in Dohoun have manure applied to them. Overall,

---

[21] Prudencio, 'Ring Management of Soils and Crops'.

throughout the sample only 12 per cent of Bwa use manure, while 34 per cent of migrant farmers do. An ANOVA analysis indicates these differences between Mossi and Bwa farmers to be statistically significant at a p<.001 level. However, when Bwa use manure they tend to use more than do migrant farmers. Table 4.6 shows the average number of sacks of manure applied to fields by village and by ethnic group. These figures indicate that more migrant farmers apply manure in Dimikuy and Dohoun than do Bwa, although Bwa in Sara use more manure than do migrants.

Table 4.6 Average manure application (in sacks/ha) by village and group for farmers applying manure to their fields

	Sara	Dimikuy	Dohoun	Total
Migrant	11.35	16.38	6.8	12.48★★
Bwa	24.49	38.93	0 27	.38
Total	16.35	19.85	6.8	

Note: ★★ Significant at p<.01

Fertilizer is generally available to village cotton-growing co-operatives *(Groupements Villageois)*. All farmers, unless they have incurred debts that they have not repaid, are eligible to borrow inputs and obtain cotton seeds from their co-operatives. Many growers have incurred debts and can no longer grow cotton. As a consequence, wealthier farmers are more heavily involved in cotton production than poorer farmers. Many Mossi migrants are also heavily involved in cotton production, but this is very much related to the time they have spent in the area. Newer migrants generally do not grow cotton, concentrating instead on millet and sorghum production. This is partly due to experience; newer migrants do not have expertise in cotton production, a risky endeavor, and are hesitant to adopt it. But it is also due to attempts by local Bwa in Dohoun to exclude migrants from the local village co-operative.

Table 4.7 Average fertilizer application (in sacks/ha) by village and group

	Sara	Dimikuy	Dohoun
Migrant	2.59	1.16	.67
Bwa	2.67	1.78	2.29

This is reflected in data regarding fertilizer use. Table 4.7 indicates the average number of sacks per hectare for the different villages and for the migrant and **Bwa** communities. The largest application of fertilizer appears to be in Sara, **where** there are no differences between the migrant and Bwa communities. In Dimikuy, migrants tend to use less fertilizer than do Bwa, but the difference is not that great. In Dohoun the difference in fertilizer application by migrant and Bwa is significant. Migrants in Dohoun have arrived more recently than in the other two villages, tend to be poorer and do not grow cotton.

Farmers in the sample had many strategies for stopping erosion on their farm fields. Some brought in stones and built stone lines along the contours of hills; others built bunds or dug canals alongside their fields to direct water away from them. Yet others planted bushes along water courses to slow down the rate of water flow. The decision to construct a bund or stone line was often related to perceived erosion, which in turn was related to whether a farm was located on a sloped field.

Table 4.8 Number of fields on which farmers made anti-erosion improvements

| | Sara*** | | Dimikuy | | Dohoun | |
	Migrant	Bwa	Migrant	Bwa	Migrant	Bwa
No	30	39	29	22	26	10
Yes	10	1	1	2	9	6

Note: *** significant at p<.01

We see the same trend for conservation structures that we saw for manure. Only 11 per cent of Bwa used anti-erosion structures on their fields, while 19 per cent of migrant farmers did. There are, however, differences by village. Table 4.8 presents the data on whether farmers make anti-erosion improvements to their fields. We can see that farmers in Dimikuy in general do not. This can be explained by the fact that Dimikuy is overwhelmingly flat. In Sara and Dohoun, which both have an abundance of sloped land, many farmers construct bunds or stone lines. In both villages, migrants are more likely to do this than are Bwa; a chi-square analysis indicates that the only significant differences between migrants and Bwa are in Sara. It appears that migrant farmers are using strategies more often than Bwa farmers to combat erosion. This could be because they have land of poorer quality, but it is more likely that they do it because they have no option to abandon a field that has erosion problems or is no longer productive. Many Bwa farmers in Sara indicated that erosion was a problem on their fields, but that they had done nothing to control it.

Finally, we examined the presence of trees on farmers' fields. These are tree species that are not deliberately planted by farmers, but are left after a field has been cleared. The number is related to the number of trees left on a plot before it is made into a field. The dominant species include *Vitellaria paradoxa* (*karité*), *Parkia biglobosa* (*néré*) and *Acacia albida*. Farmers leave these trees in their fields for their economic and ecological value. They harvest the fruits of the karité and néré trees to use in local cooking or to sell in the market. They leave acacia trees in their fields because they improve soil quality. Experiments with these tree species indicate that they do benefit soil health. They maintain or increase organic matter through leaf litter and root decay, while decreasing rates of erosion and runoff of water.[22]

[22] J. Kessler and H. Breman, 'The Potential of Agroforestry to Increase Primary Production in the Sahelian and Sudanian Zones of West Africa', *Agroforestry Systems*, 13 (1991), pp. 41–62; A. Young, *Agroforestry for Soil Conservation* (Wallingford: CAB/ICRAF, 1989).

Table 4.9 Density of trees per ha by village and ethnic group

	Sara	Dimikuy	Dohoun
Bwa	20.2	8.3	9.3
Migrant	15.9	12.5	13.5
Village Total	18.0	10.5	12.1

Table 4.9 indicates differences of tree densities in farmers' fields in all three villages disaggregated by ethnic group. Densities are highest in Sara, the village with the oldest and most intensively farmed fields. In Dimikuy and Dohoun land is still largely available. However, in both these villages there are striking differences between migrant and Bwa fields. Migrants, on average, leave more trees in their fields.

The discussion presented thus far indicates that farmers of both ethnic groups are intensifying their production. Studies of trees on farmers' fields, fertilizer use, manuring and anti-erosion structures confirm this. Farmers in Sara, the village with the most severe land scarcity, invest more in soil quality than do farmers in Dimikuy and Dohoun. Migrants, who have the least secure access to land, also make the most investments in their soil quality, applying manure and constructing conservation structures more often than the indigenous Bwa.

However, it is important not to present a picture that all migrants or all Bwa are doing this or that. Farmers from both groups are constrained in undertaking practices. Wealth, available labor, social standing within the community all affect the types of practices that an individual will undertake. Poorer farmers have difficulty gaining access to manure and fertilizer. Wealthier farmers, who use tractors and plows, leave fewer trees in their fields because they are an impediment to plowing. Newer migrants are often poorer than migrants who have been in the region for longer periods. In Dohoun, very few of the recently arrived migrants make intensive use of inputs.

To sum up, the first and most important factor in whether a farmer intensifies production appears to be land scarcity. Farmers in Sara use more of most inputs. Farmers in Dohoun use least. These aggregate statistics, however, hide a great deal of group and individual variability. Farmers were quite explicit in stating that they use these inputs to remain on a field which, if they were to let it regenerate through fallow, might be taken away from them. For that reason, Mossi farmers, especially wealthier farmers in the village of Dimikuy, used much more manure than did local Bwa.

Conclusion

Government and donor aid programs in Burkina Faso are changing institutions for allocating land. These policies have their roots in the belief that indigenous tenure systems increase land degradation because they prevent farmers from investing in soil quality. This chapter challenges this assumption by comparing the land invest-ment strategies of farmers in three villages in the cotton-growing region of southwestern Burkina Faso. In the study area, land has become more scarce; aerial photographs show that the area under cultivation has increased at the same time as the population has more than doubled. But it is unclear whether land scarcity and the resulting uncertainty over land rights have resulted in land degradation. Instead, farmers are responding to both land scarcity and what they perceive as land degradation by investing in agricultural practices that promote soil regeneration. They apply manure to their fields, construct anti-erosion structures, and implement agroforestry techniques.

The discussion investigated differences among villages, between ethnic groups and among individuals. Farmers in Sara, the village facing the most urgent problems of land scarcity, are making the most investments in soil quality of the three study villages. They use more fertilizer and leave more trees on their fields than do farmers in the other two villages, and apply manure and construct anti-erosion barriers at higher or similar rates to the other study villages. The inter-ethnic differences are striking as well. Mossi farmers, who are at greatest risk of having their land taken away from them, make the most improvements to their soil. However, the ability to invest in soil quality is influenced by economic status and social standing. Wealthier farmers are more likely to use manure than poor farmers. This is especially true for the poorer Mossi farmers, who are frequently given land of extremely poor quality and do not have the means to improve it.

Farmers improve the quality of their soils not only to increase agricultural production, but also to strengthen their rights to land. Given the choice, most farmers would prefer to improve soil quality by shifting to a new field. Now, however, with increasing land scarcity, even if farmers have the option of moving to a newer more fertile field, they are hesitant to leave their old fields for fear of abandoning the rights to them that have developed over time. Farmers therefore invest in soil quality, intensifying their production systems. By so doing, they increase their ability to stay on a piece of land and leave it fallow, essentially strengthening their land rights in an atmosphere of growing land scarcity and uncertainty.

5

Farmer Tree-Planting in Wällo, Ethiopia

DONALD CRUMMEY & ALEX WINTER-NELSON

Introduction

Popular and academic narratives about rural Ethiopia attribute environmental and social crises to population pressure and stasis in peasant farming technologies. They see the combination of 'backward' technology and rising population as the root cause of environmental degradation, manifested in deforestation. In turn, à la Malthus, they cite this population-induced deforestation as a chief contributor to declining agricultural productivity and rising famine. This chapter presents evidence from a sample of farmers in Wällo, Ethiopia, to show that, despite deep continuity in farming practices, Ethiopian peasants have innovated and responded to changes in the physical and social environment. It first raises doubts about the link between famine and deforestation by comparing changes in tree cover and experiences of famine in two regions, Wällo and Gondär. It then reveals the evidence that farm technologies have not remained static in the face of environmental change, but have been dynamic. It demonstrates that farmers have been incorporating trees into their planting activities for at least 60 years.

Our observations suggest at least three revisions in the dominant narrative linking farmers to deforestation to famine in Ethiopia. First, we characterize Ethiopia's farmers as dynamic rather than static, although subject to severe resource constraints that limit their ability to manage their physical environment. Second, while there is very little economic differentiation in rural Ethiopia, tree-planting, rather than deforestation, has been a frequent form of environmental management by farmers. Finally, changes in tree cover may have little bearing on vulnerability to famine. Our revised narrative gives more credit to farmers for managing their natural resource base, but does not suggest complacency with respect to famine or environmental stress. Our research also reveals that many households may be unable to manage the environment in the way they perceive to be best and that farmers' perceptions of optimal management do not address biodiversity loss. An implication of the paper is that governmental policies in the 1980s and 1990s, which included

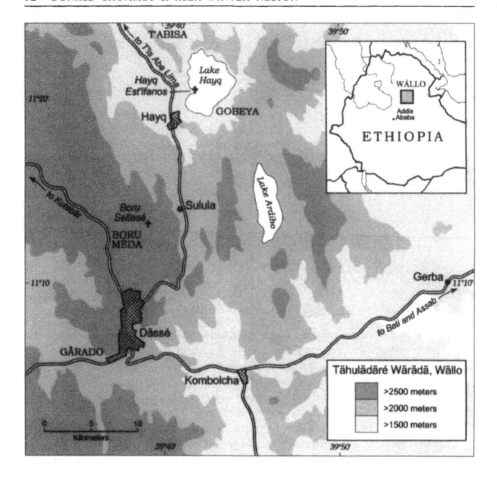

Figure 5.1 Map of Tähulädäre Wärädä, Wällo

large-scale re-settlement and extensive state control of natural resources, were based on serious misconceptions of farming practices in Wällo, and were ineffective and, quite possibly, counter-productive.

We gathered the data on which we base this paper by means of interviews with farmers, conducted within the framework of an international collaborative research project funded by the John D. and Catherine T. MacArthur Foundation under the title of 'The African Environment: Experience and Control'.[1] We were members of a larger Ethiopian research team, whose work is reflected in Chapter 9 by Dessalegn in this collection, and by a special issue of the *Journal of Ethiopian Studies*, whose articles we cite below. Our research team chose the province of Wällo[2] as the locus of its work. At the center of the great famines of 1973 and 1984, Wällo offered the team the possibility of illuminating the environmental dimension to vulnerability to famine. The Ethiopian team further chose the historic district (*awraja*) of Ambassäl,[3] and within it, the sub-district (*wäräda*) of Tähulädaré, to the north and northeast of the provincial capital of Dässé. Within Tähulädaré the team concentrated its work in three Peasant Associations – Borru, Gwobeya, and Gerba – each representing one of the three principal agro-ecological zones into which highland Ethiopians divide their landscape: *dägga*, *wäyna dägga*, and *qwolla*. (For the location of places, please see Fig. 5.1.) These categories represent farming zones distinguished primarily according to altitude, and, secondarily, by rainfall and exposure.[4] To add a further comparative dimension, we also conducted interviews in Gondär province, an area with environmental conditions and farming practices similar to those of Wällo, but without the recent experience of famine.

[1] The Introduction to this volume describes the overall structure of the MacArthur project. Like many of the other contributors to this volume, we are deeply indebted to the MacArthur Foundation for its generous support.

[2] The meaning of both 'Wällo' and 'province' require some definition. Ethiopia's administrative structure and its accompanying nomenclature have changed with each of the major changes in government which occurred in 1941, 1974 and 1991. Throughout these changes Wällo has continuously been used to refer to those lands in the Western Ethiopian highlands, which lie between the basins of the Blue Nile and the Awash. Dating from 1941, Ethiopian provincial administration has worked up from the *wäräda*, the basic administrative unit, through the *awraja*, which encompassed numerous *wäräda*, through the *keflä hagär* or *täklay gizat*, roughly 'province,' to the Ministry of the Interior in Addis Ababa. The 'province' of Wällo, as we constitute it here, corresponds to none of the actual administrative units into which the imperial or revolutionary governments divided the country, but indicates an historically constituted area, which encompassed many *awraja*, and, within them, many *wäräda*. The *wäräda* has survived all the changes in higher-level administrative organization and might reasonably be compared to 'county'. This chapter reports research conducted in three different *wäräda*, whose total population, as Table 5.1 indicates, ranged from 119,240 to 228,520.

[3] Ambassäl offered the attraction of two previous studies, which helped frame the team's work: Alemneh Dejene, *Peasants, Agrarian Socialism, and Rural Development in Ethiopia* (Boulder, CO: Westview Press, 1987); and Dessalegn Rahmato, *Famine and Survival Strategies. A Case Study from Northeast Ethiopia* (Uppsala: Scandinavian Institute of African Studies, 1991).

[4] *Dägga*, *wäyna dägga*, and *qwolla* are culturally constituted agro-ecological terms, which combine altitude, climate and soil. Their objective correlates are fairly vague. However, *dägga* may conveniently be translated as 'highland,' *qwolla* as 'lowland,' with *wäyna dägga* referring to the zone intermediate between the other two. These terms are relative, one to the other, within an agricultural landscape that ranges from about 1,200–1,300 meters above sea level to about 3,000 meters. 'Lowland' thus lies *above* 1,200 meters in altitude.

Interpretive framework

We situate our paper within the debates concerning the relationship between population growth and environmental stress. The more prevalent view today, a neo-Malthusian one, holds that population growth invariably leads to environmental degradation.[5] A minority view, most energetically articulately by Ester Boserup, holds that population pressure triggers technological and economic innovation, which mitigates stress and avoids degradation.[6] Malthus himself, and many of those who hold similar views, underestimate the capacity of people to innovate, while Boserup's position sometimes seems to hold, erroneously, that population pressure will *always* trigger the innovations necessary to avoid disaster. The *rate of innovation* would appear to be the factor determining which of these paths is more representative of a specific setting. If production technology is static, rising population on a fixed land base is ultimately unsustainable. Where the rate of technical change is sufficiently rapid, production may rise to make population growth compatible with a fixed resource base.

In Wällo, population has assuredly been growing, probably commensurate with the national figures, which indicate that rural population grew on an annual average of 2.2 per cent in the years between the two national censuses of 1984 and 1994.[7] We are not able, with the evidence at our disposal, to compare overall growth in production with overall growth in population, nor does our evidence allow us to calculate the rate of innovation, but it does clearly indicate that Wällo farmers have been innovating to an extent not previously appreciated. Establishing this fact ought, we believe, to advance the terms of the debate by focusing attention on the rate of innovation and its relative success in coping with rising population. The importance of doing so lies in the extent to which Malthusian explanations are commonly used to explain Ethiopia's vulnerability to famine.

The people who shape Ethiopian government policies, and, more widely, newspaper-reading Ethiopians, believe that the country's farmers, left alone, are incapable of innovation. Urbanized Ethiopians assume that the countryside is unchanging and always has been. Influential NGOs buttress and reinforce these views. Consequently, interventions in the Ethiopian countryside embody neo-Malthusianism assumptions. These beliefs are clearly stated in policy documents, in the scholarly literature and in articles in the national press. For example, Ethiopia's National Conservation Strategy incorporates the standard claims concerning deforestation: 87 per cent of the surface area of highland Ethiopia was 'once covered with high forests'; a figure which by 1900 had declined to 40 per cent; and by 1990 was estimated to be about 5.6 per cent. The strategy document further links deforestation to landscape degradation. By 2025, it estimates, 'the last high forests may have disappeared, … and some 10 m. ha. of former arable land in the

[5] With Malthusian overtones is James McCann, 'Ethiopia', in Michael Glantz (ed.), *Drought Follows the Plow* (Cambridge: Cambridge University Press, 1994), pp. 103–15.

[6] See, for example, Ester Boserup, 'Environment, Population, and Technology in Primitive Societies', in Donald Worster (ed.), *The Ends of the Earth. Perspectives on Modern Environmental History* (Cambridge: Cambridge University Press, 1988), pp. 23–38.

[7] Markos Ezra, *Demographic Responses to Ecological Degradation and Food Insecurity. Drought-Prone Areas in Northern Ethiopia* (Amsterdam: Thesis Publishers, 1997). For the figure in question, see pp. 243–4.

highland will be unable to produce crops'.[8] These data and the line of argument were carried forward, with little modification, into the National Programme to Combat Desertification, which claims that the country's deforestation is a main cause of land degradation and the main cause of deforestation is 'agricultural expansion, both through shifting cultivation and the spread of sedentary agriculture...'[9]

Scholars have articulated the vision of society which underlies these claims. Adhana Haile Adhana, in his historical study of famine in the provinces of Wällo and Tegray, argues that a shaping condition was the static nature of 'peasant society', which, during the years in question, went through a process of 'ever greater naturalization' – or, more simply, intensification of its historic methods of production – driven by population increase. In Adhana's conception, peasant societies are inherently incapable of innovation.[10] Michael Ståhl, one of the authors of the National Conservation Strategy, in a review of environmental degradation in Ethiopia, asserted that 'Boserup's thesis that demographic pressure spurs agricultural innovations is definitely not applicable to northern Ethiopian peasant communities in the 1990s'.[11]

So far as the press is concerned, an article in the *Ethiopian Herald* made the standard claim that population growth meant that 'marginal lands have been used for food production, thus leading to massive destruction of forest resources which in turn led to a high rate of soil erosion'.[12] The national Relief and Rehabilitation Commission, in a statement in 1995, on the tenth anniversary of the great famine, noted that, 'Traditional agricultural practices which demand the use of large areas for crop production resulted in widespread forest clearance. The lack of soil and water conservation and reforestation programs brought ecological deficiencies that eventually led to famine.'[13] And the *Herald* editorialized: 'Deforestation has especially wrought havoc on the environment leading to the present ecological imbalance.'

In short, Ethiopian elites perceive Ethiopian farmers to be, to a considerable extent, agents of their own misfortune.[14] Elites perceive farmers to be acting out the

---

[8] Adrian Wood and Michael Ståhl, *Ethiopia National Conservation Strategy. Phase I Report. Prepared for the Government of the Peoples' Democratic Republic of Ethiopia with the assistance of the IUCN* (Gland, Switzerland: IUCN, March 1990), pp. 10 and 6.

[9] Federal Democratic Republic of Ethiopia (FDRE), Environmental Protection Authority (EPA), *National Action Programme to Combat Desertification* (Addis Ababa, 1998), p. 28; See also Shibru Tedla, *Environmental Management in Ethiopia. Have the National Conservation Plans Worked?* (Addis Ababa: Organization for Social Science Research in Eastern and Southern Africa, 1998), Environmental Forum Publication Series No. 1.

[10] Adhana Haile Adhana, 'History of Selected Famines in Peasant Societies in Tigray and Wällo, Ethiopia, 1941–1974', unpublished Ph.D. dissertation, Addis Ababa University, May 1996, pp. 549–58.

[11] Michael Ståhl, 'Environmental Degradation and Political Constraints in Ethiopia', *Disasters*, 14, 2 (1990), p. 148. The best study of demography in the context of famine, Ezra, *Demographic Responses*, also has Malthusian overtones.

[12] Seifu Mahifere, 'Population: Population programmes implementation in the Amhara National Regional State', *Ethiopian Herald*, (Addis Ababa), 22 January 1997. Wällo is in the Amhara National Regional State.

[13] *Ethiopian Herald*, 11 and 12 January 1995.

[14] An explicit claim of Yeraswork Admassie, Mulugetta Abebe and Markos Ezra, *Ethiopian Highlands Reclamation Study. Report on the Sociological Survey and Sociological Considerations in Preparing a Development Strategy* (Addis Ababa: Land Use Planning and Regulatory Department, Ministry of Agriculture, 1983), p. 56.

same script as impoverished tropical farmers around the globe – in the Amazonian rainforest, the Philippine islands, and the West African Sahel. Driven by their desperate poverty they mine the soils and forests which give them life. Opinion would be one thing, but action is another. On the basis of these beliefs, the Derg, which then governed Ethiopia, implemented large-scale programs in the 1980s, resettling Wällo farmers in less densely populated, better watered parts of the country, and imposing environmental programs on those farmers who remained in Wällo: hillside enclosures, tree-planting and bunding.[15] While resettlement was abandoned with the Derg's collapse, many of the other programs remain in place and the National Action Plan to Combat Desertification indicates that government perception of the problem has not significantly changed.[16]

In contrast to this narrative of rural stagnation and decline, our informants recounted personal histories of innovation, although they also spoke eloquently of the constraints with which they had to cope. Ironically, in light of the dominant narrative of *deforestation* in Wällo, one of the principal expressions of farmer innovation has been the planting of trees and shrubs. Landscape photographs paired over a sixty-year gap indicate the success which farmers have had in placing trees on the Wällo landscape.[17] The dynamics behind farmers' tree-planting are explained through survey results presented below. Finally, contrasts between Wällo and Gondär will reveal the limited relationship between changes in tree cover and famine.

Methodology and research sites

Survey methods

Our information comes from two distinct rounds of interviewing: a household survey of three Peasant Associations, and personal narratives from elders in both Wällo and Gondär. The first round of data collection consisted of a farm management survey covering 50 households in Borru (*dägga*), 110 households in Gwobeya (*waynä dägga*), and 50 households in Gerba (*qwolla*).[18] The interviews focused on farm household demographics, decision-making, and farm management using closed and open-ended questions in a randomly selected sample. Data were analyzed to isolate elements of diversity in farm management as well as factors that could explain variation in farmer behavior. We could observe little diversity in management in Borru or Gerba, but the 107 households providing usable data from Gwobeya demonstrated marked differences in strategies, despite similarity in demographics.

Historical dimensions were added to the cross-sectional data through a survey of elder residents. With the assistance of Merera Ejjeta, Donald Crummey interviewed

---

[15] See, for example, Dessalegn Rahmato's chapter in this volume.

[16] For the National Action Plan, see note 8.

[17] Donald Crummey, 'Deforestation in Wällo: Process or Illusion?', *Journal of Ethiopian Studies*, XXXVI, 1 (1998), pp. 1–41.

[18] The survey was conducted by Sosina Asfaw of the Department of Geography of the University of Illinois between June and August of 1996.

88 individual farmers (39 women and 49 men).[19] These interviews were conducted in Amharic, but recorded in summary form in English, between April and July, 1997. Interviews were open-ended, but followed a consistent set of questions, which elicited information on personal, family, farming and environmental history. Crummey selected his informants in approximately equal numbers from each of the project's three Peasant Associations – Borru, Gwobeya and Gerba. Additional individuals were selected to cast light on a set of photographs taken in 1937 by Armando Maugini, director of Italy's Istituto Agronomico d'Oltremare. Maugini's photographs document the agricultural landscape alongside the newly opened Italian highways. In February 1997, sixty years to the day later, Crummey reproduced Maugini's photographs, establishing a visual framework for landscape transformation during the intervening years.[20] Informants, whose lives parallelled the landscape changes reflected in the 1937–97 pairs of photographs, lived in *wäyna dägga* Peasant Associations around Sulula Town (near Gwobeya) and in T'abisa, and the *qwolla* area of T'is Aba Lima.

Table 5.1 Population density by district

District name	Surface area sq. km.	Total population	Population density per sq. km.
Armac'äho	4889	122,944	25.15
Dämbeya	1271	228,520	179.80
Tähulädäré	486.53	119,240	245.08

Finally, Crummey's interviews included yet another comparative dimension with informants from two districts – Armac'äho and Dämbeya – around Gondär, hundreds of kilometers west of Dässé, where he conducted interviews in August 1997. Cultural factors and farming practices in the area were much the same as those in Wällo, but different climatic conditions prevailed. Gondär enjoys more reliable Atlantic rains falling in June-September (the *krämt* season), while the parts of Wällo on which the MacArthur project focused have better exposure to Indian Ocean moisture falling during the short *bägga* season, February through April. The Gondär and Wällo research took place at different scales. Whereas practically all of the research sites in Wällo fell within the single *wäräda* of Tähulädäré, by contrast, the two major Gondär research sites – Dämbeya and Armac'äho – were *wäräda* in their own right. Population densities vary markedly from one *wäräda* to another. Tähulädäré is the most densely populated at 245 people per square kilometer, compared with 180 per sq. km. in Dämbeya, and 25 per sq. km. in Armac'äho. (Further information on population is given in Table 5.1.)

In the area around Gondär, again with the assistance of Merera Ejjeta, Crummey interviewed 32 individual farmers, 14 women and 18 men, plus two groups

---

[19] A print-out of the database into which the record of these interviews was entered has been forwarded for deposit at the Institute of Ethiopian Studies, Addis Ababa University. The software program is Notebook II, an elegant and efficient DOS-based program.

[20] Crummey, 'Deforestation', provides an inventory of the photographs, discusses them, and reproduces five pairs.

composed one of 5, and the other of 6 men. Dämbeya is classic *wäyna dägga* country. The parts of Armac'äho, accessible to Crummey when he visited it during the rains of 1997, were described by his informants as *wäyna dägga* and *qwolla*. For the purposes of analysis we have clustered the interviews into eight groups, listed in Table 5.2.

Table 5.2 Interview groups

Cluster Name	Agro–climatic zone	Men	Women	Total
Borru	dägga	9	6	15
Gerba	qwolla	9	11	20
Gwobeya	wäyna dägga	8	8	16
Sulula	wäyna dägga	14	5	19
T'abisa	wäyna dägga	5	6	11
T'is Aba Lima	qwolla	4	3	7
Wällo subtotal		49	39	88
Armac'äho	wäyna dägga	8	6	14
Dämbeya	wäyna dägga	10	8	18
TOTAL		67	53	120

Research sites

Natural endowments differ considerably between the three main research sites at the Peasant Associations in Wällo. As already noted, *dägga* is usually translated as 'highland', *wäyna dägga* as 'moderate or temperate highland,' and *qwolla* as 'lowland'. Consider, however, that lowland here means land more than 4–5,000 feet above sea level, and low only with respect to the other zones of reference, which range up as high as 10,000 feet. In the *dägga* areas represented by Borru, the climate is cool with annual rainfall varying between 600 and 2,000 mm. Rain falls bimodally, but the short *bägga* rains, falling in the northern hemisphere Spring, have been more important agriculturally. In Gwobeya, which is representative of the *waynä dägga* zone, the climate is more temperate with rainfall ranging between 500 and 900 mm annually. Farmers in *wäyna dägga* districts like Gwobeya try to exploit both periods of rain by double-cropping. Gwobeya's soils tend to be deeper and more fertile than those of Borru and farmers frequently harvest twice a year. Like other *qwolla* areas, Gerba is much drier than the other two sites, receiving between 300 and 450 mm annually, mostly during the *krämt* season.[21] Farming in Gerba is dominated by one big harvest and this is the only region in which farmers reported intercropping in significant numbers. Annual averages mask a significant dimension to rainfall in Wällo, the fact that, from year to year, it is less reliable than the rainfall on the more westerly reaches of the plateau.

---

[21] While Gerba is sometimes described as *entirely* dependent on *krämt* farming, at the end of May farmers were busy harvesting *bälg t'éf*.

Farm practices

Farm production

The farm management survey confirmed that farm activities vary consistently over the environmental zones. In all three zones farms are small and crop production is oriented to self-provisioning far more than to market sale. All land is prepared using ox-plow technology requiring two oxen. In each of the study areas over half the households reported having at least one ox. Those with only one usually had exchange arrangements with friends or relatives to complete farm operations. In the lowest, driest area (Gerba), sorghum is the main crop, with maize, sesame, and beans (*adengwarê*) also represented. As water allows, fruit trees are grown. In the intermediate elevations (Gwobeya), *t'éf* and maize cultivation dominate, with sorghum and wheat represented. Farm sizes tend to be smaller in this zone and yields higher than in either the lowest or the highest elevations. In the highest elevations (Borru), wheat, barley, maize and beans are produced. Further descriptive statistics indicating activities in these zones are given in Table 5.3.

While yield and farm size vary across the zones, household size averages about 5.5 persons in each area; average cropped land per household member thus ranges from about 0.1 hectares in Gwobeya and Borru to 0.5 hectares in Gerba, where yields are lower. No private farms could be characterized as large, but some are exceedingly small. A quarter of the households surveyed in Gwobeya had less than .05 hectares per person, while only 3 per cent commanded over 0.3 hectares per resident household member (Table 5.4). In absolute terms this difference is slight, but it can have considerable impact on the ability to meet household consumption needs.

Extremely limited land availability was a major motivation for the policy of resettlement. In *waynä dägga* areas of Wăllo, cereal production at average yields is insufficient to meet household needs on farms with 0.15 hectares per person.[22] This suggests that unless there is substantial non-crop (or at least non-cereal) production, some 90 per cent of the farms are not sustainable. Indeed, in the sample, only 10 per cent of the farms in Borru reported that they usually have at least small amounts of cereal to sell in the market. Corresponding figures for Gwobeya and Gerba were 42 and 34 per cent. When asked whether market prices affected their decisions about what crops to plant in farm plots, only the food-deficit households responded positively. Hence, the influence of the market on farm-crop choice is through consumption needs, rather than a commercial orientation in production.

Inability to meet cereal consumption needs on many farms mandates non-cereal production. In terms of income diversification, households in the *dägga* areas historically engaged in long-distance trade during the season of the long rains, but mechanized transport and government regulations have made this practice obsolete. Remaining non-farm income sources are limited to activities such as beer-brewing, hair braiding and tool repair, which generate little income. Most households in our sample do report income from livestock or poultry, but the numbers of animals are

---

[22] Alex Winter-Nelson, 'Rural Taxation in Ethiopia, 1981-1989: A Policy Analysis Matrix Assessment for Net Consumers and Net Producers', *Food Policy*, 22 (1997), pp. 419–32.

Table 5.3 Descriptive statistics (mean values unless noted)

Location	Gerba	Gwobeya	Borru
Zone	*qwolla*	*wäyna dägga*	*dägga*
Cropped land per household (ha)	2.23	0.39	0.54
Average household size	5.25	5.50	5.66
Cropped land (ha) per household member	0.51	0.08	0.13
Main crops	sorghum	t'éf	wheat
	adengwaré	sorghum	barley
	sesame	maize	maize
% of households planting trees	83	97	86
% of households with tree lots	40	26	73
% of households with non-farm income	4	17	2
% of households with non-crop income	15	89	10
Number of large animals per household	1.7	3.33	2.33
Number of small animals per household	1.25	1.5	2.3
% of households with a cereal surplus	34	42	10
% of households whose field crop choice is influenced by market conditions	26	0	37
% of households with market gardens	15	26	42
Sample size	47	107	48

Source: Farm management survey data

Table 5.4 Percentage of households with less than X hectare per person

X	0.05	0.075	0.1	0.15	0.2	0.3	0.5	1.0	3.0
Gwobeya	25.5	56.1	83.2	97.2	98.4	100	100	100	100
Borru	14.6	35.0	56.1	77.3	91.2	98.1	100	100	100
Gerba	4.3	8.1	12.8	17.0	21.3	40.4	51.0	68.0	93.0

Source: Survey data

Table 5.5 Percentage of respondents engaged in specific farm activities

	Cereals	Livestock	Small-stock*	Trees	Market gardens
Borru	100	83	51	86	42
Gwobeya	100	96	26	97	26
Gerba	100	73	30	83	15

Note: * Two or more head.
Source: Farm management survey

Table 5.6 Stated uses for trees

Sample size		% Listing each use				
		Construction	Fuel	Sale	Fertilizer	Erosion control
Borru	40	73	78	73	23	5
Gwobeya	104	82	88	20	5	7
Gerba	39	95	95	28	3	0

Source: Farm management survey

low and this income source appears to be no longer a very reliable safety net for crop failures (Table 5.5).[23]

Crop production is highly diversified. In addition to the main cereals (*t'éf*, maize and sorghum), farms produce *ch'at*, *gésho*, coffee, vegetables, meat, fruits, sugar cane, various pulses and timber. As Table 5.5 shows, all households surveyed produced cereals, but in addition 83 to 97 per cent reported production of tree-products and 15 to 42 per cent reported that they maintained garden plots for market sale. In many cases the garden crop was a tree, often eucalyptus. The proportion of informants testifying to tree-planting varied across ecological zones. In Borru almost 75% of the households reported maintaining tree lots, while in Gerba, most trees were simply scattered around the homestead (Table 5.3). The prevalence of tree-planting suggests that trees are an integral part of the small-scale farmer's strategy. As Table 5.6 shows, farmers had multiple reasons for planting these trees. Fuel and construction were the dominant uses, but market sale was also a prominent motive. Soil erosion control was not an important reason for planting trees. Although farmers were certainly aware of this problem, they preferred to address it by other devices such as constructing water channels and terracing.

On-farm tree-planting

Farmers in all the areas in which we did research are active tree planters. We have information on tree-planting practices from 82 of the 88 respondents to the survey of elders and 201 of the respondents to the farm management survey. Of the total 283, 90 per cent (256 households) claimed to have planted trees. The response rate was identical for both surveys, although it varied by ecological zone. The most commonly planted species were eucalyptus. No fewer than 233 respondents (including 65 informants from the elders' survey) told us that they had planted eucalyptus. Other frequently planted trees were: *qulqwal* (the indigenous *Euphorbia candelabrum*); *qundo bärbärē* (the exogenous *Mentha piperita*); *shushuwa* (the exogenous *Casuarina equisetifolia*); *wanza* (the indigenous *Cordia africana*); *grar* (the indigenous *Acacia abyssinica*); and *t'ed* (the indigenous *Juniperus procera*). A full listing of species the Wällo informants had planted is given in Table 5.7.

Wällo informants also indicated that they were active growers of bushes, this,

[23] Dessalegn, *Famine and Survival Strategies*.

Table 5.7 Trees and bushes planted by informants

agulo	orange
*arorésa (Grewia mollis)	papaya
avocado	**qach'ona
**bäkralomi	*qänt'afa (either Entadopsis abyssinica, or
besanna (Croton macrostachys)	Pterolobium stellatum)
coffee	*qench'eb (Euphorbia tirucalli)
ch'at	*qulqwal (Euphorbia candelabrum)
dädäho (Euclea schimperi)	*qundo bärbäré (Mentha piperita)
**dedeya	*roman (Punica granatum)
**dokma (Syzygium guineense)	*shushuwa (Casuarina equisetifolia)
**donga (Apodytes dimidiata)	*sisal
eucalyptus	*t'ed (Juniperus procera)
*fränj t'ed	**trengo (Citrus grandis)
gésho (Rhamnus prinoides)	*wäléns (Erythrina brucei, or Erythrina
*grar (Acacia abyssinica)	abyssinica)
*gravillia	*wanza (Cordia africana)
*koba (Ensete ventricosum)	**wayra (Olea africana)
kok (Persica vulagris), the Ethiopian peach	welkefa (= makanisa = Dombeya goetzenif)
makanisa (= welkefa = Dombeya goetzenii)	**zäytun (guava)

Notes: Items unmarked were reported by both Wällo and Gondär informants
    Items marked * were reported by only Wällo informants
    Items marked ** were reported by only Gondär informants
Source: Survey of elders

generally, with an eye to the market. Here three crops predominate: coffee, ch'at (Catha edulis), and gésho (Rhamnus prinoides). Seventy-nine respondents to the elders' survey provided information on bush-growing, with 60 per cent indicating that they grew one bush or another. Only 20 per cent of the respondents to the farm management survey listed bushes among their crops, but they were not asked specifically about such plants. The most popular bush crops were ch'at, also known as qat, which has a growing market in the Horn and South Arabia, and coffee. 34 of the elder farmers and 6 respondents to the farm management survey reported growing ch'at; 2 of the elders and 27 respondents to the farm management survey grew coffee. Another 26 elders and 4 of the farm management survey respondents grew gésho, which is used to hop beer. Culture plays some role in who grows what: ch'at is culturally identified with Muslims, although increasingly Christians use it and grow it for market. Gésho, being associated with beer, is avoided by the stricter Muslims.

Changes in woody vegetation

From the foregoing it is clear that farmers are both consumers and producers of trees. Which of these two roles predominates? Has the net effect of farmers' activities been to lessen or increase woody vegetation? The interviews with the elders addressed these questions within the context of the landscape changes documented by the historic pairs of photographs. The responses of Wällo

Photo 5.1 *Dägga* landscape, Borru Peasant Association.
(Photographs 5.1–5.7 by Donald Crummey, 1997)

informants concerning overall changes in numbers of trees are summarized in Table 5.8. The chronological point of departure – the meaning of 'then' – was set by the point in time at which the informant left their father's house, usually to get married. Thus, in absolute time it varied from individual to individual, but, on average, may be taken to fall within the decade 1935–45, the decade including and following Maugini's photographs.

Our question used the Amharic term for trees, *zaffoch*, but we think informant responses probably reflect the slightly broader sense of woody vegetation. The question often followed another question in which informants were asked for their memories of useful trees and bushes. The collective inventory is an impressive one,

Table 5.8 Perceived changes in numbers of trees, Wällo

Peasant Association cluster	More trees then	More trees now	No response	Total
Borru	5	6	4	15
Gerba	18	0	2	20
Gwobeya	4	3	9	16
Sulula	9	6	4	19
T'abisa	4	3	4★	11
T'is Aba Lima	5	0	2	7
TOTAL	45	18	25	88

★ One of the T'abisa informants judged that there had been no overall change in tree numbers.
Source: Survey of elders

Photo 5.2 *Wayna Dägga* landscape, Hitacha Peasant Association, near Sulula.

Photo 5.3 *Wayna Dägga* landscape, Gwoboya Peasant Association, with Lake Häyq in the background.

Photo 5.4 Yemam Yuséf with his coffee bushes, Bädädo Peasant Association.

Photo 5.5 Wärqenäsh Täfara with her eucalyptus grove, Qorké Peasant Association.

Photo 5.6 Zäwde Gétahun with her banana tree, T'is Aba Lima Peasant Association.

Photo 5.7 Ahmäd Bäkurä with his eucalyptus grove, Gerba Peasant Association.

as is the range of uses to which its members could be turned. Nonetheless, fuel was never far from informants' minds when thinking about 'trees', and fuel often came from bushes. Forty-five of the 64 Wällo informants (70 per cent of the total), who responded to the question, judged that the number of trees (or amount of woody vegetation) had generally declined since their childhood. This leaves a substantial minority of 30 per cent deeming that there are more trees today than formerly. Gender seems not to be a factor here: 20 women made judgments and 6 (30 per cent) of them considered there to be more trees now than before.

Much more significant than gender in shaping the response of informants was their location. Informants in the *qwolla* locations of Gerba and T'is Aba Lima were unanimous in believing woody vegetation to have declined since their youth. In contrast, a plurality of informants in *dägga* Borru believed the opposite to be the case. Fittingly, informants in the intermediate *wäyna dägga* zone evinced a broader range of views. Of the 30 informants in Gwobeya, Sulula and T'abisa who offered their views, 17 (55.7 per cent) believed woody vegetation to have declined overall, 12 (40 per cent) judged the opposite and one (3.3 per cent) considered there to have been no change in the overall number of trees. These reports are fully consistent with the patterns of tree-planting given above. In areas where tree-planting has been more intensive, residents are more likely to perceive there to have been increases in tree coverage.

Received wisdom and a documentary tradition that reports frequently on the cutting down of trees would take the majority view as accurate.[24] Received wisdom infers from the known fact of rapid population growth in Wällo that there must be fewer trees. However, the view of the minority of informants confirms the impression gained from the historic pairs of photographs, that parts of Wällo under consideration have more trees on them now than they did when Maugini photographed them in 1937. Indeed, this dissenting view is as strong as the perception of tree loss in the lower altitude zones.[25] Morever, in evaluating farmers' perceptions of changes in tree cover, one has to take into account that, for at least twenty years prior to the interviews, political cadres had been promoting a 'tree loss' reading of the landscape and it is probable that, to some extent, farmers have internalized this imported reading.[26]

Historical recall implies a mixed record of vegetation change in Wällo. This is in

---

[24] For two accounts, archivally based, which stress peasant felling of trees, see Dessalegn Rahmato, 'Environmentalism and Conservation in Wällo Before the Revolution', and Bahru Zewde, 'Forests and Forest Management in Wällo in Historical Perspective', *Journal of Ethiopian Studies*, XXXI (1998), pp. 43–86 and 87–121 respectively.

[25] Unfortunately, the sites of Maugini's photographs of *qwolla* areas have proven very difficult to identify, although we have replicated some of these *qwolla* photographs and the 'before and after' comparison does not suggest dramatic changes in vegetation.

[26] Admittedly, the Maugini photographs were all taken from a major highway. A widespread view at the turn of the millennium is that roads lead to a loss of woody vegetation from the landscape. Such a dynamic is unlikely to have been at work in this instance, since the road had only been completed about a year before Maugini's trip. Such effect as the road has had is likely to have been in the opposite direction – the transmission of information which could be used for local innovation and the creation of market opportunities for wood and for bush crops like *ch'at* and coffee. This market has encouraged the planting of wood, rather than the mining of naturally propagated trees.

Table 5.9 Perceived changes in numbers of trees, Gondär

Peasant Association cluster	More trees then	More trees now	No response	Total
Armac'äho	12	1	2	15
Dämbeya	9	3	7	19
TOTAL	21	4	9	34

Source: Survey of elders

contrast to the dominant narrative concerning deforestation and famine in the region. To further explore the commonly posited link between population, deforestation and famine, we compare the recollections of residents of Gondär with those in Wällo. Table 5.9 reports the recollections of Gondär informants on the question of overall changes in woody vegetation. Twenty-five informants gave us judgments on overall changes, and of them 21 (84 per cent) believed the number of trees to have declined, while only 4 (16 per cent) believed it to have increased. Both of the Gondär areas contained *dägga* and *wäynä dägga* environments, and the informants in Armac'äho were more emphatic than were those in Dämbeya about tree loss. In general, the topography and contemporary vegetation of the two areas contrast remarkably. Dämbeya is open and rolling, close to Lake T'ana. Its soils are heavy, prone to water-logging, and it gives few signs of having once been more heavily wooded. Armac'äho, by contrast, is a rugged, broken country, dotted with woods and copses. Farmers in Dämbeya have depended on their cereal crops, which they market, for as long as they can remember, and speak fondly of the rich grazing meadows, which formerly they enjoyed. By contrast, the farmers of Armac'äho depend more heavily on tree crops, and, when they speak of lost grazing lands, they refer to uncultivated bush and woodlands. Overall, subjectively today one would judge Armac'äho the most heavily wooded of all the places in which we carried out research, yet memory is unambiguous: there are fewer trees today. Particularly eloquent on this point was informant, Mängesté Täklé, whom we interviewed in the PA of Kärkär, on 11 August. Mängesté described the youthful experience of tall canopy forest, of which little or none is evident today.

The information gleaned from informants, when we combine the responses in Wällo with those in Gondär, is significant so far as connecting 'deforestation' with famine is concerned. All our informants in Wällo had experienced famine, most recently and grievously in 1984, while *none* of the Gondär informants had done so. All the Gondär informants had learned from their parents and grandparents about the Great Famine of 1889–92, which, they insisted, had affected their areas. Quite a few informants also recalled a famine early in the Italian years, caused by locusts; but none had personally experienced it. Nor could the bulk of the Gondär informants even recall the year – 1977 in the Ethiopian calendar – in which their Wällo compatriots suffered so grievously. Admittedly, we are at several removes from historical reality. Nonetheless, at the level of perception as filtered through memory, 'deforestation' does not connect positively to famine. The next four sections of this chapter set out our evidence concerning innovation, its dynamics and some of its landscape implications. Given the importance of deforestation in established views, our principal focus is on tree-planting; we also provide evidence on innovation as indicated by market gardening.

Determinants of on-farm tree-planting

The relationships between population and tree coverage and between tree cover and vulnerability to famine are clearly more complex than Malthusian thinking would suggest. Both statistical analysis of farm management data and farmer narratives shed light on the factors that drive patterns in planting.

Statistical analyses

Despite nearly universal recognition of the uses of trees, variation exists in the intensity of planting activities, with some farmers planting only a few trees around their houses and others devoting farmland to tree-lots. Yet others report planting trees around the farm and home. Tree-lots are usually eucalyptus, but also include *t'ed* (juniper) and *grar* (acacia). Since tree production appears to have environmental and economic benefits for the region, our analysis will consider the factors that are associated with more and less intensive planting. In the first instance, it is clear that agro-ecological zone is significant. Farmers in the *dägga* areas, which are wetter, higher and more densely populated, are most likely to grow trees intensively in tree-lots. Farmers in the warmer, drier and least densely populated *qwolla* zones are

Table 5.10 Potentially explanatory variables of tree density

Category	Variable	Measurement
Demographic	Household size	Discrete
	Female-headed household	Dummy
	Members per hectare	Continuous
	Adults in household	Discrete
Wealth/Income	Large animals	Discrete
	Small animals	Discrete
	Land area	Continuous
	Cropped land per person	Continuous
	Market garden	Dummy
	Non-farm income	Dummy
	Cereals for sale	Dummy
Knowledge	Literacy	Dummy
	Age	Discrete
	Years farming in location	Discrete
	Years of education	Discrete

Source: Survey data

Table 5.11 Factors influencing tree-planting. Dependent variable: tree-lot

Consistently significant independent variables	Borru	Gwobeya	Gerba
Large animals	not significant	+ (5%)	not significant
Market garden	not significant	+ (10%)	not significant
Cereal sales	+ (10%)	not significant	not significant
Non-farm income	not significant	− (5%)	not significant
Small animals	not significant	− (5%)	not significant
Female-headed household	− (10%)	not significant	not significant
Age	− (10%)	not significant	not significant
Years of education	+ (10%)	not significant	not significant
Sample size	47	107	47
Nagelkoeke R^2*	0.46	0.45	not applicable
Prediction accuracy*	76.6%	82.25%	not applicable

*Results for the most parsimonious specification; Logit model

least likely to plant trees at all. Logit and logistic regressions were applied to determine what factors explain behavior within ecological zones.

The purpose of these regressions is to determine what variables can predict whether households in specific ecological zones engage in particular behaviors. Data from 47 *dägga* households from Borru, 107 *waynä dägga* farms from Gwobeya and 47 *qwolla* farms from Gerba were analyzed in separate sets of regressions. This exploratory analysis is not based on theory, so direction of causality can be disputed. Colinearity, endogeneity and heteroskedacity were addressed in the statistical analysis. Table 5.10 lists variables considered to be potentially explanatory in the regressions. Multiple specifications using the variables in Table 5.10 were applied and the variables that were consistently significant across specifications are presented in Tables 5.11 and 5.12. In all cases, data collected explain only a small part of the variation in farmer behavior. Unmeasured variables that may be influential certainly include micro-climatic and environmental variation and management ability. Nonetheless, statistically and conceptually significant relationships do appear in the data.

Presence of tree-lots. The most ambitious tree-planters are those who develop tree-lots. Specific variables that predicted the presence of tree-lots differed across zones, but in both Borru and Gwobeya indicators of wealth (large animals, food surpluses, years of education, market gardens) corresponded with tree-lots (Table 5.11). Similarly, poverty indicators, like the absence of an adult male and reliance on small-stock, were negatively associated with tree-lots. The negative sign on non-farm income suggests that either such activities draw labor and other resources away from tree planting or that engagement in such activities is a proxy for the degree of poverty that makes intensive tree-planting impossible. The insignificance of

Table 5.12 Factors influencing limited tree-planting. Dependent variable: few or no trees

Consistently significant independent variables	Borru	Gwobeya	Gerba
Cropped land per capita	not significant	– (5%)	not significant
Female-headed household	+ (5%)	+ (5%)	not significant
Non-farm income	not significant	+ (1%)	not significant
Food deficit	+ (10%)	not significant	not significant
Years of education	– (10%)	not significant	not significant
Sample size	47	107	47
Nagelkoeke R²	0.41	0.31	not applicable
Prediction accuracy★	78.7%	63.5%	not applicable

★Results for the most parsimonious specification; Logit model
Source: Farm management survey

farming experience and age suggests that knowledge is not the critical determinant of behavior in this case. Similarly, the insignificance of indicators of land and labor availability suggests that marginal changes in population pressure do not influence the decision to plant tree-lots. For Borru and Gwobeya, the most parsimonious specification, including only the statistically significant variables, correctly predicted the presence or absence of tree-lots in 70 and 82 per cent of the cases, respectively. Results for Gerba were insufficiently robust to be considered valid and are not reported.

Absence of trees. Whereas land availability did not predict maximum farmer adoption of tree-planting, unavailability of land does predict the opposite, minimal farmer adoption of tree-planting. As shown in Table 5.12, increased land per person reduces the probability of planting only a few or no trees in Gwobeya. Similarly, in Borru, food-deficit households are more likely than others not to plant trees. Like the relatively land-poor, female-headed households are more likely than others to plant few or no trees in both Borru and Gwobeya. Female-headed households are typically disadvantaged in terms of access to land, adult labor, and forms of institutional support. Since labor markets are restricted, it is reasonable to suspect that female-headed households are less able than others to command labor for planting and maintaining tree-lots. The presence of non-farm income in Gwobeya with a positive sign is consistent with the findings in the tree-lot regression. Farm households engaged in non-farm activities are disinclined to plant trees because they are either time or income constrained.

The pattern that emerges from the data is of widespread private tree-planting over a long period of time. However, while tree-planting is common, there are environmental and economic constraints on the degree of planting across households. In short, poverty is the main barrier. As the following personal accounts attest, this barrier can be overcome.

Selected narratives of tree-planters

The experience of the Wällo tree planter, Käbbädä Hamza of Gwobeya Peasant Association, is illuminating. At the time of the interview, on 19 July, Käbbädä was living and farming in the locality of Gubawahel. He was 68 years old and had been born some distance away in Kutabär. He was a boy herding his father's livestock when the Italians came in 1935. Just after the Italians left, Käbbädä attended school for a couple of years. He can read and write. After farming for a few years, around 1950 he moved to Borru Méda and took employment with the foreigners (Americans), who had set up a hospital there. He learned to cook and was chief gardener, but, after six years, he felt his salary was insufficient so he moved to the town of Bestima, in the hills to the east of Lake Häyq, where, around 1956, he opened a tea house and a place where he prepared and sold mead (*t'äj*). He chose Bestima, because he had two aunts and a number of friends already living there. He stayed in Bestima for about 14 years and made enough money to send his children to school. He obtained land on contract, since he had no inherited rights to land in the area, and grew vegetables which he marketed at the town of Häyq. He also bought land in four different places in Gwobeya and moved here around 1970, farming until the Revolution broke out in 1974, at which point he became a land holder on the same footing as his neighbors. Around 1980 a producers' co-operative was founded in the area and Käbbädä became its chairman. He found the experience irksome, since people did not want to co-operate. Käbbädä actually moved to this particular site, known then as Village Number 1, in 1986 as part of forcible villagization in the area.

Käbbädä did not plant trees during his years at Borru or Bestima, but has been very active in Gwobeya, where he has planted over ten different kinds of trees and bushes: eucalyptus, acacia, *shushuwa*, *tini* (?), *besanna*, foreign juniper, *koba*, *agulo*, *qulqwal*, avocado, oranges, coffee, *ch'at*, and *wanza*. 'I like trees and cultivate them', he said, 'even when I find them.' Käbbädä sells oranges, *ch'at*, potatoes, and onions. He makes enough money to buy processed feed from the market for his livestock (3 oxen, 2 cows, 3 calves, 4 sheep, and 1 donkey). Käbbädä has been married three times, to two different women, the first of whom he divorced and later re-married. Altogether he has 7 children, 3 boys and 4 girls.

Käbbädä would be an exceptional man in almost any situation in which he found himself – ambitious, curious and innovating. Landlessness proved no obstacle to him. In this respect, his experiences were parallelled by several informants in the Borru area, notably Kassaw Bäqqälä, who grew up landless, but who invested income earned from trading grains and pulses in purchasing land. We interviewed Kassaw at the locality of Gaddiso on 27 June. He was then about 61 years old. He told of how his mother's divorce and re-marriage had deprived him of any land rights he might have derived from his father and of how he traded quite long distances to earn the money to acquire land, which he bought from local land-owners. Not until the land redistribution following the 1974 Revolution did he acquire enough land to stop trading and concentrate on farming. Kassaw first started to plant trees in the late 1940s or early 1950s: eucalyptus, foreign juniper, *shushuwa*, and *qulqwal*. In fact, one of the plots which he bought he planted entirely in eucalyptus, which he used for house construction and to sell. Not all rural people could so readily overcome the challenge of landlessness, and, for a number of our informants, this was their explanation for not growing trees. Other informants, notably in the Peasant Association of T'abisa and in the village of Duré in Gerba, drew attention to the amount of water which growing trees demand, and pointed to the difficulties which the local terrain imposed on them in getting that water to their plots.

Tree-planting as an extension of farm life is not unique to Wällo, but is confirmed by the evidence of our interviews in Gondär, where farmers faced similar challenges of access to land and water and showed similar enthusiasm and persistence in planting trees and bushes. In Armac'äho 12 of our 15 informants reported planting trees, at least 17 different species. These are listed in Table 5.7. Nine informants planted bananas, 8 planted *kok* (peaches), 6 planted eucalyptus, 5 planted *besanna* (*Croton macrostachys*), 4 planted papaya, and 3 each planted *wanza* (*Cordia africana*), *wäyra* (*Olea africana*), orange and a lime-orange hybrid they called *bäkralomi*. The fruit trees are all planted for marketing. Armac'äho farmers also grew bushes, 13 out of our 15 informants. *Gésho* was most popular, being grown by 13 farmers; followed closely by coffee, favored by 11 farmers; while only 3 grew *ch'at*. To some extent this commitment to trees and bushes reflects the more broken terrain of Armac'äho, which contrasts with the sweep of land on which the farmers of Dämbeya sow their cereals.

Mängesté Täklé, the Armac'äho informant who reported the experience of canopy forest, rivaled Käbbädä Hamza of Gwobeya in his commitment to tree-planting. He may actually have planted a wider range of varieties than Käbbädä, and, in any case, his commitment to tree-growing was greater in the sense that the bulk of his livelihood came not from his fields but from his bushes and trees. Mängesté sold tree crops to buy food. Mängesté was about 67 when we interviewed him. He told us of planting 12 different kinds of trees: eucalyptus, *kok*, oranges, citron, guava, date palm, olive, *welkefa*, *besanna*, *qach'ona*, and bananas. In addition, he grew both coffee and *gésho*. Some of the trees – olive, *welkefa*, *besanna*, and *qach'ona* – he planted to shade his coffee. Like Käbbädä, Mängesté was not born where he farmed, but, in his case, in the town of Gondär. He was brought to the Kärkär district of Armac'äho as an infant by his parents and has farmed the land settled by his father since he was about 15 or 16 years old. Although he has never had trouble getting access to land, Mängesté was not secure in his access to the oxen needed for plowing until he married his fourth, and current, wife, Bällät'äch Chäkol, who brought with her 2 oxen, 2 cows, 4 sheep, and 2 donkeys.

Dämbeya farmers planted trees in similar proportions to those in Armac'äho, 16 of 19 Dämbeya informants reporting this practice. Two Dämbeya informants reported that they had never planted trees. As was the case in Wällo, eucalyptus was the overwhelming favorite, 14 farmers reporting having planted it. *Wanza* was the second most popular tree, reported by 9 farmers, with *qulqwal* and *besanna* a distant third at 2 farmers each. Bush crops were less popular in Dämbeya, a bare half of respondents, 10 of 19 reporting having planted them, while 7 respondents said they never had. *Gésho* (8) was marginally more popular than coffee (7). However, the most striking contrast with Armac'äho is to be seen with respect to fruit trees. Dämbeya farmers report planting none.[27]

---

[27] One of the participants in the group interview with which we started our work in Dämbeya did

An explanation for the different behavior seems not far to seek. Dämbeya appears historically to have been grassland and relatively treeless, and farmers face serious shortages of fuel and building materials. On the other hand, their rich, ample fields produce good returns on cereals. We asked farmers across all our research sites about the various ways in which they generated cash. In contrast to the Wällo and Armac'äho farmers, the Dämbeya farmers overwhelmingly reported field crops as their principal means, hence their lack of interest in the opportunities which fruit trees offered to generate cash.

Woodcutters

Admittedly, as we have seen, a minority of northern highland farmers have never planted trees; some lacked the necessary access to water, others had problems of access to land. We shall take up the general question of land later, but here we want to report on the two informants who identified themselves as woodcutters, people whose principal source of livelihood was bringing fuel to market. Both of these informants, one man and one woman, were coping with the challenge of landlessness.

Sheikh Mohammed Asfaw was born near Lake Häyq in the village of Demamo. His father, a weaver, died when Mohammed was 2 years old. Mohammed claims that the father had inherited land, but that it was simply too small to farm and there were too many claimants on it. He started Quranic school in Demamo and then moved to the Peasant Association of T'abisa, where we interviewed him on 24 July. He was then 65 years old. He studied at Tola in T'abisa and married there. To support himself as a Quranic student, he collected wood – bushes really – and grass and sold it in the market. His wife shared this work with him. They carried it on their shoulders. They found the fuel on Guba and Tola mountains, places which, even today, are not protected. Today the wood is all gone and olive trees, which used to grow there, can no longer be found. Four years ago, Sheikh Mohammed moved from Tola to Lägda Mäsgid, where we interviewed him. Mohammed obtained land under the Derg's land redistribution program, but has never had the capital, or maybe the skills, to work it effectively. He has never owned oxen for very long, so he lets out his land on a 50 per cent share arrangement. In 1991 he stopped trading in firewood and grass, because he felt too old and weak. He never faced government restriction.

Tärräfäch Färrädä in the Peasant Association of Abärjuda, not far from the port of Gorgora on Lake T'ana, had a very similar story to tell. Tärräfäch has been a poor woman all her life. She was 80 years old when we interviewed her. Until the Derg, she never had land in her own right, and made her living by selling firewood and brewing beer. Today the woods on which she used to draw for her firewood have been declared protected by the monks of the monastery of Mandaba. Tärräfäch does not think that the spread of eucalyptus has had an adverse effect on the market for fuelwood. Indeed, she thinks that with the price of fuelwood today, much higher than when she worked at it, one could make a better living.

[27] (cont.) report growing a variety of fruit trees, but he seems to have been quite wealthy and was really growing them as a hobby.

Other innovators: market gardeners

Just as a minority of households relied on mining forests and bush through wood-cutting, another minority was engaged in intensive production of high-value crops in market gardens. While extraction of forest products is consistent with a Malthusian scenario for environmental decay, market gardens have significantly increased production and income from small plots of land in many developing regions.[28] Distinct from small gardens for household use, these are gardens for which market conditions influence crop choice. Crops in these gardens commonly include onions, cabbages, peppers and also bush crops and trees. The farm management survey provides some indication of the factors that contribute to this form of commercialization.

Lack of land, skills and capital seems to correspond to that of woodcutting. It is less clear what factors predict intensive market production. All of the farmers in the farm management survey of Gwobeya asserted that market conditions had no bearing on their cereal cropping decisions, but 26 per cent (31 households) reported that the market influenced the garden plot crops. In Borru and Gerba, 42 and 15 per cent respectively kept crops in their gardens for market sale, while none expected to sell field crops. As seen in Table 5.13, the only variables significantly related to the presence of a market garden were indicators of wealth (large animals) and income (presence of cereals surplus). The direction of causality is unclear in this case. Indicators of education, experience, and land availability had no explanatory power for any of these variables. In Gerba and Borru, no consistent statistical relationship could be discerned between market gardens and any of the potential explanatory variables considered. Presumably, this form of intensification is associated with unmeasured ecological features and unmeasurable skill and management traits.

Table 5.13 Factors influencing presence of market gardening in Gwobeya. Dependent variable: market conditions influencing garden crop choice

Significant variables	Sign	Level of significance
Large animals	+	1%
Cereal surplus	+	3%

Notes: Nagelkoeke R²: 0.3; predictive accuracy: 77.6%; Logit model

Historical dimensions of farmer innovation

It should to be clear from the foregoing that farmers of the northern Ethiopian highlands have a keen interest in growing trees. The rhetoric and activities of governmental and non-governmental agencies suggest that such tree-planting is generally perceived to be socially and environmentally desirable. Moreover, the

[28] Robert McC. Netting, *Smallholders, Householders: Farm Families and the Ecology of Intensive, Sustainable Agriculture* (Stanford, CA: Stanford University Press, 1993).

Crummey-Maugini photographs and the memories of residents outside of the *qwolla* areas, imply that farmers' tree-planting has been substantial enough to affect the landscape. But important questions remain unanswered. In particular, was tree-planting a strategy originating from farmers themselves, or was this interest in trees created by government agents? The implications for farmers' capacity for innovation are considerable. If all the innovations which we document can be attributed to the 1980s and early 1990s, then they might reasonably be viewed as the result of government interventions dating back to the 'Green Revolution' campaign starting in 1979 and to the government and NGO interventions of the latter half of the 1980s directed to the underlying causes of the famine of 1984–5. Tree-planting is currently supported by government tree nurseries, which provide seedlings, some free, others at very low cost. The network of nurseries was established by the previous government starting around 1980, building, in turn, on the first forestry school and nurseries set up by Haile Selassie's government.

The chronology of tree-planting indicates that the practice pre-dates government initiatives. Eucalyptus was introduced to Ethiopia early in the twentieth century by Emperor Menilek. Thereafter, government actions had little or nothing to do with the origins of tree-planting on the part of the farmers of the northern Ethiopian highlands. Tree-planting was a strategic decision reached by individual farmers and farm families at different times in different places, but it was a decision that some made early in their farming careers, and it was an option passed on to some of them by their parents.[29] Eucalyptus was known in the countryside of Wällo and Dämbeya before the Italian invasion of 1935, but it was not widely adopted. Farmers in Wällo described first seeing it around important centers like Borru. Few informants mentioned Dässé as the place where they first saw the tree, perhaps because it had already been adopted in the secondary centers of the region. It was certainly well established in Dässé by 1935, according to Italian sources.[30]

Information on dates of initial planting is provided in Table 5.14. Of the 82 tree-planters in Wällo, 64 provided information about when they began to plant trees, or planted particular varieties. Four of them started under the Italians; another 11 began during the 1940s; another 5 mentioned the 1950s; and 2 the 1960s. Twenty-seven attributed the start of their tree-planting activities to the reign of Haile Selassie, which, in this context, means the years running from 1941 to 1974, with the probability that the bulk of them meant the 1950s and 1960s. By contrast, only 15 farmers said that they first began to plant trees under the Derg: i.e. following the Revolution of 1974. Thus the ratio of those planting trees before the Revolution compared with those who started after is more than 3 to 1.

An even more dramatic ratio emerges from the information from Dämbeya in Gondär, although, admittedly, from a much smaller pool of informants. Fifteen of the 16 Dämbeya tree planters provided us with chronological information. Thirteen of them started before the Revolution (1 pre-Italian; 1, 1940s; 2, 1950s; 9, Haile Selassie), 2 after. The Armac'äho story follows the same pattern. Eleven of the

---

[29] When we asked a 60-year-old informant, Mohammed Ali, of Borru, what was the first eucalyptus tree he had ever seen, he gestured dramatically to the stump on which he was sitting, planted by his father.

[30] See here, in addition to the Maugini photographs, the account by Alessandro Lusana, 'L'Uollo Orientale,' *Gli Annali dell'Africa Italiana*, II, 2 (1939), pp. 477–515.

Table 5.14 Period of initial tree-planting

Date	Wällo	Dämbeya	Armac'äho
Italians or earlier	4	1	0
1940s	11	1	0
1950s	5	2	0
1960s	2	0	0
Haile Selassie (1941–74)	27	9	10
Derg (1974–91)	15	2	1
TOTAL	64	15	11

Source: Survey of elders

Armac'äho tree-planters provided chronological information, and, of them, 10 first planted trees during the reign of Haile Selassie.

Discussion

Farmer resource management: capacity and constraints

In contrast to the view of a tradition-bound peasantry entrenched in unsustainable farm practices, survey data and farmers' narratives demonstrate that resource-poor peasant farmers are modifying and developing their production systems. The farmers of Wällo are not passively caught in a trap of increasing land scarcity, declining land quality and intensified poverty and hunger. The data also indicate variation in their ability to innovate, despite the narrow band of resource differentiation. For example, woodcutting remains a source of livelihood for a handful of people with the most limited resources of land and capital, while the development of market gardens and tree plantations is dominated by the more successful, entrepreneurial farmers.

Overall we gained the impression that, poor and hard-pressed as they are, the farmers of northern Ethiopia are not desperate, nor have they forsaken sound farming practices. On the contrary, they have a truly astonishing range of knowledge of soils, seeds and seasons. Their sense of themselves, and of the value of their culture and institutions, was quite intact. They expressed themselves clearly, drawing on a rich store of metaphors. Their lives are stories of innovation, adaptation and adjustment, and, if the wider forces working on them are not always clear to them, they face those forces with resolve, a sense of irony and humor and an admirable capacity for work.

Not surprisingly, since participation in the market constitutes only a small percentage of household economic activity, cropping patterns reveal that households engage in a food-first production strategy, which limits the scope for innovation.[31]

---

[31] A 1983/4 survey indicated that Wällo farmers marketed 20 per cent of their sorghum and horse beans and 16 per cent of their *t'éf* and lentils. They marketed only 5 per cent of the remainder of their crops. See Alex Winter-Nelson, 'Rural Taxation in Ethiopia, 1981–1989: A Policy Analysis Matrix Assessment for Net Producers and Net Consumers,' *Food Policy*, 22 (1997), pp. 419–32.

Outside of what is grown for home consumption, the spectrum of crops is not wide and the market served is local and only occasionally regional or national. The need to ensure household food self-sufficiency before engaging in other farm activities is likely to be a barrier to innovation, especially when farms are too small to allow market production beyond household needs. The findings that households with less land per capita are likely to grow few or no trees and that food-surplus households are more likely to keep market gardens indicate that peasant farms with greater food security are more able to innovate. Because there is little apparent room for innovation within food crop production, the food-first strategy is probably unsustainable. However, in the absence of reliable food- and cash-crop markets, households are forced to strive for food self-sufficiency.

Farm management data and farmer narratives reveal that the less poor farmers have been more able to innovate and plant trees on a large scale, and that their expanded production is of goods with market value. Hence, one can conclude that past government policies that have impoverished farmers through taxation or inhibited commodity markets have probably undermined tree-planting and rural development generally. If food markets were improved, the reduced need for a food-first strategy would allow more variation in cropping choice. Similarly, improved cash-crop markets would broaden production options. Greater opportunities for non-farm employment could generate further resources for farm investment.

The absence of strong commodity marketing institutions constitutes a glaring difference between Wällo and areas where farmers have successfully managed increasing land scarcity, such as Machakos, in Kenya.[32] So, too, does the absence of non-farm income. In the case of Machakos, money earned in Nairobi was invested in commercial farming. In Wällo government intervention could promote market development through better legal protection of merchants, market information, finance for traders, and market infrastructure. Meanwhile, credit or grants to households that are willing but unable to move beyond cereal production could enhance tree-planting and the production of other high-value crops. In the absence of a more enabling economic framework, and the emergence of increased opportunities for both market participation and cash income, it is likely that many households will remain unable to adopt sustainable, intensive practices like commercial gardening and wood-production.

Clearly, access to land is an issue in the ability to engage in tree-planting and other innovations. But a focus on resettlement and tenure reform may be misleading. We find little evidence that the form of land tenure has influenced farm practices, although the land redistributions of 1975 and 1997 have constrained land accumulation and preserved the independence of households who are unable to innovate. The Revolutionary land reform of 1975 is generally given high marks for equity, but low marks for encouraging productivity. Legally, land remains state property, administered by local Peasant Associations and subject to alienation, redistribution, and externally generated policy interventions. This system has been much criticized by Ethiopian scholars, by NGOs and by Western government development agencies, all of whom argue that greater security of tenure is necessary

[32] Mary Tiffen, Michael Mortimer and Francis Gichuki (eds), *More People, Less Erosion: Environmental Recovery in Kenya* (Chichester: Wiley & Sons, 1994).

to encourage farmers to invest in land. Outside of government circles, there is widespread opposition to the continued state ownership of agricultural land, a provision written into the 1995 federal constitution.[33] However, the evidence reviewed here suggests that farmers have spent less time worrying about these issues than they have dedicated energy to coping with the circumstances in which they found themselves. Neither of the very different tenure regimes, which obtained under Haile Selassie and the Derg, seems to have deterred farmers from planting trees and making other investments in their land. Whether alternative tenure arrangements would have *increased* their investments is uncertain, but investment seems to depend on a far more complicated matrix of factors than simply land tenure.[34]

Farmers have not been the only tree-planters in Wällo. Governments are also responsible for the increase in forest cover between 1937 and 1997 documented in the Crummey-Maugini photographs. The Derg used its ownership of land to enclose hilltops and hillsides and to nationalize existing woodlots. It embarked on huge tree-planting schemes and it implemented policies, which radically restricted, where they did not halt altogether, local exploitation of woods for grazing and fuel. The government which replaced it has kept those policies in place. But a close observation of the landscape also reveals intricate patterns of tree-planting, and tree nurture, which can only be attributed to local farmers.[35]

Perceptions of environmental stress

We do not want to end with simplistic optimism. In Wällo afforestation by farmers and governments has not prevented environmental losses. Farmers feel acutely that the land is becoming 'narrow' – that population growth is reducing the amount of land available to each farming family. They experience this narrowness in many ways, not least in crises in obtaining fuel and grazing. Their memories of famine are vivid and painful. But growing land scarcity has not triggered a downward spiral of deforestation, soil erosion, yield losses and hunger.[36] Hunger remains real – in contrast to Gondär – and land scarcity is a constraint on farmers, but the links from population to trees to hunger are not supported by our data.

Both despite and because of farmer initiatives, the environment is changing. Northern Ethiopian farmers do not make light of environmental change. How much they have internalized the rhetoric presented to them by a succession of political cadres, how much the perception arises from their own experience, and how much older people always view the past as a landscape of loss, is impossible to

[33] See the contributions by Dessalegn Rahmato and his collaborators: Dessalegn Rahmato (ed.), *Land Tenure and Land Policy in Ethiopia after the Derg. Proceedings of the Second Workshop of the Land Tenure Project* (Trondheim: University of Trondheim, Centre for Environment and Development, 1994), Working Papers on Ethiopian Development, No. 8.

[34] T. Bassett and D. Crummey (eds), *Land in African Agrarian Systems* (Madison: University of Wisconsin Press, 1993).

[35] The relative area coverage of farmer-planted trees as opposed to government plantations awaits a careful reading of aerial photographs, which are available for the area from 1957 and 1986 and, less promisingly, of the data generated by remote sensing.

[36] For a revisionist view on soil erosion, see Belay Tegene, 'Indigenous Soil Knowledge and Fertility Management Practices of the South Wällo Highlands', *J. Eth. St.*, XXXI, 1 (1998), pp. 123-58.

say, but a landscape of loss is how most of our elderly informants view it; not a landscape stripped of vitality, nor denuded of value or meaning, but one, nonetheless, impoverished from the one they knew as younger people. Farmer initiatives and government programs have combined greatly to increase the number of eucalyptus trees in the landscape. While Wällo is far from becoming a eucalyptus desert, informants do speak of dwindling overall numbers of the principal indigenous tree species – junipers, olives and acacias – and of the smaller sizes today of the individual members of those species, although they do not indicate any absolute species loss within the last 60 years.[37] Nevertheless, farmers have planted trees and bushes – for fuel, construction and marketable fruits and stimulants. Government efforts would be better directed to finding solutions to the grazing crisis and to arresting the decline of junipers, olives and acacias.

[37] Observations supported, in general, by Sebsebe Demissew, 'A Study of the Vegetation and Floristic Composition of Southern Wällo, Ethiopia,' *Journal of Ethiopian Studies*, XXXI, 1 (1998), pp. 159–92; see especially the comments on pp. 159, 186–7, and 189.

6

The Wild Vegetation Cover of Western Burkina Faso
Colonial Policy & Post-Colonial Development

MAHIR ŞAUL, JEAN-MARIE OUADBA
& OUETIAN BOGNOUNOU

This chapter presents an overview of the evolution of the wild vegetation cover in western Burkina Faso from the early colonial period to the present day, highlighting the effect on vegetation of government policies and of recent agricultural practices.[1] Commercial development since 1960 has had a massive impact on the flora. To provide a reference point for this transformation and explore its sources we start our analysis at the beginning of the colonial era. We then describe a number of projects and larger-scale commercial ventures that have left their mark on the countryside. In Burkina Faso agricultural change is still strongly connected to smallholder farming. We therefore investigate the popular appeal of new forms of production such as tree plantations and their effect on wild plants. The example of western Burkina is instructive in the light of recent challenges to the widespread belief in a straight and narrow path leading to increasing deforestation in the West African savanna. Our survey confirms the contradictory and inconclusive character of human/environment relations, and the unforeseeable twists and turns which mark their development. We describe how recent commercial expansion leads in directions that are unexpected on the basis of a narrow demographical approach, and unlike what is reported for other places in West Africa. We suspect, however, that what we observe in Burkina is not unique.

The Ecological Setting

Our project area lies in the southern half of western Burkina Faso and is

---

[1] Project members Jean-Baptiste Kiéthega and Christophe Dya Sanou, both of the University of Ouagadougou, have contributed substantially to our research and to the information provided in this chapter. Kiéthega has presented the results of his many years of research in extensive form in his monumental doctoral thesis, *La métallurgie lourde du fer au Burkina Faso*, 2 vols., Université de Paris 1, 1996. Sanou is continuing his erosion experiments in the Péni station and will present his findings separately in the form of articles and a doctoral thesis in preparation.

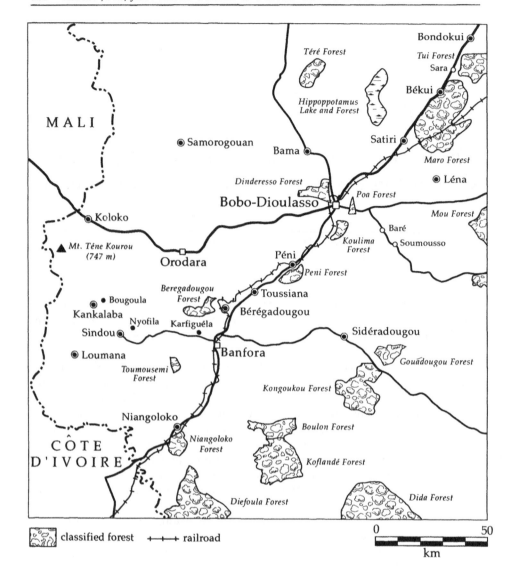

Figure 6.1   Project area and research sites

spontaneously a Sudanian orchard savanna conditioned by fire.[2] Average yearly rainfall ranges from 1100 to 1000 mm in a summer rainy season of 180 to 200 days. Predominant in the flora are the species *Burkea africana* Hook., *Detarium microcarpum* Guill. and Perr., *Trichlia emetica* Vahl., *Hymenocardia acida* Tul., *Ostryoderris stuhlmannii* (Taub.) Dunn ex. Harms, and *Pteleopsis subrosea* Engl. and Diels (Group 2 of the Vegetation Map of Burkina Faso).[3] The Tagouara plateau in the westernmost part of the project area, where two of our sites are located, is a highland over 700 m above sea level containing the headwaters of the Mouhoun (Black Volta), which flows north, and the Comoe, which flows south. East and south of the Banfora cliff the plateau drops to 500 or 400 m above sea level.

Fringing forests (*galeries forestières*) and thick dry forests are important components of the environment of the project area. The fringing forests are of unusual ecological importance. They are situated on the banks of streams or rivers that carry water throughout the year, and are delicate systems that can quickly go into irreversible decline. The rich flora that are characteristic of these forests, and the serious risks that farm clearings present to them, have long been understood.[4] In some respects the fringing forests are like branches of Congo-Guinea vegetation that extend northward following the favorable environment of the river valleys, rather than a more luxurious version of the surrounding Sudanian vegetation. Of the 60 inventoried species of the Mouhoun fringing forests, 63 per cent belong to the Sudanian-Zambesian belt, 27 per cent to the Congo-Guinea belt, and 10 per cent are other species.[5]

The fringing forests are of greater ecological importance than would be expected from the small surface area that they occupy, because they contribute to plant diversity and protect the hydrographic system. The largest rivers of Burkina Faso and Côte d'Ivoire have their sources in our project area: the Mouhoun (Black Volta), Comoe, Leraba, and some tributaries of the Bani River, a major tributary of the Niger. The forests protect the banks of these rivers from erosion, maintain a humid microenvironment that reduces evaporation losses, improve water quality, and yield the most valuable timber trees.

The river valleys where these forests are to be found were not populated within our study period, as they were infested with the two disease vectors of this region, the tsetse fly, which transmits trypanosomiasis (sleeping sickness), and the simulium fly, which transmits onchocercosis (river blindness). In the 1970s and 1980s the

---

[2] For the vegetation we use the terms 'wild' and 'spontaneous' as synonyms – the latter term being borrowed from the authors writing in French – to indicate plants that are not planted deliberately. Since the presence and distribution of these plants in the landscape are significantly affected by human activities, the commonly encountered expression 'natural vegetation' is not adequate for them.

[3] J. Fontès, A. Diallo, J.A.Compaoré, *Carte de la végétation naturelle et de l'occupation du sol, Burkina Faso*, Institut de la Carte Internationale de la Végétation, Université Paul Sabatier de Toulouse and Institut du Développement Rural, Université de Ouagadougou (Toulouse, 1994); J. Fontès and S. Guinko, *Carte de la végétation et de l'occupation du sol du Burkina Faso: Notice explicative* (Toulouse, 1995), p. 33.

[4] L. Bégué, 'Contribution à l'étude de la végétation forestière de la Haute-Côte d'Ivoire', *Bulletin du Comité d'Etudes Historiques et Scientifiques de l'Afrique Occidentale Française*, Série B, no. 4 (1937).

[5] E. G. Bonkoungou, 'Inventaire et analyse biogéographique de la flore des galéries forestières de la Volta Noire en Haute Volta', *Notes et Documents Voltaïques*, 15, 1–2, pp. 64–83.

forests were systematically sprayed, in an international campaign to eradicate these diseases.[6] They were also cut down or thinned, as part of the treatment, without serious consideration of the ecological consequences. However, the onchocercosis program did set up a monitoring procedure to ensure that the applications of insecticides did not seriously perturb the fresh water system. These zones, now declared free of disease, are coveted by organized bodies and private individuals. Among the organizations dedicated to settling them and opening them to agriculture, are the AVV (Aménagement des Vallées des Volta) and the Fronts Pionniers du Sud and du sud-Ouest, to which we shall return later.

The Colonial Period

The French colonial government shaped the successor nation-state in its baseline economic structure, but the effect was uneven when we consider specific domains. The direct colonial impact on changes in the vegetation cover was slight. Some extractive activities led to the transformation and impoverishment of the flora, but they were not of a magnitude and intensity to make a major difference. The same can be said of conservation efforts, which gained momentum in the later decades of the period. A more substantial impact on the flora was mediated by changes in farm life, which themselves were linked to administrative policies which set the rural areas on their course, but the more significant transformations came only after independence was won in 1960.

Persistent throughout the colonial and post-colonial periods combined and giving them coherence is the evolution of farm life towards greater commercialization. From the type of farming in which only surpluses were sold agriculture moved towards increasingly greater use of purchased inputs, including labor. The correlate of this sequence relevant to this chapter was the play between the expansion of areas under cultivation and fallow land management. After independence export-crop production and the extraordinary growth of grain farming for national and regional markets accelerated the pace of this dynamic. However, the wild trees were not replaced only with a few domesticated cultivars from annual grasses and leguminous species; instead, in our project area, agriculture turned towards fruit orchard 'agroforestry'. The following sections of this chapter are directed to a description and analysis of these developments, showing the range of variation in the different parts of our project area, and assessing the importance of local particularities. They conclude by showing how crucially fruit orchards are integrated into the farming pattern in the project area, and bringing out the implication for fallow land and wild vegetation.

Colonial research. The beginning of the twentieth century in what is now Western Burkina Faso was marked by the disruption of nearly three decades of un-interrupted war. The 1890s witnessed major confrontations between the forces of Sikasso, Samori, and the French, and an acceleration of smaller conflicts between villages plus heightened banditry. Even more destructive to the local economy were

---

[6] In Burkina Faso this action was carried out by the Service Général d'Hygiène et de Prophylaxie (SGHMP).

the French colonial occupation and efforts to establish a colonial administration in the first decade of the twentieth century. The aggregate result of these events was loss of population and wealth, and a retrenchment of farmed areas. Political crisis dragged on into the second decade of the century. There were growing movements of insubordination in the southern and western parts of the *cercle* (administrative district) of Bobo-Dioulasso, and then the great anti-colonial war of 1915–16, which spilled into the central and northern parts of the same *cercle* and was suppressed with destruction of property and mortality of disastrous proportions.

As a result of these misfortunes, when the early colonial observers penned the records that constitute the horizon of our own research, they were witnessing an economy that had severely regressed in a relatively brief period of time. What they described in static terms or as an incremental development of farming was instead a landscape that was returning to a state of relative non-use, following a period of more intensive occupation only a few decades earlier. Under the coercive colonial peace that was imposed after World War I the population returned only slowly to its former level, and the rural economy started on its course of development as a colonial 'backwater'.

Compared with other colonial powers, France was late in conducting systematic surveys of natural resources and economic potential and in formulating a conservation policy. We do not find in French West Africa the environmental 'conservation lobby' of the British colonial establishment, nor the energetic coalition of local forces that opposed such conservation efforts.[7] The late pre-colonial explorations of L. G. Binger and L.-P. Monteil, in 1888 and 1890 respectively, give some indications of the landscape, some vignettes of occasionally prosperous agricultural economies and busy trade life, but they are far from providing a coherent picture of the environment. The gathering of information by men of science starts with August Chevalier, a botanist who in 1898/99 was included as a researcher in the conquering columns.[8] His mission was followed by that of Emile Perrot in 1927/8, and of Bégué in 1936.[9] These missions, while few and far between, give us precise information on the natural vegetation, land use, farm production, plants and their uses, the constitution of a classified forest, and the mosaic of ethnic groups (a list of more than thirty) in what is now western Burkina Faso.

Colonial agronomic research started almost as soon as botanical exploration. Starting in 1901, the administrators introduced fruit trees such as mango (*Mangifera indica*), citrus and pineapple, and spread species that were not foreign, such as bananas and papayas, to areas where they were scarce or non-existent. These initiatives were carried out in an authoritarian manner. Villagers were ordered to plant and maintain fruit trees, especially in the roadside colonial compounds dotting the countryside for travelling officials to spend the night. The first agronomic research and experimentation station in what is now western Burkina Faso was

---

[7] For the cases of the Gold Coast and Nigeria, which contrast sharply with colonial Upper Volta, see R. Grove and T. Falola, 'Chiefs, Boundaries, and Sacred Woodlands: Early Nationalism and the Defeat of Colonial Conservationism in the Gold Coast and Nigeria, 1870–1916', *African Economic History*, 24 (1996), pp. 1–23.

[8] A. Chevalier, 'Mon exploration botanique du Soudan français', *Bulletin du Muséum d'Histoire Naturelle*, 5 (1900), pp. 248–53.

[9] Bégué, 'Contribution'.

established in 1904 in Banfora. By 1908 this station had a ten-hectare-wide plantation near the waterfalls of Karfiguela. Here attempts were made to create *Landolphia* (vine rubber) and *ceara* (*Manihot glaziovii*) plantations. Other programs of tree-planting in the villages were under the supervision of canton chiefs. Later, additional stations in Niangoloko and Farako-Ba were opened. These institutions assisted in the introduction of species such as Teck and Neem, and also the success-ful propagation of new fruit trees, which revolutionized farmer strategies in organizing production, a topic to which we shall return in the section on 'Mango and other fruit orchards'.

Colonial policy and development. The most important colonial policies to affect the environment fell under the notorious expression *mise-en-valeur.* This term can roughly be translated as development, but includes in its meaning the idea that resources were wasted in the hands of the local population, either for lack of investment or because the locals were indifferent to intensifying production. The ideal against which this failure was registered was not always spelled out clearly. Sometimes it meant the potential for a higher level of production, desirable both for the well-being of the population being administered and for greater government revenue. At other times foremost in the mind of the *mise-en-valeur* planners was sustaining the imports of the metropole. These different objectives intertwined, with the most prosperous colonies having the strongest export regimes, generating most local revenue from taxes on foreign trade.

Western Burkina Faso, under the various names it assumed in the different phases of its colonial existence, was both export-poor and poor in revenue. *Mise-en-valeur* here therefore meant primarily the encouragement of export crops, and secondarily large public work projects. Furthermore, exports meant commodities that could be exported to France, restricting the list of development crops to peanuts, sesame, and cotton. The older export products of the region, most importantly cattle and iron tools to the forest region, and secondarily dyed cotton cloth, were only of interregional significance and were not encouraged. The production of these older commodities did come into conflict with major colonial policies, but they were simply ignored rather than being actively suppressed.

Other important aspects of colonial policy were the struggle to spread the use of minted currency at the expense of the generalized use of cowrie shells and supporting barter-like exchanges, and the sustained effort to increase direct taxation in the form of the head tax. Again these two were connected, and together determined how this region was integrated with the rest of the colonial economy. The development of the grain market to supply the growing cities, for example, cannot be understood without taking account of these two colonial policies. In the pre-colonial period there had been a large group of non-producers (warriors, Islamic clerics, traders), probably comparable in size to the urban population of the colonial period, but nevertheless grain markets were small compared with the active commerce in other commodities. The growth of food production for sale and the development of a network of grain traders were here typically colonial developments with important consequences for agriculture, and therefore for the environment. Monetary transformation and taxation underlay these developments.

The abuse of human resources in the colonial period amounted to a kind of 'mining'. It included, first, forced labor and various corvées, and then, in addition, coerced migration of workers to develop the forest region of Côte d'Ivoire and the inner delta of the Niger. An unintended consequence of these policies was a parallel stream of voluntary migration to the neighboring British colony of the Gold Coast, which further drained the work force.

The guiding thread for the rest of this section and the following one can be given here in a few sentences. All things considered, the colonial period forced the population into a regime of intensified production. Besides its disproportionate human costs, this intensification showed poor long-term returns. The total colonial effect on spontaneous plant cover and the other elements of the environment is difficult to determine with precision, but does not appear to have been drastic except for the near extinction of large wild animals, particularly mammals, the combined result of an increase in the available means of destruction.[10] Specific policies and developments brought contradictory positive and negative effects.

In terms of public projects, the most salient undertaking of the colonial government was the establishment and maintenance of a road network. Initially, the primary purpose of this network was administrative and military rather than commercial. Today it appears quite humble. But observers have pointed out that it was quite lavish relative to the technology and capital resources made available for its construction, or in terms of the uses to which the roads were put.[11] Until almost the end of the colonial period the network was built and maintained exclusively with requisitioned manual labor. In the 1920s the decision was taken to join the region to the ports of Côte d'Ivoire, and the road network was further expanded. River transportation was gradually abandoned in favor of cars as these became available, thus increasing the commercial value of the roads. The roads were generally lined with trees. But the transportation policies had another impact on forest resources that was barely veiled by this cosmetic improvement. We shall proceed here in reverse chronological order following the order of magnitude of the impact.

The most capital-intensive transportation project of the colonial period was the Abidjan-Niger railroad. The railroad came to the Mouhoun region late; it reached Bobo-Dioulasso in 1934 and Ouagadougou only in the 1950s.[12] It was a project which absorbed tremendous amounts of labor and foreign capital, and completed the re-articulation of the commercial flows of the region southwards, while cutting off the bend of the Mouhoun from the Niger basin of which it had previously been a part. The steam engines consumed a large volume of fuel, which, in the absence

---

[10] Of the large mammals that disappeared only the elephants have made a comeback in Burkina Faso in recent decades, following the world-wide conservation efforts focusing on the banning of ivory.

[11] In 1930 the colony of Upper Volta had 2,192 kilometers of motor roads, about twice as long as in the twice as large French Soudan, and more than four times the roads of Côte d'Ivoire which was roughly the same size. At that time there were only 71 cars and 222 trucks in the entire colony, while the French Soudan had almost five times as many cars and more than twice as many trucks. The comparison with Côte d'Ivoire is even more unfavorable. See R. Delavignette, *Afrique Occidentale Française* (Paris, 1931), pp. 211, 212. Forced labor for road construction contributed to repeated anti-colonial uprisings.

[12] The project was never completed in its original design. Under the regime of Captain Thomas Sankara the tracks were extended as far as Kaya, with the ultimate aim of reaching the rich Tambao manganese deposits, which does not appear likely to be realized soon.

of mineral coal deposits, had to be supplied in the form of wood cut along the railroad tracks.

The station at Péni, which is one of our research sites, had a particular place in supplying wood to the trains coming from the south. The engines needed extra fuel to climb the steep slope between Bérégadougou and Péni before reaching the plateau of Bobo-Dioulasso and there was a two kilometer radius of wood cutting around this station. In this zone the productivity of ligneous plants is estimated to be only 16 cubic meters per hectare, with a growth rate ranging from 0.6 to 1.66 cubic meters per hectare. Steam engine demand led to the creation of the first classified domains in western Burkina – the classified forest of Péni, covering 1,200 hectares, in 1942.

The consumption of wood fuel had already started with automobile roads. Before the introduction of diesel engines, there were trucks that burned charcoal or wood. Locally called *gazo* (gazogene lorries), they absorbed large quantities of wood. In addition, the Péni region supplied the city of Bobo-Dioulasso with fuel. The demand was strong because of companies such as CFAO, CFCI, and the military camp, which was one of the largest in French West Africa. The city was the hub of industrial development in the region, and most of it was driven by wood or wood derivatives. For a while there was also a sawmill in Banfora and selective extraction of timber woods.

It is simply too difficult to evaluate quantitatively how important these pressures on the local plant cover were, and how much of a change they represented compared with the previous situation, because the industrial use of firewood did not start with technologies made available in the colonial period. Pre-colonial production had already intensively used selected hardwoods as a source of energy, the primary users being the blacksmiths engaged in smelting operations. In the project area there were extremely important iron production and tool-making centers, and the region as a whole was an iron exporter until scrap metal of European origin made this line of commerce superfluous. The extraction of iron ore, conducted on a scale that was capable of transforming the composition of local flora very quickly, continued well into the colonial period. [13]

Ironwork is the occupation of endogamous groups, but these groups do not constitute a culturally and socially homogeneous category throughout the region. Blacksmiths form small clusters that contrast with the agricultural population, but also with each other, ethnically as well as in their work techniques. The presence of blacksmith groups in certain areas and villages, but not in others, is the result of contingent historical factors. In the research area, major smelting sites existed near Sara in Bwa country, around Péni, and on the Tagouara plateau. In these places, and in others too, local blacksmiths still continue to fashion farming tools in their forges, and supply almost all the tools that the farmers need, including some replacement parts and repair work for the newly introduced ox-drawn plows. In contrast, smelting operations stopped around 1950, as the tools started to be made using imported or scrap metal.

[13] J.-B. Kiéthega, 'La metallurgie lourde du fer au Burkina Faso', Doctorat d'Etat thesis, Université de Paris I, 1996. Elisée Coulibaly, 'Savoir et savoir-faire des anciens métallurgistes: Recherches préliminaires sur les procédés en sidérurgie directe dans le Bwamu (Burkina Faso-Mali)', doctoral thesis, Université de Paris I, 1997.

When the smelting industry was still alive, its impact on the flora was specific because charcoal was produced from a limited number of woods of high caloric value. The impressive mounds of slag left behind in old smelting sites give a rough measure of the huge quantities of charcoal used.[14] The principal species used for smelting were *Prosopis africana* Guill. and Perr., *Burkea africana*, *Parinari polyandra* Benth., *Swartzia madagascariensis* Desv., and *Hymenocardia acida*, which grow mostly on valley bottoms, and *Afrormosia laxiflora* Harms, which is a tree of the escarpment woods with ironstone concretion outcrops.[15] As iron ore is very common in this area, the location of iron-producing blacksmiths was probably more dependent on the availability of these trees and their rate of exhaustion than on any other factor. Today the high-caloric species sought after by smelters are still rare in the landscape. The end of smelting activity in mid-century lifted this particular pressure on these species, at the same time as it was replaced by colonial industrial and transportation needs and as the government stepped up its effort to classify forest domains.

Forest plantation efforts were virtually null in the colonial period. The classified forest action was limited to forbidding access to the local population, and to modest fire protection. Underlying this policy were arguments of the botanists that the orchard savanna was the product of human activity, and if left undisturbed it would revert in the long run to the assumed climax of dense dry forest.[16] In the 1940s the fuel needs of the growing cities increased concern for conservation, and classified domains were created around urban centers expected to expand in the future. The archipelago of classified forests that can be observed on any tourist map today was achieved at the expense of the population of the immediate area. With each decree many villagers lost their farmland, and the conflict between the putative economic advantages at the national level of having classified domains and the local populations' loss of resources when they were excluded from these domains was rarely addressed, either in the colonial period or in the days of independence that followed.

Unplanted Tree Crops from Colonial Times to the Present

The harvesting of wood for fuel was not the only colonial interest in forest products. Even earlier, the harvesting of another forest commodity shaped the colonial economy. Before even completing the military occupation, French officials in the Volta region seem to have decided that the major export commodity of the new territories was going to be rubber obtained from the spontaneously growing vine *Landolphia heudelotii* A. DC. France imported all its rubber from foreign markets outside its control and the economic potential of this plant had already

---

[14] The furnace models are very different, but one firing of a modest furnace that produced 15 or 20 kg of iron necessitated at least 25 cu.m. of charcoal.

[15] Coulibaly, 'Savoir et savoir faire', also lists *Terminalia macroptera* Guill. and Perr. and *Cassia sieberiana*, DC.

[16] Empirical research does not always justify this assumption. In a study carried out on a zone fenced and protected from fire after long years of agricultural use (in Sapone, in the central Mose plain) Ouadba found that the vegetation evolved toward perennial grasses rather than trees. J.-M. Ouadba, 'Note sur les caractéristiques de la végétation ligneuse et herbacée d'une jachère protégée en zone soudanienne dégradée', in C. Floret and G. Serpantié (eds), *La jachère en Afrique de l'Ouest* (Paris: ORSTOM, 1993).

been identified in eastern Senegal at the end of the nineteenth century. Rubber exports had started before the French occupation. The African leader Samori had this product gathered, and sold it to help finance his firearm purchases.[17]

The *Landolphia* vine was fairly abundant in the southern and western parts of the Mouhoun, and in the first years of the occupation, the latex was promoted as a means of paying the head tax in kind. From 1900 to 1901 tax payments in the form of latex in the *cercle* of Bobo-Dioulasso doubled, although the overall level of production was still very modest. Soon afterwards private traders started to complain that the administrative policy of accepting taxes in the form of rubber latex deprived them of a market. The collection of taxes in kind ended in 1908, but the payment of tax from the proceeds of rubber sales continued. The gathering of rubber fitted the production schedules and internal division of labor of households. The latex was gathered by dependent young men and women in the non-farming season, relieving the heads of household from having to make substantial changes in their farms to pay the taxes, allowing them to make the junior members contribute without having to give them more autonomy.

The harvesting of rubber from wild vines proved disastrous for the species.[18] After exceeding 22 tons in 1903 and reaching 29 tons in 1904, rubber sales stagnated and soon began to fall.[19] The growth of rubber production then stagnated and soon began to fall. The damage done to the vines when they were incised frequently killed them, and the spontaneous stock rapidly started to disappear. In addition, in some places the population started to destroy the vines deliberately. They realized that the head tax had become a permanent imposition, demanding increasingly more hours of work to pay rising taxes, and they reasoned that without the *Landolphia heudelotii* vines most people could not be made to pay very much.

In 1910 rubber production went into full crisis because of the combination of falling local production and deteriorating world prices, as rubber from *hevea* plantations in other parts of the tropics entered the market. Between 1910 and 1913 the harvesting of rubber was officially prohibited in order to allow the natural stock to recuperate. In 1914 harvesting was permitted again and production reached a ceiling of 80 tons in 1920. The administration tried to establish *Landolphia* and imported *ceara* plantations to promote a different production strategy. However, the *ceara* did not do well and the artificial stands of *Landolphia*, while not totally unsuccessful, did not prove to be commercially viable. Eventually the gathering of rubber went down to nothing because merchants lost interest in it, although it picked up briefly during World War II, disappearing again after the war.

A few other forest products became export commodities in the colonial period. The most important was *shea* 'butter', known in the francophone world as *karité*, an oil which is in solid state at normal temperatures. It is obtained from the kernels of

---

[17] Yves Person, *Samori: Une révolution dyula* (Dakar: IFAN, 2 vols., 1968), vol. 2, pp. 929, 931, 938, 941.

[18] In the late nineteenth century in Gabon rubber plants had already been almost exterminated under export pressure. Alfred Molony, *Sketch of the Forestry of West Africa* (1887), p. 90, quoted by Grove and Falola, 'Chief Boundaries', pp. 5–9.

[19] M. Şaul, 'L'économie formelle du cercle de Bobo-Dioulasso, 1899–1930', Paper presented to the colloquium, 'Le Cercle de Bobo-Dioulasso dans le Processus d'Intégration des Colonies Ouest Africaines', Bobo-Dioulasso, December 1997.

the tree *Butyrospermum paradoxum* (Gaertn. f., synonym: *Vitellaria paradoxa*) Hepper, of which there are vast stands in western Burkina.[20] This oil is the principal fat in the local diet and also has very important cosmetic and medicinal uses. Since pre-colonial times some of it entered local market exchanges. In the early colonial period, *shea* oil, while highly visible in the *cercle* statistics, was not regulated by government or targeted for development. In the 1920s production estimates varied between 100 and 200 tons per year. But *shea* oil found uses in the European chocolate, cosmetic, and soap industries, and became a significant export commodity after 1924. In the early 1970s *shea* oil and kernels together rose to become the third foreign-exchange earner for the country, although their share has been going down ever since.

The other important tree product is the African locust bean, which like *shea* nuts are gathered from unplanted trees (*Parkia biglobosa* Benth. and Hook). These seeds are processed into a condiment (*sumbala*, a kind of strong vegetable cheese) and constitute one of the highest value-per-weight commodities of the rural economy. The pods provide another important food product, the yellow sweet powder that is sold in the markets and enters into many kitchen preparations, as well as some other secondary products.[21] While overseas export commerce found no use for these seeds, they are of great regional commercial importance. Note that although the *shea* and the African locust trees were not deliberately planted until very recently, farmers did protect them. Yields of the trees improve when they are incorporated into a farm, and their very propagation and survival depend on the farming and fallow sequence. Farmers spare these trees when they clear new farms.[22] When a sorghum field has, as is commonly the case in our project area, 5 to 10 *shea* trees standing, they bring in a revenue of about 3,000 CFA francs per hectare. Two or three locust trees in the same hectare bring in 5,000 CFA francs. At the same time, their shade reduces the grain crop yield by 50 to 70 per cent, and in the case of a locust tree this shade covers up to 200 sq.m.[23] If the total area of the farm cannot be extended, the farm owner may fell some of these trees, and thus forgo their valuable produce, in order to obtain larger grain harvests. This system of 'careful balance' has been dubbed the traditional agroforestry. It provided the model for the fruit tree orchards, the new agroforestry, which we shall describe in the section on 'Mango and other orchards'.

Another forest commodity was *kapok*, which is produced by two unplanted species, *Ceiba pentandra* (L.) Gaertn. and *Bombax costatum* Pellegr. and Vuillet. In

---

[20] Burkina Faso as a whole is a *shea* tree heaven. The savanna woodlands dominated by *shea* trees are estimated to amount to 6.5 million hectares, about a quarter of the total land area: J.J. Kessler and C. Geerling, *Profil environnemental du Burkina Faso* (Wageningen, 1994), p. 44.

[21] E. G. Bonkoungou, 'Monographie du Néré, *Parkia biglobosa* (Jacq.) Benth., espèce agroforestière à usages multiples', (Ougadougou: Centre National de la Recherche Scientifique et Technologique, Institut de Recherche en Biologie et Ecologie Tropicale (IRBET), 1987).

[22] In the village of Dolékha in northern Côte d'Ivoire a study found that between 1962 and 1993 the parkland with shea and locust bean trees had become much richer following the expansion of the farmed area for cotton and cereals. L. Bernard, M. Oualbadet, Ouattara Nklo and R. Peltier, 'Parcs agroforestiers dans un terroir soudanien: Cas du village Dolékha au nord de la Côte d'Ivoire', *Bois et Forets des Tropiques*, 244 (1995), pp. 25–42.

[23] J. J. Kessler, 'Agroforestry in Burkina Faso: A Careful Balance', *International Agricultural Science*, vol. 63, pp. 4–5.

Photo 6.1a,b,c *Parkia biglobosa* (néré or dawadawa): the tree, the flower, and a woman carrying her harvest of beans.
(Photographs 6.1–6.3 by M. Şaul)

Photo 6.2 *Butyrospermum paradoxum* (shea or karité). The caterpillar (*situmu* in Jula, *kpiye* in Bobo) that thrives on the tree in the rainy season is gathered and eaten.

1920 the *kapok* production of the *cercle* of Bobo-Dioulasso was estimated at 10 tons. While *kapok* is now known mostly as a stuffing for pillows and upholstery, it used to have industrial uses. The international demand for these latter needs subsided over time and the commodity lost its visibility in foreign trade.

The cases of *Butyrospermum p.*, *Parkia b.* and *kapok* contrast with that of the *Landolphia* vine, because their commercialization has been environmentally benign. Despite the tremendous trade interest in the first two in particular, their intensive harvesting did not lead to the destruction of the trees. On the contrary, their role in 'agroforestry' intensified and production figures went up, even when the agronomic establishment was indifferent to them. The different impacts of commercialization resulted from the different ways in which the tree crops became part of the rural economy. The gathering and harvesting of *shea* nuts and locust beans was one of the economic activities of women. It was motivated by the contractual obligation of women to provide condiments within the household and by their desire to generate autonomous income. In both cases most community members were supportive of these objectives, and there was widespread interest in protecting the trees and resolving potential conflicts by regulating their use. Rubber, by contrast, was externally imposed and it provided few benefits as long it was tied to fiscal obligations. The gathering of the latex fell upon the dependent household members, and this situation exacerbated intra-household tensions. Few people were motivated to safeguard the survival of the plants, while, on the contrary, some stood to benefit from their disappearance.

Very recently, in one of the many startling turns that tree-planting fervor is taking in Burkina Faso, there is a trend to make *Parkia biglobosa* a plantation tree, including it in the jump from the traditional agroforestry to the new one. Aspiring plantation owners are purchasing the seeds in large quantities and experimenting with different techniques of planting them over parcels of several hectares, even though the rate of failure in establishing the seedlings is high because the process has not yet been mastered. The pioneers for this trend seem to be urbanites, as we shall see later in the case of fruit trees as well.

Peanuts and Cotton from the Colonial Period to the Present

The colonial era inaugurated several trends of agricultural innovation, especially involving industrial crops, primarily oil crops – peanuts and sesame – and cotton. Cotton has grown in importance up to the present and has clearly discernible impacts on land use. Farmers have also transferred some of the technology of cotton-growing to cereal-growing. These developments, separately and in interaction with each other, have worked simultaneously to expand the area under cultivation and to eliminate fallowing. This, in turn, has placed increasing pressure on spontaneous vegetation.

Peanut production started to grow between 1928 and 1932, in response to the world economic crisis. Peanuts were produced as a subsidiary crop and fitted in a particular way into the crop rotation pattern. They were sometimes intercropped in cereal fields, where they also served as a ground cover. They could also be grown in separate parcels, and this has increasingly become the case in the last few decades. The advantage of peanuts is that they require little weeding, and are thus not very demanding of labor.

In the colonial period, peanut areas went from 55,000 hectares, producing 25,000 tons in 1937 to 110,000 hectares producing 46,000 tons in 1942. An oil press was opened by the Compagnie Française de la Côte d'Ivoire, in Kiribina near Banfora. Colonial intensification programs primarily involved the extension of cultivated surfaces and not the increase of yields per unit area. Such programs might therefore have led to higher rates of clearing new land. But because peanuts were always subsidiary to grain cultivation in this region, this did not happen.

Since independence the area under peanuts has effectively increased. Until the 1990s the area under peanuts was larger than that under cotton, which made it the largest non-cereal crop. With the strides that cotton has made in recent years this is no longer true. In the project area peanut parcels are commonly located on old fields near the village that have been retired from cereal cultivation. This makes the initial preparation an easy step, although the parcels are now often ploughed before planting and manured or fertilized later. Peanuts thus extend the useful life of old fields and simultaneously prevent them from going back into fallow.

We have already noted that colonial planners did not always distinguish between increasing food production for the well-being of the population and increasing exportable crops. The encouragement of peanuts illustrates this confusion. Senegal had demonstrated the success of peanuts as a major export crop since the end of the nineteenth century, but the authorities also promoted peanuts for food security.

The encouragement of peanuts was part of a broader strategy of promoting subsoil crops, in order to limit the periodic ravages of locusts. A report from 1900, when the war of colonial occupation was still raging, reads: 'Serious propaganda has been carried out in favor of the cultivation of manioc, sweet potatoes, peanuts, cowpeas, harvests which will help people live when their farms have been destroyed by locusts.'[24] A report four years later reads similarly: 'We constantly exhort the natives to increase their underground crops, to extend their farms of tobacco and cotton'.[25] These 'exhortations' had no impact in extending aggregate farm area.

Sesame occupied smaller surfaces and fits, in the region of Bobo-Dioulasso, at the opposite end of the crop rotation cycle. Many people plant it during the first year in which they clear a new parcel, to hold the ground until the second year when grains are planted after a more thorough cleaning.

Cotton is by far the most important industrial crop of Burkina. It had a turbulent history in the colonial period. Only after independence, and especially in the past two decades, did it emerge as a great success story. In 1997 the surface area devoted to cotton production was around 325,000 hectares, over 9 per cent of all farmed land in the country. No other crop besides food grains comes even close to having this kind of centrality in farming, and there can be no doubt that the history of cotton is an important part of the evolution of plant cover in the country.

The French colonial authorities systematically promoted the production of cotton by compulsion. The drive started in 1902 when the textile industry in France was in crisis because of speculation in the United States, then the world's leading cotton producer. However, this initiative had no impact until after World War I. With the constitution of Upper Volta as a separate colony in 1919, the government decided that the production of cotton for export was going to be its main vocation.

In our project area some farmers had produced cotton in pre-colonial times for the local weaving industry, but it was not grown everywhere, and was at best only a subsidiary crop. Neither the quantities of cotton that the producers offered nor the high prices that the weavers paid for it allowed this cotton to become an export commodity. When the colonial government started to promote cotton for export, the farmers failed to respond with enthusiasm, and the administration resorted to coercion. Between 1924 and 1929 cotton was produced in fields supervised by colonial chiefs and agents of the administration. Production went from about 300 tons in 1923–4 to 3,528 tons in 1924–5 and 6,238 tons in 1925–6.[26] In that year, in order to increase production more rapidly, the administration enjoined the creation of joint village fields with sizes proportional to the population, 4 hectares per 100 people. The measure backfired and cotton production took a downward turn, never again in the colonial period reaching the peak of 1926. Cotton exports fell to insignificant levels after 1930. The government ended up withdrawing the coercive measures. Nevertheless in 1928, during the export cotton 'crisis', the representative

[24] Archives Nationales Côte d'Ivoire, 5EE 21 Cercle de Bobo-Dioulasso, *Rapport Administratif: Agriculture*, 2e trimestre 1900.

[25] Archives Nationales Côte d'Ivoire, 5EE 21 Cercle de Bobo-Dioulasso, *Rapport Agricole*, 1er trimestre 1904.

[26] T. Hartog, 'La culture du coton dans l'Ouest voltaïque', *Cahiers du L.U.T.O.*, 2 (1981), pp. 75–107; and A. Schwartz, 'Brève histoire de la culture du coton au Burkina Faso', *Découvertes du Burkina*, vol. 1 (1993), pp. 207–37.

of French textile interests in the colony claimed that 25,000 tons of cotton were being produced and sold to the local weaving industry.[27] While these rhetorical figures may have been exaggerated, cotton production to supply local weaving did continue after the forced export crop regime collapsed.[28]

The production of cotton for export gained new life after independence, entering a spiral of sustained growth in the 1970s. Several organizations offered new incentive packages premised on the principle of voluntary participation, and cotton production reached new records in every decade. Most of this production was in the western part of the country. The high points were: more than 35,000 tons in 1970: about 77,000 tons in 1980; 115,000 tons in 1986; and nearly 190,000 tons in 1991. Now grown exclusively in single stands, the surface area under cotton went from about 50,000 hectares in the mid-1960s to 185,000 hectares in 1991. After that year, however, export production fell rapidly for two years in a row, down in 1993 to the unexpectedly low level of 116,000 tons. The setback was triggered by an infestation by a caterpillar, *Heliotis armigera*. The crisis was hard to overcome despite highly favorable conditions, the devaluation of the CFA franc in 1994, and a phenomenal increase in world cotton prices. In 1997 export cotton production reached a new high, 334,000 tons – 75 per cent above the previous high of 1991. The sales accounted for half of all the foreign-exchange earnings of the country. Despite this new triumph, cotton remains very sensitive to price fluctuations and dependent on the smooth operation of the purchasing network and credit programs.

Growth went together with an impressive increase in yields, from about 300 kg of grain cotton per hectare in the 1960s to about a 1,000 kg in the 1980s. Well-to-do, skilled farmers, who have several hectares under cotton, now obtain yields of more than 1700 kg per hectare. These yield increases were achieved by using chemical fertilizers and insecticides, more experience and better laboring techniques, and the adoption of animal traction. There is even a growing number of tractors purchased not only through credit programs but also with private savings.[29] Worries about the various long-term effects of fertilizer use are occasionally voiced in different quarters. They range from increasing acidity of the soils to its impact on drinking water and aquatic life, but these voices are muffled in the euphoria of higher household incomes and national export revenues.

Cotton is grown in a range of patterns. Today many farmers still have only one or two hectares of it and for them it is simply another crop in the rotation cycle. The cotton parcel is located on the oldest part of the farm, where sorghum or maize had been grown for several successive years. In these parcels the clearing of the spontaneous vegetation is very advanced. The stems of the felled trees are now

---

[27] M. C. Henry, 'De la naissance à la remise en question d'un métier: encadreur', 3e cycle doctoral thesis, EHESS, 1988, quoted by Schwartz, 'Brève histoire', pp. 215–16.

[28] For the example of the Katiali region in northern Côte d'Ivoire where the importance of the local weaving industry contributed to the failure of colonial cotton export policies, see T. J. Bassett, 'The Uncaptured Corvée: Cotton in Côte d'Ivoire, 1912–1946', in A. Isaacman and R. Roberts (eds), *Cotton, Colonialism and Social History in Sub-Saharan Africa* (Portsmouth, NH and London: Heinemann and James Currey, 1995), pp. 247–67.

[29] For a case study conducted in the large cotton producer village of Boho-Kari, which is near our Sara-Hantiaye site and where a good number of farmers own tractors, see Philipe Tersiguel, *Le pari du tracteur: La modernisation de l'agriculture cotonnière au Burkina Faso* (Paris: ORSTOM, 1995).

mostly rotted away, the deeper roots have mostly disappeared and there is no sucker regrowth. Even trees left standing in the early years of the farm, have been gradually cleared away to get more sun and achieve a smoother plowing surface. Farmers say that at this stage the soil of the parcel also has better qualities for the cotton plant. Once the cotton is finished, however, in the following year most farmers put the parcel back under sorghum or maize, for a new cycle of cereals that may last a few years. The intervening year of cotton stops the progress of parasite and weed infestation, which is the main reason for abandoning cereal fields, and the residual effect of the cotton fertilizer helps subsequent cereal yields. While this is happening, in the second year the cotton parcel is moved to a new location on the farm that has also been farmed for many years, and on the third year to yet another location, and so on. The cotton parcel also benefits in protection against parasites from being moved around on the old grain parcels of the farm. This practice rejuvenates the farm for both cotton and cereals and has made possible an extension of the uninterrupted use of farm sites. A few farm histories reveal that, within a farm covering 8 or 10 hectares of suitable land, farmers who follow this schedule can put the entire surface under cultivation continuously for twenty years or more. As more and more people do this, the older practice based on a sequence of five years of farming followed by at least fifteen years of fallow is becoming a thing of the past. This intensification involves not the shortening of the fallow period, but instead more years of uninterrupted field use with a new crop rotation.

The second pattern for cotton growing is large farms of 6 hectares or more. In these farms cotton has become the main production activity. Some cotton farmers have 18 hectares under cotton, and buy some of the food that they need for their family. For the first time, cotton expansion is responsible for an increase in the total cultivated area of the country and is entering into competition with food grains. So far, cotton growth has not obstructed food grain production, which has also increased. The expansion of both cotton and cereal farms, however, is changing the plant cover because there is less fallow land and fewer of the plants characteristic of that transition.

Cereal farming remains very important. Of Burkina's farm area 85 per cent is thought to be under food cereals. However, between 1963 and 1988 the total farmed area in the country increased by approximately 60 per cent.[30] This is probably still below the rate of population growth over the same period, but considering that about 35 per cent of the population is less than ten years of age, it is most likely above the rate of increase of the adult population. If one also notes the gains in yields – 50 to 60 per cent in cereals and quadrupling in cotton – the tremendous growth of total agricultural output in the country is beyond any doubt. The western part of the country including our project area is responsible for much of this increase. The expansion of cotton, the most reliably documented crop, indicates that the growth of cereals is no longer the engine of the overall expansion of farming, either in surface area or in crop value.

The principal cereals of the savanna used to be red and white sorghum and millet (*Pennisetum*). These remain important in many places, yet in some parts of western Burkina Faso maize has become more important than all three of the older cereals

---

[30] This figure has been calculated from Kessler and Geerling, *Profil environnemental*, Table 6, p. 32.

put together. In the old days these cereals were intercropped, with each other and with legumes, vegetables, and other kitchen crops, practices which are now old-fashioned and small-scale. The spread of plows has made it more important for the farmer to have a clear flat surface with which to work. Fields are gradually cleared more thoroughly, and fewer trees are left standing.

The use of fertilizer, but not insecticides, has also become almost universal, especially with maize parcels, despite the withdrawal of all subsidies for chemical fertilizers since 1987. This fertilizer is often obtained from the cotton agency on credit, and has become an incentive to add a small parcel of cotton to the farm. The indebtedness following the cotton setback of 1991 arose from the fact that fertilizer purchases from the cotton agency were much larger than what was needed for growing cotton. Medium-size farm owners obtained fertilizer for their grain fields on cotton credit, intending to pay it back with the proceeds of their small cotton parcel. The cotton failure of 1991 demonstrated that this strategy may end up in large debts that burden the farmers for many subsequent years following just one bad cotton year.

From statistics concerning the nation in its entirety it may be surmised that unused land is abundant in Burkina Faso, but this impression is misleading for particular areas. Land shortage in the central part of the country has been acknowleged for a long time, but now the same is true for the fast-growing areas of the west, although they have lower population densities. Many localities in our project area are experiencing land shortage. Fallow land is becoming very scarce, swamp rice is invading riparian zones, and the culturally and ecologically significant distinction between 'bush' and 'farm' is eroding. Part of the reason is that the expansion – whether in cereals, cotton, or fruit orchards – is driven by the desire for market sales. Therefore areas that are accessible by car are filling up with farms at a rate totally out of proportion to their population growth. The standard demographic measures of farm potential and development are becoming now less and less relevant. Not all of these features of the rural dynamism of Burkina Faso are due to the development efforts of the past decades, but these efforts did nurture them, in both their positive and not so positive aspects.

The rural development efforts of the independence period evolved in a 'multilateral' context. This is different from the colonial period, when the 'commandant' stood for the expert interpreting the directives of the 'ministère', and development was a matter of administration. However, the new situation has not encouraged greater local autonomy. Agricultural and environmental research has been carried out by expatriate experts, working in organizations funded by multilateral agencies, and projects remain initiated and guided by external centers of decision-making. The capital funds and the philosophies have their origins not in the region, not even in the national state, but in international donors and constituencies. After 1983, with a new political orientation in the country, a partial attempt was made to change this situation by spelling out national research and policy priorities. A Land Act (Réorganisation Agraire et Foncière) and an Environmental Code were enacted, and the scientific research institutions, which are independent from universities in Burkina, were reorganized. The extent to which these changes alter the extrovert nature of research and development efforts remains to be seen.

Major Development Projects of the Recent Decades

So far we have dealt with private farms, mostly owned by smallholders, a large number of whom still have as their main farming target feeding their own household, even if they are being imperceptibly drawn into a deeper engagement with the market. This section reviews the other major development undertakings since independence: plantation agriculture, settlement of river valleys, dam construction, internal migration, and growth in cattle population. They had non-trivial social consequences and contributed to the reduction in fallow land, which is transforming the spontaneous vegetation cover.

The first important case is a large-scale industrial sugar cane reduction enterprise in the Bérégadougou plain, established in 1972. It was run by the Comoe Sugar Company (SOSUCO) and covered 10,000 hectares. There is no other agricultural operation of comparable size in the country. The plain was emptied of its occupants, water was piped in from the upper reaches of the Comoe, and a high-yield irrigated perimeter of permanent cultivation was put in place. The under-taking had a major impact on the hydrographic system of the area. The expropria-tions, monetary compensations, and resettlement of the population had social consequences that we know only from anecdotal evidence. At the same time, the fields and the sugar factory and distillery created many jobs, especially for women.

A different initiative but with similarly broad environmental and social consequences was the Sourou project which was planned to bring irrigated agriculture to the Sourou river valley. In the 1980s this project became a hot political topic, when the government decided to allocate part of the project area to unemployed young university graduates. The unemployment of people with advanced degrees is new and is perceived as an acute problem. Until recently the country had the reverse problem, a shortage of people with higher education. Overshooting in the opposite direction came suddenly and accelerated quickly. Just as the investments made since the 1960s in higher education started to pay off, producing a large number of graduates, civil service employment began to shrink under structural adjustment programs. The private sector provides less than 10 per cent of non-agricultural jobs in the country. The Sourou project was offered as a solution, and as a way of seeding a new dynamic entrepreneurial class of educated farmers. This vision has not yet been translated into reality, while the policy has been criticized in principle and in terms of its suitability for its stated goals.

An earlier, more ambitious resettlement and agricultural development project was the Volta Valleys project, a national sequel to the international effort to eradicate river diseases in a group of neighboring countries. In the late 1970s the Volta Valleys Authority (AVV) received large amounts of multilateral development funding.[31] It did not achieve its grand original objectives, but it did effectively open up the valleys to agriculture, not only with official settlers, but also by building roads and social services and bringing in its wake an inflow of independent farm operators. Crowded communities now lived in places that until recently had been very sparsely populated. Movement to the Sourou and Volta valleys accounts for a

---

[31] For the AVV, see D. E. McMillan, *Sahel Visions: Planned Settlement and River Blindness Control in Burkina Faso* (Tucson: University of Arizona Press, 1995).

good part of the expansion of farmed areas in the country. In the process, large areas of fringing and dense dry forests were cleared, sometimes with heavy machinery. In the Bondokui region between 1952 and 1981 the fringing forests may have shrunk by 70 per cent. During the same period farmed surfaces went up by 200 per cent, basically at the expense of the vegetation of the lower parts of the catena.[32]

The construction of dams – for hydroelectric power, irrigation, or the supply of water to the cities – is another major development theme. The dams are of varying scale and scattered throughout the country. In 1987 there were 714 of them with a total water holding capacity of 400 million cubic meters. Most of these are small-scale and support only modest irrigated perimeters. The irrigation dams sometimes flood more good land than they make available, but some of the vegetable-growing schemes around them have been minor success stories. Another serendipitous result has been a greater supply of fish. For example, fresh fish has entered the daily diet of the inhabitants of Ouagadougou in a way that was not anticipated twenty-five years ago, and the experience has been similar, albeit at a smaller scale, elsewhere. Endogamous fish professionals, migrating from the Niger valley in Mali, have now established themselves in Burkina. Our project area now includes one of the largest recent projects, the Nyofila hydroelectric power plant, located on the hills of one of the feeding branches of the Leraba. The works consist of two reservoirs, the upper one powering the plant and the lower one collecting the water for irrigation.

Two other social and economic processes had important consequences for the wild vegetation cover. One is the stream of migrants from the lower-rainfall northern parts of Burkina Faso to its eastern, southern, and western parts, including our project area. For the past three decades, this demographic movement has been one of the principal themes of the social science and development literature in the country. The flows started in the late colonial period, but accelerated during the drought years of the early 1970s, and are an important component of an impressive growth of cereal and cotton production in the receiving area (see the chapter by Leslie Gray in this volume). There is now a literature alarming experts about the negative consequences of this migration. The warnings are based on different considerations, including the destruction of fallow land, but mostly stress the land-wasteful cultural disposition of the migrants themselves. However, population density and farmed area are increasing everywhere, and the shrinkage of fallow land, as we have seen, is a common consequence. Naturally all of this is happening at a faster rate where agricultural potential is greatest, namely, in the west. Migration should be seen not as a special environmental threat, but as part of the general transformation of farming practices.

The second process involves cattle, which we have not taken up anywhere else. Our project area now includes numerous large herds mostly owned by wealthy Fulbe proprietors who have them herded by young Fulbe migrants brought by them from the north. The herds of western Burkina Faso had been heavily reduced during the colonial occupation and in the first decades of the century, by expropriation, taxation, and plunder. They were reconstituted in the 1970s,

---

[32] J. L. Devineau and G. Serpentié, 'Paysages végétaux et systèmes agraires au Burkina Faso', in M. Pouget (ed.), *Caractérisation et suivi des milieux terrestres en régions arides et tropicales* (Paris: ORSTOM, 1991), pp. 373–83.

primarily from stock moved down from the Sahelian regions during drought years. Other developments have also contributed to the increase of cattle in our project area. Numerous plow acquisition credit programs since the 1960s encouraged village farmers to purchase animals. The monies from grain and cotton sales allowed some well-to-do farmers to invest in oxen, cows and calves. Farmers take care of these animals as part of their household resources. The prevailing entrepreneurial spirit motivates many such farmers to look after these animals within the household instead of entrusting them to Fulbe herd owners as in the past.

The combined result is a large increase in the number of cattle. Since the 1970s animals on the hoof and animal products have been consistently the country's second or third largest source of foreign exchange. To measure the effect of grazing on the evolution of the flora was not one of our project aims. Nonetheless, grazing is related to agricultural change. Some herders have manure contracts with farmers, in the dry season bringing their animals to the empty fields to feed on the stalks in return for the benefit of the dung left behind. Those villagers who own animals pen them similarly in their own fields. The increased manure has contributed to the lengthening of the period of uninterrupted use of field sites. Thus the presence of cattle has become integral to the move to a more permanent farming style. At the same time, large herds owned by outsiders are causing a growing number of conflicts between farmers and herders because of crop damage.

What is the effect of larger herds on the vegetation? With many more animals to graze, there is more stress on grasses, although animals do not eat only grass. Tree leaves are an important part of their diet, especially in certain months of the year, and in our project area several choice fodder species, *Pterocarpus erinaceus* poir., *Cajunus Kerstingii* Harms, and *Khaya Senegalensis* (Desr.) A. Juss., are both depleted and protected by the wandering herders.[33] One outcome of the increased presence of cattle may be conditions more favorable to tree growth in the grazed bush savanna (see Chapter 3 in this volume). Casual observations by local people indicate a greater density of shrubs and trees in the bush compared to the past. Thus, on the one hand, most things seem to conspire to reduce the fallow land successions and to produce a more tightly domesticated landscape, with farms, orchards, and less plant diversity. On the other hand, the development of herding seems to be creating conditions favorable to denser dry woodlands in between the farms.

Land Use and Cover Between 1952 and 1983

The trends described so far are evident in the evolution of farm and fallow land in six project sites over a period of about thirty years, as documented by two sets of aerial photographs taken in the 1950s and 1980s. The areas that we mapped are small, and therefore cannot provide conclusive evidence for the commentary that accompanies them. But the illustrations do show variation between sites and allow a more detailed discussion of some of the practices transforming land use.

---

[33] For an exhaustive empirical study of tree lopping by Fulbe herders in western Burkina Faso, see S. Petit, 'Environnement, conduite des troupeaux et usage de l'arbre chez les agropasteurs de l'ouest burkinabe', Doctoral thesis, 2 vols., Université d'Orléans, 2000.

Table 6.1 Land use in the selected sites

	Sara-Hantiaye		Bare		Soumousso		Péni		Bougoula		Kankalaba	
	1952	1981	1952	1981	1952	1981	1956	1983	1956	1983	1956	1983
Farmed	3.30	12.36	10.61	28.75	1.33	33.17	44.28	28.87	35.77	36.16	18.08	11.79
Fallow	3.80	2.38	3.10	4.00	2.21	11.50	9.55	9.12	5.11	3.14	1.57	4.71
Unfarmed	91.00	84.60	82.53	65.05	92.92	52.23	46.17	62.01	59.12	60.70	80.35	83.50
Riverine	1.90	0.66	3.76	2.20	3.54	3.10						

Figure 6.2a Land use in Sara–Hantiaye, 1952

Figure 6.2b Land use in Sara–Hantiaye, 1981

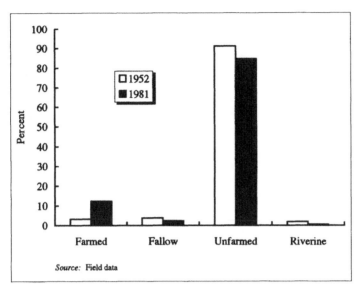

Figure 6.2c Land use in Sara-Hantiaye, 1952 and 1981

The six sites are Sara-Hantiaye, Bare, Soumousso, Péni, Bougoula and Kankalaba. They are scattered through western Burkina Faso in zones with different agricultural histories and topographies. They vary in terms of demography and migration status, and in the ethnic characteristics of the population, which is heterogeneous. Sara-Hantiaye, a Bwa village in the cotton region, is the northern-most site. The next two sites, Bare and Soumousso, are near each other and show the contrasting and yet similar impact of current developments on two situations starting from different historical baselines. The photographs in all three cases come from the coverage of 1952 and 1981. The fourth site, Péni, shows a mix of the characteristics of the other sites. The final two sites, Bougoula and Kankalaba, are clustered in the Tagouara plateau, which is hilly, remote from, and until recently not well connected to, the national center. Over the past three decades the Tagouara plateau became the site of agricultural innovation combining cereal and fruit orchard cultivation, which has raised incomes and spurred the interest of national elites. The coverage for Péni and the last two sites comes from the photographs of 1956 and 1983.

We present the maps resulting from the photographs in Figures 6.2 to 6.7 and summarize the data in Table 6.1. The maps separate the surface area into four categories: farmed, fallow, unfarmed, and riverine forest, and the bar charts aggregate the proportion of area in each category.[34] Farmed area refers to stretches of land with a mix of crops, including the trees that have been spared during clearing for the farm. Fallow land refers to former farm sites, recognizable on the

---

[34] Cf. J. R. Anderson, E. E. Hardy, J. J. Roach, R. E. Witmer, *A Land Use and Land Cover Classification System for Use with Remote Sensor Data*, USGS Professional Paper No. 964. (Washington, DC: USGS, 1976).

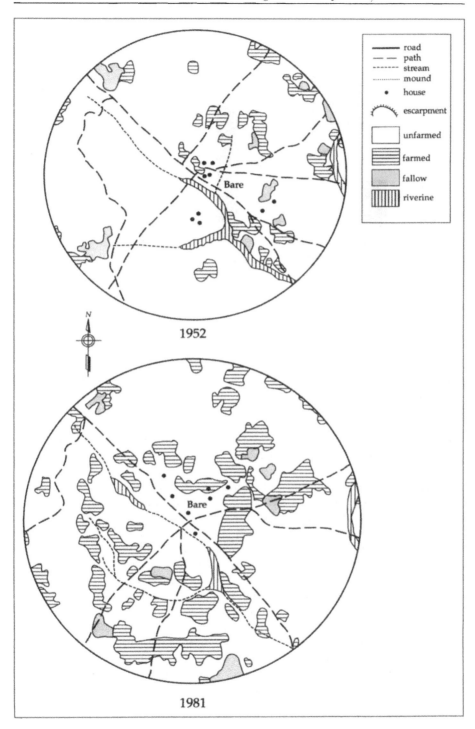

Figure 6.3a Land use in Bare, 1952 and 1981

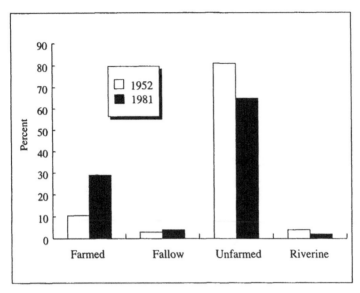

Figure 6.3b Land use in Bare, 1952 and 1981

photographs by their artificial shape with a lower density of plants, in different stages of succession toward savanna woodland. Unfarmed area refers to what is not recognizable as farm or fallow land. It includes both areas that support little vegetation, and more extensive areas that are wooded or bushy. Within the general unfarmed area, we single out riverine forest. With the spread of fruit orchards, distinguishing plantations from woody 'unfarmed' areas presents technical difficulties. This has a bearing on our commentary.

Sara-Hantiaye. Our northernmost site, Sara, is a large Bwa village located north of the Maro classified forest. In the 1952 map it appears as a well established agricultural community, with approximately equal amounts of land under cropping and fallow, and plenty of uncleared bush (Figure 6.2a). In 1981, there is about a fourfold expansion of cultivated areas, while fallow and uncultivated bush go slightly down (Figure 6.2b). Riverine forests have also decreased as the stream banks have come under cultivation. This village is still well endowed with land and has plenty of uncultivated bush, although some of it is hilly and difficult to farm.

The two maps cover an area that is more than twice the size of the area covered by the maps of the other cases and allow us to take note of the effect of migration on agricultural expansion. While the cropped areas of the village have increased at the expense of fallow land, one can see in Figure 6.2b that most of this increase is due to the colonization of a separate bloc to the west of the village, on territory that was empty in 1952. This new bloc has farm areas three times larger and fallow areas five times larger than the village itself. Migrants who have arrived from the north and who are mostly Mose occupy this new zone. They grow primarily cotton, sorghum, and peanuts. They also provide most of the labor for the forest

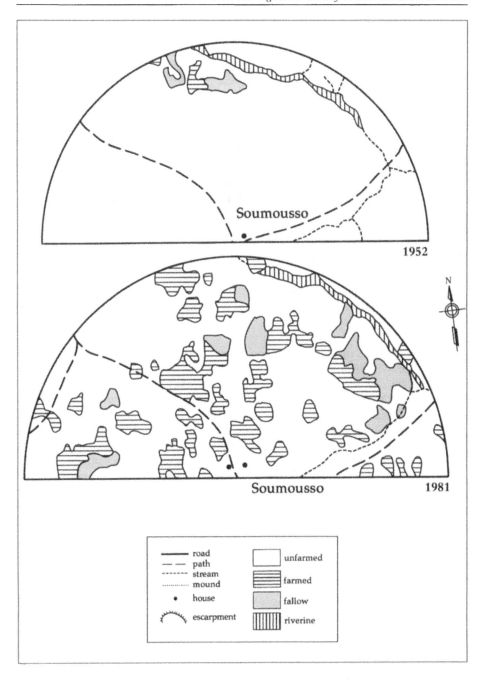

Figure 6.4a Land use in Soumousso, 1952 and 1981

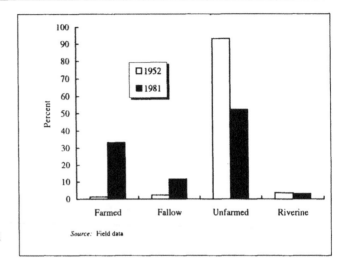

Figure 6.4b Land use
in Soumousso, 1952 and
1981

maintenance and development work carried out by a government agency in nearby forests. The area farmed by the migrants is separated from the village by a ridge of uncultivated hills.

Bare-Soumousso. The maps of Bare cover an area with a radius of about 3 km, which represents only a portion of the territory on which the villagers have their farms. Some families in this old and well populated village (with an approximate population of 2000) possess rights to land at large distances, and the farms of some villagers included in a survey conducted in 1983 were 10 or 12 km south of the village.[35] These areas were very old fallow lands that had become shady woodland where fire could not penetrate. Young members of the families were clearing farms there to reassert customary rights and protect them against claims from outsiders.

Nonetheless, the maps in Figure 6.3 provide a sense of the general trend of development in farmed area from 1952 to 1981. Farmed area has tripled within the radius shown, most of the extension being to the south of the village, where most village families have land rights. The riverine plant cover has nearly disappeared between these two dates, except for two locations that have been spared for ritual activities. There were mango tree plantations near the core area of the village, and cashew tree (*Anacardium occidentale* L.) plantations established by neighboring villagers for commercial purposes to the north of the mapped area which do not appear in this figure. The planting of trees, both of the fruit variety and of the timber species propagated by extension services, has accelerated since 1981, and plantations can now be found far from the habitation area.

The village of Soumousso (Figure 6.4) illustrates the most dramatic impact of migration on the extension of farmed surfaces. Until the 1960s this was a small hamlet with a few families of Tiefo origin. Since then it has received a large influx

---

[35] For agriculture in this village see M. Şaul, 'Farm production in Bare, Burkina Faso: the technical and cultural framework of diversity', in G. Dupré (ed.), *Savoirs paysans et développement* (Paris, 1991), pp. 301–29.

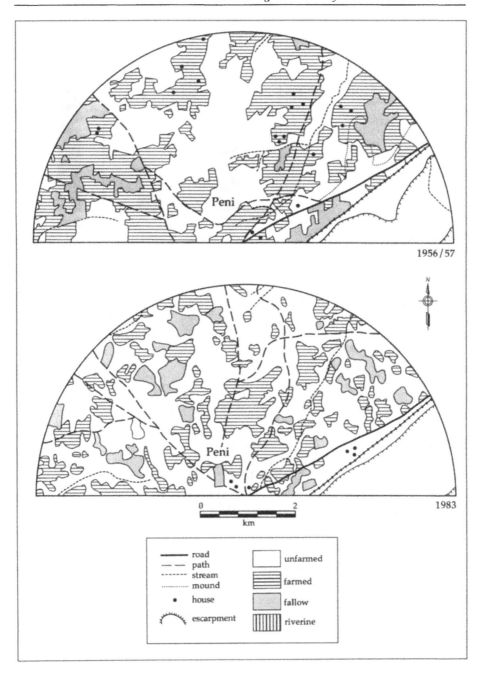

Figure 6.5a Land use in Péni, 1956/7 and 1983

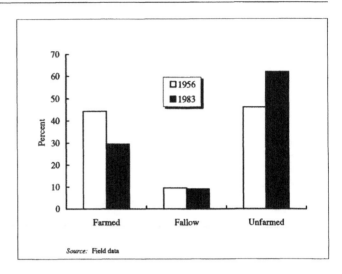

Figure 6.5b Land use
in Péni, 1952 and 1983

of Mose settlers to whom the original inhabitants have lent their lands. Thus both inhabited area and farms started to expand. In the 1951 map the surroundings of the village are almost empty, except for a few fields in the northern edge of the map, which were farmed by the inhabitants of Bare. In 1981 35 per cent of the mapped area is under cultivation, a proportion larger than that of Bare. Most of the farms are under sorghum, millet and maize, and the village sells a large quantity of grain to merchants. The land greed of the settlers has caused conflicts between them and the inhabitants of Bare, as well as tensions among land controlling groups in Bare.

Péni. The maps of this village, representing an area with a radius of about 4 km, seem to show a reduction of farm area from 1956 to 1983 (Figure 6.5). However, if the 1956 farm areas had been abandoned, by 1983 one would expect to find in their place a larger area under fallow land. That this is not the case hints at one of the interpretative problems that we encountered. In fact, ground observations show that much of the area appearing as bush in the 1983 map is under mango orchards. Starting in the mid-1960s this village, with many others to its west, started systematically to convert cereal fields into mango orchards by a cropping pattern that we discuss below. The village is on a national highway and has attracted many settlers from near and far. Besides the local and neighboring populations of Tiefo, Sambla, and Jula origin, the ethnic mix includes Bobo, Bwa, Lobi, Mose, and Fulbe settlers. The mango orchard business has brought large revenues to households, and vast areas around the core of the village and far into the bush are under mango groves. This is the most important effect of new farming practices on the evolution of the vegetation in this village and the far western zone that is the Tagouara plateau.

Bougoula and Kankalaba. These two sites on the Tagouara plateau differ from each other and from previous cases, and exhibit certain important features of this part of the country. The maps of Bougoula show farm areas slightly larger in 1983 than in

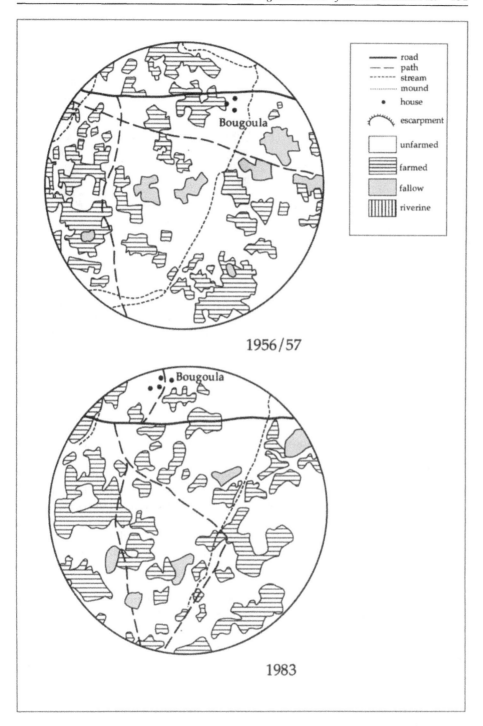

Figure 6.6a Land use in Bougoula, 1956/7 and 1983

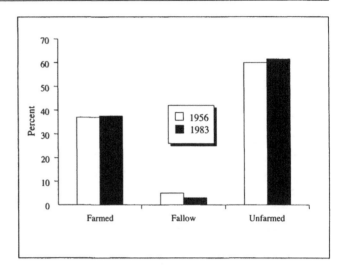

Figure 6.6b Land use
in Bougoula, 1956/7
and 1983

1956 (Figure 6.6a). Note that this village is very intensively farmed and the
proportion of farm to fallow land, which was already high in 1956, is even higher
now. Kankalaba shows an increase in fallow land between 1956 and 1983 (Figure
6.6b). The village territory of Kankalaba is extraordinarily hilly. People here
adopted farming techniques unlike those anywhere else, using special tools and
practices such as terracing. The difficulties of farming have motivated the
inhabitants to go to distant areas in the region and request farm sites from other
villages with less hilly surfaces. The topography of the village also discourages
expansion of production by making it difficult of access for trucks and cars. The
combination of these factors pushed this village into agricultural recession, showing
the uneven effect of developments.

The Tagouara plateau is undergoing a phenomenal development of fruit tree
plantations, which is not revealed in our mapped cases. Crucial in this development
was the proximity of Orodara. Continuing the early colonial efforts started in
Banfora, in the early 1970s this town became the primary node for the spread of
mango, citrus, and other fruit tree seedlings and related technical knowledge.
Farmers around the town perfected the transition from cereal farm to orchard, a
practice that is now universal. No family in this area farms only food crops as was
the case in earlier days.

Another plant that is undergoing tremendous expansion around Banfora and in
the Tagouara plateau is the *ban* palm (*Borassus aethiopum* Mart.). Its abundant sap can
be tapped, ferments quickly, and is sold and consumed as palm wine. In this region
it has always been planted from seeds, but with the growth of the market the
plantations have become much larger and more numerous. The mature plant can be
tapped several times during its short life, and with a useful life of a few years it stands
half-way between annual crops and longer-term fruit trees. In many villages one
now finds farmers who own plantations of 1,000 or more of these palm trees.

One final development concerning the Tagouara plateau deserves mention. The
profitability of combining food crops and orchards has brought the plateau to the

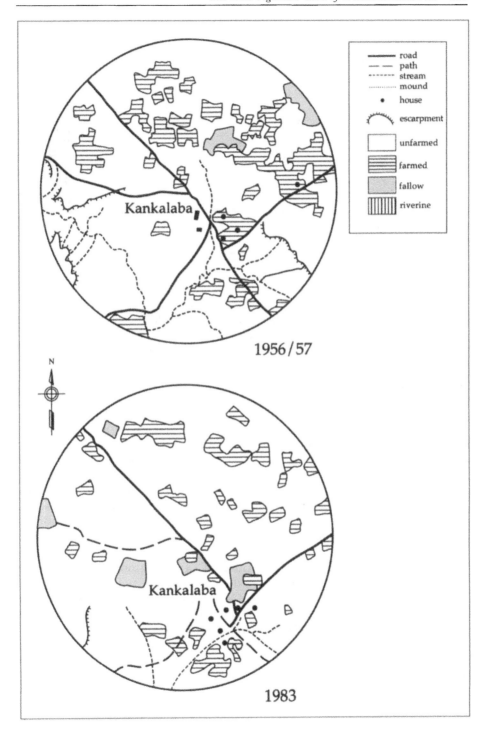

Figure 6.7a Land use in Kankalaba, 1956/7 and 1983

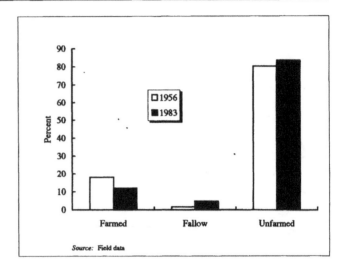

Figure 6.7b Land use in Kankalaba, 1956 and 1983

Source: Field data

attention of the national political elite. The trend started with civil servants in the middle rungs of the administrative hierarchy – retired soldiers, schoolteachers, and all kinds of salaried personnel, who had connections to this area and access to land. Now individuals occupying the highest political positions in the country have joined in the development. They secure tracts of more than 100 hectares, farming them in the pattern that has become common: clearing and cereal cultivation in the early years, while planting tree seedlings which will transform the farm gradually into a plantation. They use heavy machinery as well as hiring workers with old-fashioned hoes; they also throw giant work parties with bands of musicians and huge quantities of food. These large farms-on-their-way-to-becoming-plantations are a true mixture of forms, techniques, technology, one could almost say eras, all with a view to producing profits. The Tagouara plateau is not the only part of the country where one observes such a development, but the special interest of this area is evident in the infrastructure that has been built for it. Since the politically prominent people have considerable decision-making power, the Tagouara plateau is now endowed with some of the best roads in the country. It is also being electrified.

In the Tagouara plateau the combination of expanding orchards, wine palm plantations, annual crops, the interest of the national elite, and hilly landscape is producing an acutely felt land scarcity. Besides the disappearance of fallow land, this scarcity is provoking new strategies and conflicts to preserve land rights, which affect relations between the lenders and borrowers of farm sites.

Mango and Other Orchards

The spread of fruit tree plantations in complementarity with grain production is the most noteworthy new development of western Burkina Faso. It is the cultural practice which influences the spontaneous vegetation cover most significantly. If the trend continues at the pace it has done in the past two decades, it will change the distribution of plant species on a large scale, substituting domesticated foreign trees for the grasses and tree species of the fallow and woodlands. In this section we relate the origins of this trend, show its commercial scale and profitability, then describe how plantations are established, and consider some of the potential consequences for plant diversity.

Fruit tree plantations in Burkina Faso are mostly the work of farming households. Orchards multiplied when household heads followed the example of a few innovators. Orchards started in the 1950s with people of the cities – civil servants, retirees, and merchants – who obtained the seedlings from colonial agronomic stations. This did not result in the concentration of plantations in the hands of a few heavily capitalized ventures. Household farms, on a whole range of scales, replicated the example.

Exogenous varieties of fruit trees entered Burkina in the inter-war period, in the administrative style characteristic of that period. Orders were issued for each household to plant seedlings. Before long, a few people from the privileged colonial stratum – merchants and the Mose chiefs of Ouagadougou – noticed the income possibilities that the orchards offered. In the central plain of the country mango trees do better in depressions or river banks with high humidity, and the Mossi chiefs have customary rights to such land, which, until then, had been of little use. Other political factors were linked to decolonization. In the late 1950s, the Mossi chiefs came into conflict with the new cadres who had graduated from the colonial schools and who constituted the native intelligentsia that was going to take the country into independence. The chiefs were quickly stripped of many of their official duties, and, now facing loss of status, allied themselves to the Muslim merchant community of the capital to go into such ventures as commercial agriculture, meat production for export to the neighboring coastal countries, and transportation.[36] The Moro Naba Kougri, the most important Mose chief, had orchards and gardens planted, and shipped fruits and vegetables to Côte d'Ivoire.[37] Others quickly followed. In the 1970s René Dumont observed in Banfora and Bobo-Dioulasso plantations of fruit trees, mostly mango 'planted especially by those who have money: people of the cities, civil servants, the military... a new category of absentee owners who have land "allocated" to them by customary authorities...'[38]

Mangoes do marvelously well in western Burkina Faso; papayas and guavas do reasonably well; and for citrus, western Burkina is one of the few places in the region where they can be produced without irrigation. The popularity of mango is also related to its inherent properties as a fruit. It is sweet and therefore high in calories. The fruits come just before the rainy season sets in, a time when most heads of production units are reluctant to bring out the grain stored for the farming season and tend to let women and other dependent members fend for themselves.

---

[36] Catherine Some, 'Sociologie du pouvoir militaire: Le cas de la Haute-Volta', 3e cycle Doctoral thesis, C.A.E.N., Bordeaux University I, 1979.

[37] E. P. Skinner, 'The changing status of the 'Emperor of the Mossi' under colonial rule and since independence', in M. Crowder and O. Ikime (eds), *West African Chiefs under Colonial Rule and Independence* (New York and Ile-Ife, 1970), pp. 98–123.

[38] René Dumont, 'En Haute-Volta, une paysannerie à demi affamé', Chapter 5 in *Paysans écrasés, terres massacrées* (Paris, 1978).

A large quantity of mangoes risk rotting in this period of glut and become food for pigs, the husbandry of which also became popular at the same time in the villages. Despite phenomenal sustained increase in production since the 1970s, the price of mangoes did not drop until the 1990s, as the market expanded at the same pace, within the country and through exports to neighboring countries as well as to the luxury markets in Europe.

In the 1970s Orodara's Regional Development Organization (ORD, supplanted now by the CRPA) became the center for the introduction of new varieties, the production and sale of seedlings, and training. Truckloads, not only of fruit but also of seedlings, filled the highways out of Orodara during the rainy season. In the early 1980s this region was already said to be developing a food-grain deficit. No one in the region thought this a misfortune, though. The prosperity brought by the fruits was such that the development was seen as a boon.

In the early 1980s one good grafted tree could produce a crop worth at farm-gate prices more than 2,000 CFAF. Incomes of 4,000 or 5,000 CFAF per tree were common. A plantation of about 10 hectares with 2,800 trees on it, two thirds mangoes and the rest citrus and guava, easily produced net yearly incomes in the range of 4–5 million CFAF – more than $10,000. For this kind of return, the initial investment was relatively minor. A 10-hectare plantation would be beyond the means of most households, but at a much smaller but still very profitable scale, planting is within reach of the majority of village people in western Burkina, and the investment can be spread over many years.

Fruit trees are planted in two distinct patterns. The first was more common in the past. Some farmers plant a few trees within the walls of their compound or near their house. This initial investment seems negligible, but growing the seedlings to mature trees is difficult. The young trees have to be protected from roaming animals and children, and they do not always survive the dry season. Adding one or two trees a year, it is possible to develop a small stock over time, but in this way the planting can-not go beyond the miniscule stage, because the farmer soon runs out of space. When there are only a few trees, most of the fruit is simply eaten by family members. Now that the trees are almost always of the good grafted varieties, the crop may be boun-tiful and valuable. There are farmers who allocate one or two good trees to a spouse for cash income. The monetary returns are rarely negligible even at this amateurish scale, but to establish a true plantation a farmer has to turn to the bush fields.

Plantations in the bush represent the second pattern of planting. This is the way the first merchants, civil servants, chiefs and retirees established their plantations and it has become universal among village farmers in the surroundings of Orodara and in the Tagouara plateau, is common around Péni, and is spreading elsewhere in the area of Bobo-Dioulasso. The operation starts with the clearing of a fallow field for cereal cultivation, in the conventional manner. The farm can be more thoroughly cleared and plowed for a second year of cereal cultivation. In the months of high rainfall, July and August, part or all of the farm is planted with fruit tree seedlings between the stalks of the maturing grain. In subsequent years the farm is enlarged by clearing adjacent areas, more land being brought under cereal cultivation. The planting of tree seedlings follows that of cereal cultivation with one or two years' delay. The pace of this expansion depends on the means of the farmer. For a typical successful household, a new farm site starts with one or two hectares, and then expands for a few years with the addition of half or one hectare every year, stabilizing at about 4 or 5 hectares. The loose crop rotation schedule varies with the nature of the soil and other contingent factors.

Previously, portions of the farm that had been farmed for five or six years were abandoned while new areas were cleared ahead. Thus the farm location shifted, until a limit, such as a stream or the boundary of another farm, was reached. Today there is no longer movement, simply accumulation by expansion. Before it is even time to abandon the field as fallow, the seedlings have grown into trees and the cereal farm has transmuted into an orchard. Under the shade grains no longer grow. Some farmers plant shade-resistant peanuts, earth peas (*Voandzeia subterranea* (L.)DC.), or cowpeas (*Vigna unguiculata* (L.)Walp.) under the young trees, but annual agriculture stops completely when the trees are fully grown and producing to capacity. Many farmers continue to plow under them after this point and spread fertilizer to assist tree growth and fruit production. Once the entire tract is planted with trees and they have become mature, the household can only produce cereals by opening a new farm in a different location.

A plantation starts bringing in monetary returns after four or five years, and this revenue can finance the further expansion of the plantation. Most farmers manage their plantation with a keen sense of investment. Besides plowing and fertilizer use, many owners hire a permanent worker to live on and look after the plantation. In the Péni area farmers who had planted their first trees in the late 1960s remember that they had to water the seedlings at great pains in the drought years of the early 1970s, an unheard-of practice. In 1984 when late-bearing varieties became available, many people around Orodara and the Tagouara plateau cut down their oldest trees and planted the new varieties instead, thus forgoing a few years' worth of production to obtain better prices in the future.

A head of household who has engaged in this course stops only because he has run out of land. Within a period of 10 or 15 years a successful farmer can go through 15 hectares, plant it all with trees, and no longer have land suitable for yearly crops. Consequently, a difference is now appearing between the old founding families of the villages, who have ritually expressed rights over plenty of land, and others, more recent immigrants or outsiders, who have borrowed land for farming. The borrowing of land for grain farming was widespread and did not involve substantial payments of rent.[39] However, extended plantations do not provide a good reason to ask for more farm land, and the land-owning households are now careful to preserve their own future access to the few still unused sites. There are contradictory processes at work. On the one hand, land belonging to groups is being transformed into plantations controlled by specific individual members who have planted them with trees. On the other hand, groups are keen to assert their collective rights in the face of outsiders, resulting in the 'hardening' of corporate bodies around the land. And trees are now planted not only for revenue but also to maintain a foot in the land against lineage companions, and to forestall future demands by potential borrowers and by long-standing borrowers who want to avoid the risk of eviction. Young farmers, who discover that the family farm,

---

[39] M. Şaul, 'Land custom in Bare', in Bassett and Crummey, *Land in African Agrarian Systems*, pp. 75–100; *idem*. 'Money and land tenure as factors in farm size differentiation in Burkina Faso', in R. E. Downs and S. P. Reyna (eds), *Land and Society in Contemporary Africa* (Hanover, NH: 1988), pp. 243–79.

Photo 6.3 Mature mango trees in the process of being severely lopped in order to rejuvenate them and open the field under them to cropping (Péni).

which they had taken for granted, is located on borrowed land, plant trees to secure their continued access, an action generating very complicated conflicts in which moral sentiments, religious values, and the possibility of legal and administrative intervention by the state pile up to produce inconsistent outcomes.

The new commercial vocation of the Tagouara plateau can also be seen in that it has also become one of the major cotton-growing areas, adding to the shortage of land. In the Kenedougou province, of which the Tagouara plateau is part, almost 67 per cent of the farms are under cotton, in rotation with maize.[40]

This version of land grab, and the resulting land shortage, are stimulating new practices. The trees themselves may soon become a rotation crop. When mango trees age they start to bear smaller crops. Pruning can rejuvenate the tree and restore the yields. It is common in the project area for farmers not only to prune the old trees, but also every so many years to lop them severely, leaving only the main trunks standing. Such drastic treatment does not kill the tree in this area. After reducing the tree to the stem, the part of the plantation where this is done can again be used to plant an annual crop, a legume, or even a cereal crop, for one or two years, before the trees grow branches and start bearing fruit again. Thus there is an

[40] A. Schwartz, 'Que faut-il penser de la régression de la production cotonnière au Burkina Faso depuis la campagne record de 1990-1991 et de mesures de relance proposées en 1995?', Rapport de Mission, 1996 (mimeo), p. 17.

evolving pattern of fruit trees as the main perennial crop, supplemented with periodic cereal or pulse production on the same plot.

On the Tagouara plateau land shortage is reaching a point where some farmers are contemplating felling their oldest fruit trees altogether and bringing a portion of their farm back under grains, only to start the process soon again with seedlings of newly available and more attractive varieties of fruit trees. If such periodic clearing of orchards becomes generalized it will constitute a partial return to the old farm–clearing system, but without the succession of woodland fallow. Now the wild fallow flora are replaced with planted trees, the land is exploited more heavily than under spontaneous trees, and there is a concurrent loss of diversity in the managed area.

The contribution of fallow land to plant diversity in our project area was studied as part of our investigation in the Péni site, where our colleague Sanou Dya put in place erosion parcels. The research uncovered the richness of plant life in fallow plots and the extraordinary regeneration power of the wild species. We inventoried some 60 species, ligneous and herbaceous, on these study parcels. Seven species dominated: *Daniellia oliveri* (Rolfe) Hutch. and Dalz. (237 shoots/new plants), *Butyrospermum paradoxum* (shea, 121 shoots/new plants), *Hymenocardia acida* (66 shoots/new plants), *Crossopteryx febrifuga* Benth. (66 shoots/new plants), *Trichilia emetica* (45 shoots/new plants), *Swartzia madagascariensis* (42 shoots/new plants), *Vitex diversifolia* Bak. (26 shoots/new plants). This is the cover that is becoming increasingly scarcer as the countryside becomes shaded by huge mango trees.

Conclusion

In this chapter we have described the long development that started in the colonial period with requisitions, taxation, and the building of roads and a few agronomic stations, and led to the commercial expansion of agriculture and orchards in recent decades. This expansion explains why a large proportion of arable land, previously under fallow, came under planted species. The shortage of farm sites is no longer exclusively a reality of the central plains of Burkina Faso, but also of the western part of the country where our project was situated, and it can no longer be attributed to high population densities. The shrinking of fallow land in western Burkina is driven by the expansion of commerce rather than by the necessity of feeding the members of farm households, and is due to the growth of cotton production, grain farming for market sales, and even more importantly fruit plantations.

The lessons of the colonial period are instructive for understanding some of the changes that are occurring in western Burkina, even though the processes involved and the scale of the transformation are different today. The direct impact of both the extractive and conservation efforts of the colonial government on the environment was limited and inconsistent. We noted, for example, the destructive effect of rubber collection on the *Landolphia* vines, whereas the commercialization of *shea* nuts and African locust beans proceeded without disturbing the local conventions of conservation and production. We traced these different outcomes to the fact that rubber-gathering was an external imposition and exacerbated intra-household tensions, whereas *shea* and locust beans remained within the well-established lines of intra-household specialization that committed the members to co-operate. The use of wood as fuel for transportation and industry, and subsequent decisions to declare classified domains, had mutually correcting effects on the vegetation cover, even though both developments stressed the farming capacity of some local communities. In the case of the fuel use of wood, it is difficult to assess if the colonial period represented a more intensive rate of depletion when compared with the pre-colonial industrial activities.

The most important legacy of the colonial period with regard to the vegetation cover was the commercialization set in motion by the growing urban demand for cereals, the export of industrial crops, and the initial stimulus given to orchard production. All of these tendencies became more pronounced after independence in 1960. These developments combined have resulted in a reduction of the area under fallow, and a more domesticated plant cover in the environment. Many spontaneous plants, both ligneous and herbaceous, which have their habitat in the arable land that is being occupied at an accelerating rate by agri- and arbori-culture, are reduced in number. Increased herding activity qualifies the impact of this process. More cattle reduce the mass of grasses, encouraging shrub and small tree proliferation and contributing to the emergence of a sharper contrast between the areas under planted species and the corridors left uncultivated between them.

We stress that agricultural expansion is not leading exclusively to the spread of farm cereals, pulses, and cotton. In a manner reminiscent of what happened in the forest regions of West Africa at an earlier date with coffee and cocoa, in western Burkina fallow land is being supplanted by trees of a limited number of species chosen for plantations. The choice for these plantations is now widening from imported fruit trees to include valuable savanna species such as *Parkia biglobosa* as well as the *ban* palm. Overall, it is clear that it would be inaccurate to characterize this shrinking of fallow land as deforestation. although it may be true that the result is impoverishment in the number of individuals of many wild species compared with an earlier period, even if the species themselves are not yet disappearing even locally.

Western Burkina represents in many respects a case of successful agricultural development. The initiatives of peasant farmers have created desirable increases in income and total production. This is also happening without creating a mass of paupers in rural areas and without worsening beyond remedy income distribution at the national level. At the same time, the rich spontaneous flora of the savanna is being replaced with monotonous planted orchards and grain and cotton farms. Conventional wisdom holds that the wild savanna was the complex result of centuries of human activity. Recent developments are introducing in a very short timespan a much higher degree of artificiality in this older pattern of human-environment interaction.

PART FOUR

Pastoral Ecologies

7

Rethinking Interdisciplinary Paradigms
& the Political Ecology of Pastoralism in East Africa

PETER D. LITTLE

Introduction

Interdisciplinary research programs on land use and environmental change in sub-Saharan Africa increased considerably during the 1980s and 1990s. Research was conducted throughout Eastern and Southern Africa and many of these efforts focused on savanna areas inhabited by pastoral and agropastoral populations.[1] Empirical findings from this body of work challenged orthodox assumptions about the relationship between pastoral population growth and environmental change; the resiliency and 'patchiness' of savanna ecologies;[2] and the capacity of local institutions to regulate resource use. They also pointed to concerns about the fundamental politics of pastoral resource use, an issue that had emerged in the early 1980s.[3] In terms of theory, two broad bodies of work became especially appealing to social scientists and to a small number of range ecologists: the 'rangelands at disequilibrium'[4] and the 'political ecology' approaches. A growing number of scholars

---

[1] See N. O. J. Abel and P. M. Blaikie, *Land Degradation, Stocking Rates and Conservation Policies in the Communal Rangelands of Botswana and Zimbabwe*, ODI Pastoral Development Network Paper 29a. (London: Overseas Development Institute, 1990); James E. Ellis and David M. Swift, 'Stability of African Pastoral Ecosystems: Alternate Paradigms and Implications for Development,' *Journal of Range Management*, 41, 6 (1988), pp. 450–9; Melissa Leach and Robin Mearns (eds), *The Lie of the Land: Challenging Received Wisdom on the African Environment* (Oxford: James Currey, 1996); Ian Scoones (ed.), *Living With Uncertainty: New Directions in Pastoral Development in Africa* (London: Intermediate Technology Publications, 1996); Katherine Homewood and W. A. Rogers, *Maasailand Ecology: Pastoralist Development and Wildlife Conservation in Ngorongoro, Tanzania* (Cambridge: Cambridge University Press, 1991).

[2] Ian Scoones, *Patch Use by Cattle in Dryland Zimbabwe: Farmer Knowledge and Ecological Theory*, ODI Pastoral Development Network Paper 28b (London: Overseas Development Institute, 1989).

[3] Anders Hjort, 'A Critique of "Ecological" Models of Pastoral Land Use,' *Nomadic Peoples*, 10(1982), pp. 11–27; Peter D. Little, 'Social Differentiation and Pastoralist Sedentarization in Northern Kenya,' *Africa*, 55 (1985), pp. 242–61; Peter D. Little and Michael M. Horowitz (eds), *Lands at Risk in the Third World: Local-Level Perspectives* (Boulder, CO: Westview Press, 1987).

[4] This school has also been referred to as the 'New Range Ecology:' Emery Roe, L. Hutsinger and K.

(including some in this volume) are familiar with both schools; and both have implications for what factors are privileged in interdisciplinary research programs.

In this chapter I wish to revisit some of my own and others' work in these different theoretical genres and to reassess some of the trade-offs that social scientists, including anthropologists, accept when working within interdisciplinary research programs on environmental change. It is not meant to reach definitive conclusions, but to raise issues that remain theoretically and empirically unresolved despite the wealth of increased knowledge about the politics and ecology of local resource systems. The chapter questions whether we are asking too much of current interdisciplinary models, and social scientists in particular. It goes full circle to query whether or not the 'political ecology' approach, as reflected in some recent work,[5] has departed so much from addressing the ecology side of the equation that one is left only with political economy, or at best a revisionist cultural ecology. The proliferation of work under the rubric of 'political ecology' calls for a re-evaluation of what this concept means, especially if it is to be invoked in interdisciplinary studies of environmental change.[6] In contrast to recent castigations of political ecology,[7] I suggest a simplified approach that offers a way of linking the political and the ecological as was originally envisioned in the 1970s and 1980s. Case studies of pastoral resource use and ecological change in Marsabit, Kenya and the Lower Jubba region, southern Somalia are presented to demonstrate why a return to Eric Wolf's basic issue of resource distribution must remain central to a political ecology approach.[8] They also illustrate why recent work in other interdisciplinary paradigms might prove helpful in meeting the challenge of its critics.

Theories and Models

Recent theoretical advances in the 'ecology of disequilibrium' school are relevant to understandings of the complex relationships between politics, human agency, and savanna habitats. They also serve as wonderful entry points for interdisciplinary studies. For example, dryland parts of the African continent, including savannas, are subject to large fluctuations in rainfall and sustained droughts from one year to the next, and have seen major political changes in resource access for many groups, especially pastoralists. Climatic data collected over the past decade reveal that most grassland zones of Africa are inherently unstable and, therefore, attempts to adjust conditions to some notion of 'stability' violate the natural order and are in themselves destabilizing.[9] These findings particularly confront the 'carrying capacity'

[4] (cont.) Labnow, 'High-Reliability Pastoralism Versus Risk-Averse Pastoralism', *Journal of Environment and Development*, 7, 4 (1998), p. 387.

[5] Arturo Escobar, 'After Nature: Steps to an Antiessentialist Political Ecology', *Current Anthropology*, 40, 1 (1999), pp. 1–30.

[6] Raymond L. Bryant and Sinead Bailey, *Third World Political Ecology* (London: Routledge, 1997).

[7] Andrew P. Vayda and Bradley B. Walters, 'Against Political Ecology', *Human Ecology*, 27, 1 (1999), pp. 167–79.

[8] Eric Wolf, 'Ownership and Political Ecology', *Anthropological Quarterly*, 45 (1972), pp. 201–5.

[9] Roy Behnke, I. Scoones and C. Kerven (eds), *Range Ecology at Disequilibrium: New Models of Natural Variability and Pastoral Adaptation in African Savannas* (London: Overseas Development Institute, 1993).

concept for its failure to recognize the variability and patchiness of savanna ecologies and for its requirement of quantifying a process as dynamic as the feeding habits of different animal species.[10] Moreover, contrary to conventional wisdom, dry savanna regions 'are intrinsically "resilient" compared with more "stable" ecosystems',[11] because they must deal with such climatic perturbations and instability. They are not the inherently 'fragile' ecologies that much of the literature portrays,[12] although, as with any ecosystem, they can be made fragile through inappropriate development and land-use practices. The latter would include settlement projects that encourage excessive population densities and cultivation in zones better suited to extensive livestock systems.

Because of the high variability of climate and unpredictable drought events in dry savanna ecosystems, many of the concepts developed in the temperate zones fail to explain the dynamics of these highly variable ecologies. Many of the primary indicators, including carrying capacity and 'climax' vegetation stages, which have been used to measure the productivity of different ecosystems, are of limited use in these dynamic ecosystems.[13] Ecologists in Africa and Australia have designated these highly variable environments as disequilibrium ecosystems, to distinguish them from ecosystems where climate patterns are generally reliable enough for resident plant and animal populations to reach some sort of equilibrium.[14] The 'state and transition' model, where transition and randomness rather than equilibrium are norms, is helpful in describing such environments. Yet, most plans for conserving biodiversity and rangelands (for example, national parks and biosphere reserves) are still based on equilibrium theory and invoke notions of carrying capacity and average stocking rates to preserve an 'undisturbed wilderness', or to reverse perceived problems of land degradation. In both cases, African herders frequently resist politically, either subtly or forcefully, these efforts to restrict their access to key grazing and water resources.

The 'new range ecology' (disequilibrium) is more consistent with African models of environmental change, which have never excluded anthropogenic disturbances nor pursued ecological equilibrium as an objective. It is also appealing to those social scientists who study the complexity of herder decision-making. For

---

[10] Scoones, *Patch Use by Cattle*; G. Bartels, G. Perrier and B. Norton, 'The applicability of the carrying capacity concept in Africa: a comment on the thesis of de Leeuw and Tothill', in R. Cincotta, C. Gay and G. Perrier (eds), *New Concepts in International Rangeland Development: Theories and Applications* (Logan, UT: Department of Range Science, Utah State University, 1991) pp. 25–32; P. N. de Leeuw and J.C. Tothill, *The Concept of Rangeland Carrying Capacity in Sub-Saharan Africa – Myth or Reality?*, ODI Pastoral Development Network Paper 29b (London: Overseas Development Institute, 1990).

[11] Abel and Blaikie, *Land Degradation*, p. 20.

[12] Kenneth Hare, 'The Making of Deserts: Climate, Ecology, and Society', *Economic Geography*, 53 (1977), pp. 332–45; UNEP (United Nations Environment Program), *Status of Desertification and Implementation of the UN Plan of Action to Combat Desertification* (Nairobi: UNEP, 1991).

[13] P. Little, 'The social context of land degradation ('desertification') in dry regions', in L. Arizpe, P. Stone and D. Major (eds), *Population and Environment: Rethinking the Debate* (Boulder, CO: Westview Press, 1994), pp. 209–51.

[14] M. Westoby, B. Walker and I. Noy-Meir, 'Opportunistic Management for Rangelands Not at Equilibrium', *Journal of Range Management*, 42, 4 (1989), pp. 266–74.; Behnke *et al.*, *Range Ecology at Disequilibrium*.

example, herd management strategies of East African pastoralists, which have always been the bane of conservationists, assume drought, some degree of range degradation, and fire (burning) as norms, and have never tried to pursue ideas of carrying capacity or equilibrium. The ecology-of-disequilibrium perspective also adds considerably to our understanding of the 'overgrazing' or desertification debate, another global issue that is centered on African pastoralism and savanna landscapes.[15] It suggests that we know far less about what a degraded or overgrazed rangeland looks like, in terms of plant cover, than earlier work had led us to believe. According to this new line of thinking, a degraded or desertified parcel of land may have little to do with many of the standard vegetation indicators that had previously been used.

The political ecology approach, in turn, is relevant to this discussion; indeed, often the same scholars have an intellectual stake in both schools.[16] Political ecology can be a useful framework for weaving together different disciplines and has contributed considerably towards understandings of the social and political processes underlying resource use in savanna areas. Particular areas of politico-ecological research in savanna zones include: (i) herder and farmer conflicts; (ii) the emergence of absentee herd ownership and waged herders; and (iii) the effects of state policies on local institutions. Notwithstanding the proliferation of literature on political ecology in the 1990s, its popularity has led to a certain dilution in focus and theory. A perusal of the literature reveals ideas and concepts borrowed from theories as diverse as Marxism, neo-classical economics, critical theory, and feminism, and topics ranging from social movements and environmental discourse to land degradation and environmental values. Some scholars now claim that it marries cultural ecology with political economy (the latter a concept that is increasingly devoid of theory), while others say it includes any phenomenon where the state – even world systems – incorporate or impinge on local ecological and social systems.[17] More recent uses of the concept invoke a constructivist approach that examines how beliefs and ideology construct what we see as environment.[18] Finally, while some social scientists clearly prefer the open-ended nature of the concept,[19] others prefer a theoretical tightening and a more radical interpretation.[20]

There are, of course, some benefits to maintaining a broad interpretation of political ecology, and there is little doubt that this has resulted in important research and insights. What I propose here, however, is that, in the context of inter-disciplinary studies of ecological change, political ecology should revisit its original premises which helped to define what it is; orient field research; and selectively integrate concepts from different disciplines. In my opinion, there are three

---

[15] See Little, 'Social context of land degradation'.
[16] For example, Bassett, Behnke, Horowitz, Leach, Scoones, and Swift.
[17] James B. Greenberg and Thomas K. Park, 'Political Ecology', *Journal of Political Ecology*, 1 (1994), pp. 1–12.
[18] See Arturo Escobar, 'Constructing nature: elements for a post-structuralist political ecology', in R. Peet and M. Watts (eds), *Liberation Ecologies* (London: Routledge, 1996), pp. 46–68.
[19] Raymond L. Bryant, 'Political Ecology: An Emerging Research Agenda in Third World Studies', *Political Geography*, 11, 1 (1992), pp. 12–36.
[20] Richard Peet and Michael Watts, 'Introduction: Development Theory and Environment in an Age of Market Triumphalism', *Economic Geography*, 69, 3 (1993), pp. 227–53.

elements that define political ecology and that need to be re-emphasized: access, allocation (or management), and ecological impacts and processes (see Table 7.1). It starts with the political question of access, but all of the three elements in Table 7.1 are both causative and affected by each other. Relationships involving land (including the resources on it, such as water) are enmeshed in the political question of access, but in some pastoral areas one can gain access to lands through ownership of livestock, which, as recent work has shown, is increasingly skewed and is controlled by non-herders. The latter also raises a political issue, since many non-herders are civil servants, merchants, and others who may benefit from favorable policies of the government. Increased disparities in the distribution of wealth, in turn, have strong implications for how resources are conserved or abused in these areas. In short, while one can debate about how to define 'access' and 'resources,' there is little question that a political ecology approach – as Wolf implied more than 25 years ago – should first query access to resources. Among the three elements in Table 7.1 the social science disciplines have probably devoted the most energy to addressing the 'access' question.[21]

Table 7.1 The basic elements of a political ecology approach

Political	Social and economic	Ecological
1. Access to resources	2. Allocation and management of resources	3. The physical impacts and the feedback to 1 and 2

The resource allocation (or management) dimension of the model has also received considerable attention from social scientists. Culture and social structure are obvious components of resource management; for instance, culture-specific divisions of labor (for example, based on gender or age), community-based institutions, and social norms all play key roles in how resources are allocated and managed. Economists also devote considerable efforts to quantifying and modeling this aspect of political ecology and increasingly acknowledge that certain segments of producers or classes have differential access to resources which, in turn, affects the economics of resource allocation.[22]

The final element of the political ecology approach in Table 7.1 is the physical environment itself. Particular attention is given to the ecological or physical impacts of elements (i) and (ii) in the model, and the ways in which ecological processes, both short- and long-term, interact with and affect them. It is here where interdisciplinary collaboration with the natural sciences is most needed, but where collaboration has been particularly spotty. Because of this shortcoming many social science findings about the effects of 'access' and 'allocation' on the environment

---

[21] Jesse C. Ribot, 'Theorizing Access: Forest Profits along Senegal's Charcoal Commodity Chain', *Development and Change*, 29 (1998), pp. 307–41.
[22] Frank Ellis, 'Household Strategies and Rural Livelihood Diversification', *Journal of Development Studies*, 35, 1 (1998) pp. 1–38.

remain hypotheses to be empirically tested. While research may result in important social science insights and findings, a political ecology that does not address the physical environment falls short of the paradigm that many of us were so excited about in the 1980s.[23] In short, the politics of access may drive environmental change, but without ecological data and interdisciplinary collaboration the promise of a political ecology remains unfulfilled.[24] For this reason, I feel it is timely to seek interdisciplinary collaboration; and the 'ecology of disequilibrium' model holds the potential of this in the context of African pastoralism.

This appraisal has been an over-simplified discussion of interdisciplinary research and theory. Yet, as those who have worked on interdisciplinary studies know so well, simple conceptual frameworks and basic research questions – where some agreement across disciplines can be reached – is where most collaboration begins. Another advantage for interdisciplinary research is that the model suggests where the social science (elements (i)and (ii)) and natural science disciplines (element (iii)) should concentrate their efforts. Recent research in South Africa explores the relationship between access, local management institutions, and grasssland ecosystems and points to the kind of political ecology that integrates both social science and ecological data.[25] In this case the research program draws on concepts from both the disequilibrium ('new range ecology') and political ecology schools and shows empirically how local vegetation cover over the past 60 years has been affected by social and political pressures which have resulted in 'intensive land pressure and stocking rates'.[26]

Case Studies from East Africa

In this section of the chapter, two different case studies of pastoral resource use are analyzed. It begins with a research and development project in Marsabit, Kenya, which is then followed by a study that the author conducted in southern Somalia. The application of a political ecology framework – as elaborated in the previous section – is attempted to point both to its complexities as well as the difficulties that arise when researchers are not working on the basis of common concepts and problems.

The Integrated Project in Arid Lands (IPAL), Marsabit, Kenya

The IPAL research in Marsabit is one of the most detailed, long-term studies of land

---

[23] Little and Horowitz, *Lands at Risk in the Third World.*

[24] In my own case, understandings of element (iii) (ecology) in the context of dryland ecosystems have gone about as far as possible without additional training as a range ecologist or agronomist.

[25] Thembela Kepe and Ian Scoones, 'Creating Grasslands: Social Institutions and Environmental Change in Mkambati Area, South Africa', *Human Ecology*, 27, 1 (1999), pp. 29–53.

[26] Kepe and Scoones, 'Creating Grasslands', p. 40. The Kepe and Scoones' study is a wonderful example of political ecology that is interdisciplinary and balances the social and ecological dimensions of environmental change. Ironically, it appears in the same issue of *Human Ecology* that includes Vayda and Walters' condemning critique of political ecology (Vayda and Walters, 'Against Political Ecology').

degradation in a dry region of eastern Africa.[27] Initiated in 1977 under the direction of Hugh Lamprey, an avid believer in 'desertification', the IPAL study of Marsabit District, Kenya was a multidisciplinary effort by anthropologists, geologists, historians, ecologists, range management specialists, and hydrologists.[28] It generated masses of biophysical and socioeconomic data and resulted in numerous reports and publications (see, for example, the bibliographic list in IPAL, *Integrated Resource Assessment*). Even in this interdisciplinary project, however, a synthesis of the social and biological was difficult to achieve, often resulting in conflicting positions being taken by members of the same research team. Reasons for this were that the social science research was never treated equally with the work of the natural sciences,[29] and the study did not elaborate a common conceptual framework. The study spanned a period of approximately eight years and was strongly motivated by the twofold premise that desertification was measurable and that it was occurring in northern Kenya. The project, which was part of UNESCO's 'Man and the Biosphere Program', received substantial amounts of funding from UNEP and often reported its findings in UNEP's *Desertification Control Bulletin*.[30]

The social science research carried out by IPAL was competent and resulted in several important publications on social change and land management that had important implications for a political ecology of pastoralism.[31] They clearly address

[27] IPAL (Integrated Project in Arid Lands), *Integrated Resource Assessment and Management Plan for Western Marsabit, Northern Kenya* (Nairobi: 2 vols., UNESCO, 1984); and Elliot Fratkin, *Surviving Drought and Development* (Boulder, CO: Westview Press, 1991).

[28] Hugh Lamprey, 'Report on the Desert Encroachment Reconnaissance in Northern Sudan, 21st October to 10 November, 1975', UNESCO/UNDP, mimeo.

[29] Michael O'Leary, 'Drought and change among northern Kenya nomadic pastoralists: the case of the Rendille and Gabra', in Gísli Pálsson (ed.), *From Water to World-making: African Models and Arid Lands* (Uppsala: Scandinavian Institute of African Studies, 1990), pp. 151–74; *idem.*, 'Ecological Villains or Economic Victims: The Case of the Rendille of Northern Kenya', *Desertification Control Bulletin*, 11 (1984), pp. 17–21; IPAL, *Integrated Resource Assessment and Management Plan for Western Marsabit*; Hugh Lamprey and H. Yussuf, 'Pastoral and Desert Encroachment in Northern Kenya', *Ambio*, 10, 2 (1981).

[30] See O'Leary, 'Ecological Villains or Economic Victims'. Few intellectual debates about the environment are marked with more ambiguity, scientific 'self-righteousness', and ideology than the debate about land degradation in dry regions or, what is more commonly called, 'desertification'. First noted as a problem in Africa in the 1930s (see, for example, E. P. Stebbing, 'The Encroaching Sahara: The Threat to the West African Colonies', *Geographical Journal*, DXXXV, 5 (1935), pp. 506–24), desertification emerged as a major environmental issue in the 1970s with the publicity surrounding the prolonged Sahelian drought and the speculation regarding its causes and ecological effects (Michael H. Glantz, *Desertification: Environmental Degradation in and Around Arid Lands* (Boulder, CO: Westview Press, 1977); Erik Eckholm and Lester Brown, *Spreading Deserts: The Hand of Man* (Washington, DC: Worldwatch Institute, 1977), Worldwatch Paper No. 13. The convening of a major UN conference on desertification in the mid-1970s (UNCOD (United Nations Conference on Desertification), *Report of the United Nations Conference on Desertification, Nairobi, 29 August–9 September 1977* (Nairobi: United Nations Environment Programme, 1977)), the establishment during the same period of a research and policy unit at the United Nations Environment Program (the 'Desertification Branch'), and the increased availability of funding for research and projects on desertification heightened awareness of it among the general and scientific communities and propelled it into the center of the environmental dialogue.

[31] See Michael O'Leary, The *Economics of Pastoralism in Northern Kenya: The Rendille and the Gabra*, Integrated Project in Arid Lands, Technical Report Number F-3 (Nairobi: UNESCO, 1985); and *idem.*, 'Drought and change' (Nairobi: UNESCO, 1985).

issues of access and resource management and provide detailed data on land, labor, and income relationships, including information on tenure institutions, local resource practices and knowledge, labor organization and migration, and herder and gender differentiation. The work of O'Leary and Sobania, both of them social scientists who worked for IPAL, place Rendille and Gabbra pastoralism in its larger political, economic, and historical contexts, showing how colonial grazing policies restricted population movements and how the Rendille and Gabbra have recently entered wage labor markets to supplement their incomes.[32] While the natural science research of IPAL often claimed that desertification was widespread in northern Kenya and caused by local management practices,[33] the project's own social scientists cautioned against such sweeping generalizations. Thus, while O'Leary admits to the strong presence of degradation around large settlements, he does not attribute it to herder practice.[34] Instead, he argues that these problems are the result of faulty policies that encourage sedentarization, impoverish large segments of herders, and restrict access to distant ranges.[35] In short, there was a strong 'disconnect' between the social science and ecological research that left the social scientists with only a partial picture of the area's political ecology.

However, there are certain social and political processes in the area that had clear environmental effects. For example, herders in Marsabit increasingly pursue a sedentary form of pastoralism that overuses areas around settlements but leaves large parts of the range underutilized.[36] It is estimated that about 45 per cent of herders in the district reside around four major population settlements, and that large parts of the surrounding range areas are underutilized, including 11 per cent of usable rangeland that for more than three years contained no domestic livestock at all.[37] While not explicitly addressed in the study, at least part of the problem is due to insecurity (a political problem) that makes some distant range areas inaccessible. Nonetheless, the settlements around which a large number of herders dwell are pockets of economic and ecological impoverishment, marked by increased economic differentiation, dependence on food aid, and fragmented domestic units headed by widows and divorced women. What seems clear in this case is that

[32] Michael O'Leary, 'Drought and change'; idem., 'Ecological Villains or Economic Victims; idem., 'Changing responses to drought in Northern Kenya: The Rendille and Gabra livestock producers', in Paul Baxter with Richard Hogg (eds), *Property, Poverty and People: Changing Rights in Property and Problems of Pastoral Development.* (Manchester: Department of Social Anthropology and International Development Centre, University of Manchester, 1989), pp. 57–79; and N. Sobania, 'Background History of the Mt. Kulal Region of Kenya', Integrated Project in Arid Lands', Technical Report Number A-2 (Nairobi: UNESCO, 1979b).

[33] Lamprey and Yussuf, 'Pastoral and Desert Encroachment', IPAL, *Integrated Resource Assessment.*

[34] O'Leary, 'Ecological Villains or Economic Victims.' Rendille herders recognize the existence of overgrazing, but note that it is a reversible process, and they can identify several overgrazed areas that have recovered after being left fallow (O'Leary, *The Economics of Pastoralism*, p. 81).

[35] O'Leary, 'Ecological Villains or Economic Victims'; and idem., 'Changing responses'.

[36] IPAL, *Integrated Resource Assessment*. Insecurity is another reason why herders of Marsabit now prefer to reside in large settlements. For a recent analysis of the relationship between political insecurity and environmental degradation (including desertification), see Anders Hjort af Ornas and M. A. Mohammed Salih (eds), *Ecology and Politics: Environmental Stress and Security in Africa* (Uppsala: Scandinavian Institute of African Studies, 1989).

[37] O'Leary, 'Ecological Villains or Economic Victims', p. 17.

environmental problems are not caused by customary land use that relies on mobility, but by social, political, and economic processes that restrict movement and lead to the overutilization of certain zones. In the case of the IPAL project area, this is further aggravated by the recent tendency of wealthy and well-connected ranchers, who often are residents of local towns, to acquire private land titles in the more favorable range zones and to fence 'immense areas' of communal grazing grounds.[38]

In spite of the relatively lengthy period, from 1977 to 1985, of data collection on environmental processes, the ecological arguments of IPAL are far from convincing. None of the ecologists subscribed to models based on disequilibrium, but instead were systems ecologists who relied on equilibrium-based, functionalist models. While IPAL's environmental studies rate much of Marsabit District as degraded or under threat of degradation, the project's own data show that most degradation has been in the vicinity of a few large settlements and water points, an observation that is hardly novel. Even in the case of settlements, IPAL admits that much of the damage is probably reversible:

> The present situation is a complete absence of vegetation around the settlements in an ever-increasing circle, which, at Korr, is estimated as having a 9 km radius.... Some degree of regeneration has proved possible by the IPAL enclosures in Korr and Kargi. The successional trend appears to be from annual grasses and herbs, which are the first colonizers, to dwarf shrubs with, hopefully, after long periods of protection, a return to the original vegetation.[39]

The IPAL scientists, surprisingly, rely heavily on subjective indicators of range condition:

> The range condition for the range sites contained here is a *subjective judgement* [emphasis added] based on the following attributes: soil stability, composition of desirable and non-desirable plants, bare ground and litter cover and the state of erosion. It is based on four categories: excellent, good, fair and poor.[40]

With such general categories and with their own perceptions of 'desirable' plants and erosion, it is not surprising that the IPAL ecologists found much of the area to be degraded: 'With the exception of one mountain range type, none of the range types had a condition better than "fair", indicating the seriousness of the degradation of the grazing resource condition in the study area' (*ibid.*). The range specialists see a lack of sound management strategies by the herders (which is not a problem, according to IPAL's social science work)[41] as the major cause of land degradation.

In sum, although IPAL's work provides important information on range ecology and pastoral practice, the socio-economic and biophysical components of the study are not well integrated, which results in contradictory statements about the causes of land degradation. The project involved different disciplines but was *not interdisciplinary*. Because their program was funded in part by organizations (for example,

[38] Gunter Schlee, 'Traditional pastoralists – land use strategies', in H. Schwartz, S. Shaabani and D. Walther (eds), *Marsabit District: Range Management Handbook of Kenya* (Nairobi: Ministry of Livestock Development, 1991), Vol. II, No. 1, p. 154.
[39] IPAL, *Integrated Resource Assessment*, p. 227.
[40] IPAL, *Integrated Resource Assessment*, p. 106.
[41] See O'Leary, *The Economics of Pastoralism'*.

UNEP) with a vested interest in the reality of 'desertification',[42] and because they did not work with a consistent definition of what degradation entailed, IPAL seemed willing to assume degradation in cases where its own data and analysis indicated otherwise.

More recent ecological research in the same region, sponsored by Kenya/ UNEP, reveals similar inconsistencies and contradictions regarding the extent of degradation. Compare, for example, these two quotations from the same document:

> The general picture for Marsabit is that no significant degradation occurred during the 16 years [1956–72] except for Logologo and a little for Illaut.... Because the degradation was apparent only at Illaut and Logologo, it would not be fair to derive a rate and generalize it to the whole study area.[43]
>
> The results of this study show that desertification is a major problem in the study areas [also including Baringo, Kenya]. The main forms of desertification identified were soil and vegetation degradation. The soils are being degraded through water and wind erosion, while vegetation is through tree and shrub cutting and overgrazing.[44]

As in the IPAL study, this research raises doubts but, again, when the environmental data are inconclusive, desertification is assumed.

An even more recent study contradicts both the IPAL and the Kenya/UNEP findings, by suggesting that land degradation in Marsabit is not a problem: 'Most of the district's rangelands are in good condition.... This primarily means that the rangelands are not degraded ... less than 1% of the district's rangelands are in poor condition'.[45] Regarding this last assessment, conducted by the Government of Kenya with support from German technical assistance (GTZ), Herlocker and Walther are not willing to claim 'desertification' when serious questions remain. Considered together, the three studies raise troubling questions about the

Table 7.2 Comparison of rangeland classification, Marsabit District, Kenya

Study	Classification		
	Poor	Fair	Good
IPAL/UNEP	>70%	20%	<5%
GTZ/KREMU	<1%	<<<99%>>>	

Based on: IPAL, *Integrated Resource Assessment*; Herlocker and Walther, 'Condition of Marsabit range lands'.

[42] For a good discussion of the 'institutionalization' of the desertification debate in development bureaucracies, see Andrew Warren and C. T. Agnew, *An Assessment of Desertification and Land Degradation in Arid and Semi-Arid Areas* (London: International Institute for Environment and Development, 1988); and Brian Spooner, 'Desertification: the historical significance', in Rebecca Huss-Ashmore and Solomon H. Katz (eds), *African Food Systems in Crisis. Part One: Microperspectives* (New York: Gordon and Breech Science Publishers, 1989), pp. 111–62.

[43] Richard S. Odingo, 'Report of the Kenya Pilot Study (FP/6201-87-04) Using the FAO/UNEP Methodology for Assessment and Mapping of Desertification', in *Desertification Revisited: Proceedings of an Ad-Hoc Consultative Committee on the Assessment of Desertification* (Nairobi: UNEP, 1990), p. 152.

[44] *Ibid.*, p. 157.

[45] Dennis Herlocker and Dierk Walther, 'Condition of Marsabit range lands', in Schwartz *et al.*, *Marsabit District*, p. 52.

measurement and interpretation of degradation even when 'hard' evidence is offered (see Table 7.2).

Case study: herder response to drought and conflict in southern Somalia

The data for this case study were collected under two different research projects: a study of herder households and marketing in 1986–8 in the Lower Jubba Region (prior to the recent civil war) and a study of regional and cross-border livestock trade between southern Somalia and Kenya in the summers of 1995 and 1996.[46] The analysis is only suggestive of environmental processes but available data show that there are differences from the Marsabit case. In comparison with Marsabit, Somali herders of the Lower Jubba Region have generally fared better during recent droughts and have avoided the kind of sedentarization and social and ecological impoverishment discussed earlier. With few exceptions there is little localized environmental degradation, and human population and livestock are relatively evenly distributed in the region.[47] Even the civil strife of the 1990s has not had a major disastrous impact on the region's livestock sector, an outcome that many organizations and practitioners might challenge.[48] The reasons for the resilience of Somali pastoralism – especially in comparison with similar systems in northern Kenya – are the flexibility/mobility of the system and the relatively favorable access to key grazing and water resources that herders maintain. By contrast, on the Kenyan side access to vital resources is eroding because of the encroachment of alternative land uses (settlements, irrigated agriculture, game parks and reserves, and hydropower projects) into dry-season reserves.

The Lower Jubba Region is a particularly patchy environment of wetlands and localized water points which borders on Kenya (see Figure 7.1). Key grazing resources tend to be around seasonally flooded pans (the Lag areas of Jira and Dera and the coastal plains), swamps (the Lorian swamp of northeastern Kenya), and river valleys (the Tana and Jubba River valleys). Rangeland vegetation in the area includes such trees and shrubs as *Salvadora persica, Acacia nilotica, A. tortilis*, and *A. zanzibarica*, and such grasses as *Sporobolus helvolus, Cenchrus cilaris*, and *Echinochloa hapoclada*. In seasonally flooded zones of the basin, grasses 'form tall dense swards'

---

[46] Research for this case study has been supported during the past 12 years by the 'Settlement and Resource Systems Analysis' (SARSA) grant, Institute for Development Anthropology, the Office of Foreign Disasters Assistance, USAID, and the Social Science Research Council, USA. Opinions expressed in the chapter are those of the author and should not be attributed to either of the above funding agencies.

[47] See ARD (Associates in Rural Development), *Jubba Environmental and Socioeconomic Studies: Final Report* (Burlington, VT: Associates in Rural Development, 1989); and Murray Watson, 'Aerial Surveys of Livestock Populations in the Dry Season, Jubba Valley', unpublished report, 1987.

[48] See EC/FAO (European Commission/Food and Agriculture Organization), *Somalia: Livestock Export Market Study* (Nairobi: EC, 1995). For example, one local NGO, the United Somali Sahil Professionals' Association (UNISOPA), estimated that the Lower Jubba area had lost about two-thirds of its cattle since 1991 as a result of drought, war, and disease (UNISOPA, *Report on Relief, Rehabilitation, and Resettlement Schemes, Juba-Land, Somalia* (Mombasa, Kenya: 1993), p. 2). I would strongly question these figures, as would others. My assessment here does not include the effects of the recent (1997–8) floods and outbreaks of livestock disease in the region.

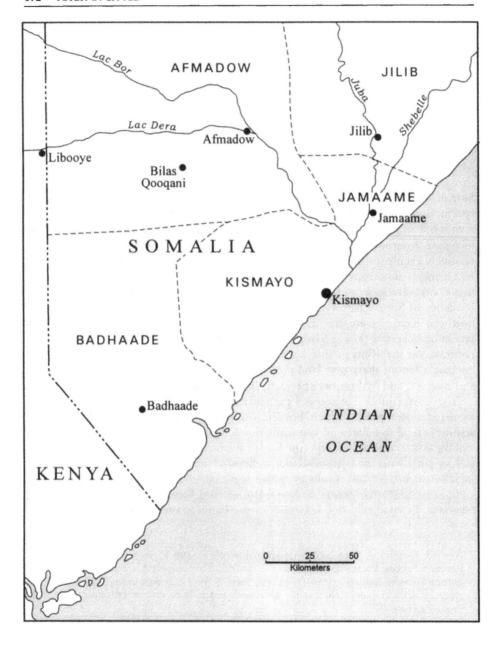

Figure 7.1 Map of Lower Jubba Region, Somalia

with annual fodder production in excess of 10 metric tons per hectare.[49] The most productive of these patchy zones is the Descheeg Waamo, which supports cattle densities in excess of 150 per sq. km during the long dry season.[50] It is the productivity of these areas – which on the Kenyan side are already increasingly cut off to herding populations – that determines whether or not herders can survive harsh dry seasons and droughts without massive livestock losses. Local groups of elders regulate the use of these pastures during the dry seasons.

Localized droughts are very common in the Lower Jubba Region, forcing herders to adjust their normal grazing patterns every 3 to 4 years. During the past 20 years full-blown regional and/or national droughts have occurred about every eight years, while localized droughts occur more frequently. A localized drought means that at least some parts of the region or nearby accessible locations have experienced at least some rainfall, while other sites have not. A regional drought, in turn, means that the long rains (*gu* season) have completely failed in the entire region; while a national drought means that the long rains have failed throughout the country. During April to July 1996 the long rains of central and northern Garissa District on the Kenyan side failed almost completely, with areas receiving about 30 per cent of normal rainfall. Fortunately, the early rains in April were quite favorable across the border in Somalia and were above average in the southern portion of Garissa District during the later part of the season (May–June). Thus, herders in the locally affected zones were able to move their animals across the border into Somalia or southward towards the coast (the latter area being in the tsetse fly zone). During June 1996 several Kenyan herders moved their cattle as far as the Kismayo coast, where rainfall was good but conflict endemic. That herders moved their animals to the Kismayo coastal zone, a tense area of conflict between the region's two major clans,[51] is indicative of the flexibility of grazing strategies and reciprocal grazing rights even in times of hostility.

In southern Somalia the prolonged dry season of 1987 turned into a regional drought affecting most of the Lower Jubba Region. Yet, because of accessibility and mobility herders were able to move animals to neighboring regions in the Shebelli river basin (Somalia) and/or to parts of north-eastern Kenya. Herd migrations in the *haga* season (June to August) of that year exceeded 200 kilometers in many cases, a distance that is very uncommon for pastoralists in Marsabit, Kenya. Because of their inaccessibility to former dry-season grazing areas in river basins and highland zones, Kenyan herders are constrained during droughts. Once again, herd movements in southern Somalia were into areas under the control of different clans, but reciprocal grazing rights were respected.

In terms of livestock distribution, the ownership of both camels and cattle in the region is highly skewed and influences how herders acquire access to resources. Among a sample of 88 herders, the richest 12.5 per cent of camel herders own 70 per cent of the camels in the region, while almost half the herders own no camels at all. Inequities are further demonstrated by the fact that one household alone controls more than 25 per cent of total camels in the sample. The ownership of

[49] Ian Deshmukh, 'Terrestial ecology baseline studies', in *JESS Final Report* (Burlington, VT: Associates in Rural Development, 1989), p. A-38.

[50] Watson, 'Aerial Surveys of Livestock Populations'.

[51] The Ogadeen and Harti.

Table 7.3 Distance of grazing migrations of livestock by season: Afmadow herders (km)

Season	Cattle	Camels	Sheep/goats	All
Long dry season	70.22	34.50	61.18	67.32
Long wet season	60.29	135.00	49.89	61.34
Short dry season	44.00	19.00	40.22	42.38
Short wet season	37.64	14.50	29.56	35.40
Average all seasons	53.49	50.75	46.05	52.09

cattle in the lower Jubba is not as skewed as noted above, but considerable discrepancies do exist. For example, 12.5 per cent of households in the sample own 39 per cent of total cattle, with the bottom 50 per cent of cattle herders owning approximately 15 per cent of the herd. The inequities displayed in cattle ownership in the region are not unusual for pastoral economies.[52]

Analysis of grazing patterns by livestock species and season reveals several important trends about access, allocation, and ecology (see Table 7.3). First, it is revealed that on average the distances of seasonal herd movements vary considerably according to livestock type, but in general movements are greatest for camels and cattle, followed by small stock. For Afmadow herders movements of camel herds are greater than for cattle during the long wet season (*gu* season), when camels move well away from the seasonally flooded pastures around the Jubba river and Lags Dera and Jira (see Figure 7.1). When critical grazing zones are flooded, access is cut off for all herders, rich and poor. During the long dry season camels are not moved nearly as far as cattle, since they are able to use browse species away from the rivers and main water points. At that time small stock and cattle herds are moved an average distance of 70.22 and 61.18 km, while camels are moved only about 34.50 km. In short, camels tend to avoid the normal seasonal grazing areas of cattle and small stock, which in the case of Afmadow tend to center around the Jubba Valley and the seasonally flooded plains near the Somalia/Kenya border. Seasonal distances of small ruminant movements approximate seasonal cattle migrations for part of the year, except that they tend to remain with the main family residence whenever they move.

The data illustrate how distances of cattle and small stock movements are greatest in the long dry season, followed in importance by the long wet season. In the case of Afmadow herders, the distances of cattle migrations during the long dry and wet seasons are 70.22 and 60.29 km, respectively, while they are only 44 and 37.64 km during the other two seasons of the year. While the distances are smaller, the timing and rhythm of small ruminant movements generally follow this same model of seasonal variation. It should be noted, however, that, except during extreme droughts, certain grazing zones became restricted to members of certain clans.

What effect does access to livestock and labor have on herd management strategies? Why do certain herders remain in the Jubba Valley area despite the

---

[52] John W. Sutter, 'Cattle and Inequality: Herd Size Differences and Pastoral Production Among the Fulani of Northeastern Senegal', *Africa*, 57 (1987), pp. 196–217; Little, 'Social Differentiation and Pastoralist Sedentarization'.

presence of tsetse flies and other hazards, especially during the long wet season? By differentiating herd movements according to a wealth indicator like cattle ownership, it is possible to show how access to seasonal pastures differs for relatively poor, labor-constrained families. Table 7.4 indicates that the average size of cattle herds in Afmadow is 74.74 head per herding unit, but that those herders who mainly remain in the Jubba Valley during the long rains own considerably fewer. If we look at cattle ownership among those herders who use the Jubba Valley during the long rains, most of them own between 6 and 48 cattle. For example, among herders who use three of the most heavily utilized Jubba locations – Boka, Descheeg Wamo, and Tingadud[53] – in the wet season, average cattle herds are 6, 44, and 48.50, respectively. By contrast, those herders who use areas at Fafudun, Jira, Boji,[54] and Hagar – all wet season grazing areas located more than 80+ km from the Jubba Valley and well outside the tsetse fly belt – average cattle herds are 165, 67, 75, and 172, respectively. In short, those herders who remain in the Jubba area during the wet season own cattle herds that are below the general average.

Those herders with less livestock wealth also have smaller households. With fewer household members to draw on for herding tasks, they are more constrained by labor than wealthier households. While the average household size in the region is 9.07 members, it is less than 7.5 members for domestic units with herds of less than 50 cattle. Thus, households which control smaller numbers of cattle usually do not move as far as other herding units. The average number of cattle owned was 63.07 head for households with migrations of less than 100 km, while it was 103 cattle for units which migrated an average of more than 100 km per dry season. Clearly those households with large herds and additional family members have a greater capacity than other families to utilize distant, more productive pastures.

In 1987 the long rains in the Lower Jubba area had consisted of only a few showers early in the wet season, with a combined amount of about 20 mm. Confronted with dim prospects for improvement, several Afmadow herders considered grazing options outside the Lower Jubba Region. The pastures of the Jubba Valley had been exhausted and movement to Kenya was difficult because of the lack of surface water. Although Afmadow herders are from a different clan than residents of the lower Shebelli region (a distance of about 200 km), they nonetheless sent a group of young men and elders to the region to negotiate a grazing agreement. The long rains had been adequate there and water and pastures were plentiful. Despite a certain degree of tension between the different clans, permission was granted for several Jubba families to graze their animals in the lower Shebelli. When I re-interviewed herders who had gone on the long trek at the end of 1987 – the wealthier had hired trucks for the move – very few reported unusual levels of cattle mortality, although there had been virtually no rain for about 9 months. Movements of 150+ km during prolonged dry seasons or droughts are not unusual in the Lower Jubba Region, but are highly atypical of Kenyan herders. The mobility of herders of the Lower Jubba and their access to distant pastures outside their home regions are key reasons why they are more successful than Kenyan herders during droughts.

---

[53] It should be noted that Boka and Tingadud are located alongside the Descheeg Wamo, although herders often distinguish them from the latter.

[54] Boji is actually located in Lag Jira, although it is distinguished by herders as a separate grazing area.

Table 7.4 Average number of cattle owned by Afmadow herders at different grazing zones, (long wet season)

Grazing location	Average cattle owned
Afmadow	47.14
Boji	75.00
Boka, Jubba Valley (JV)	6.00
Borsanga, JV	39.00
Buale, JV	41.50
Descheeg Waamo, JV	44.50
Diif	70.00
Fafadum	165.00
Gomese	50.00
Hagar	172.00
Jira	67.20
Bilas Qooqani	95.67
Tingadud, JV	48.50
Tortora	133.50
Wareers	362.00
ALL	74.74

In contrast to the Marsabit study, ecological data collection for the Lower Jubba area was not on a large-scale basis. Elsewhere I have presented some of the data on pastures and water resources that exist for the area;[55] these were collected as part of a large inventory undertaken in the late 1980s.[56] With the continued conflict in the area, such research is currently impossible, and thus, while I can hypothesize about some of the impacts of access and resource management on the environment, they remain speculative. The key resource 'patches' in the zone seem to have undergone little change in recent years, as evidenced by the fact that grazing patterns remain similar and livestock weights and conditions are very good.[57]

Conclusion

Both the Kenyan and Somali studies highlight the importance of understanding what is meant by access and the ways in which power and wealth determine who can graze 'where' and 'when'. These processes have important ecological impacts, as evident in the Marsabit case where impoverished herders are restricted to degraded grazing zones around settlements, and in the Somalia case where labor-constrained herders have less mobility and opportunity to graze both wet and dry season zones. In the Marsabit case, the role that the state plays in facilitating changes

[55] Peter D. Little, 'Traders, Brokers, and Market "Crisis" in Southern Somalia', *Africa*, 62, 1 (1992), pp. 94–124.

[56] See ARD, *Jubba Environmental and Socioeconomic Studies*.

[57] See P.D. Little, *Cross Border Cattle Trade and Food Security in the Kenya/Somalia Borderlands* (Binghamton, NY: Institute for Development Anthropology, 1996).

in access to land and other resources requires examination; in stateless Somalia the issue is irrelevant. In the Kenyan example, the actions of the state and development agencies give rise to environmental problems, since they encourage unfavorable land/people ratios and excessive use of restricted lands. On the other hand, conflict-ridden southern Somalia suffers less from localized degradation since herder mobility and access remain favorable.

This chapter has called for a return to the basic elements of a political ecology approach that highlights resource access and allocation, on the one hand, and ecology, on the other. Access is shown to be a political issue, while ecology requires an appreciation of biological processes. Both sides of the equation need to be addressed, and environmental data should be as carefully scrutinized as social and political data. This was shown to be the case in the Marsabit example, where environmental data and interpretations were highly suspect. This means making the physical scientist accountable to the same standards of data analysis and quality and logic of argument by which any scholar (social scientist or other) should be assessed.

Advantages of the political ecology approach elaborated in this chapter are that it suggests a division of labor among disciplines, as well as incorporating insights from new thinking in ecological theory, such as the 'disequilibrium' school. It narrows political ecology to a framework that integrates three fundamental aspects: access, allocation, and ecology. Studies can weigh certain of these factors more than others, but political ecology should incorporate elements of each. This framework is evaluated in the context of two East African case studies involving pastoralists. The mobility and favorable access that Somali pastoralists maintain allow them to adjust to unstable physical and political environments without serious ecological consequences. The environments of the Lower Jubba and Marsabit are both highly variable but, unlike the Somali herders, mobility and access are more constrained among Marsabit pastoralists, where localized degradation is taking place.

The pendulum in political ecology has probably swung too much towards the political and, in turn, has downplayed the importance of ecology. In this sense, Vayda and Walters are correct in their assessment that much of what is called political ecology has very little to do with ecology.[58] This does not mean, however, an abandonment of the concept, with a retreat to a naive 'isolationism' that removes external political processes from studies of environmental change. Analytically convenient as this might be, it would strongly contradict current political and social realities around the world, even in relatively isolated areas of East Africa. Instead, what is called for here is the renewal of a political ecology that integrates social, political, and ecological processes, and which recognizes that the last has to be a part of the approach. When one looks at some of the excellent current research on environmental change in Africa, there are grounds for optimism that an integrated political ecology is possible. [59]

---

[58] Vayda and Walters, 'Against Political Economy'.

[59] Behnke *et al.*, *Range Ecology at Disequilibrium*; Scoones, *Living With Uncertainty*; Leach and Mearns, *The Lie of the Land*.

8

Ecological 'Crisis' & Resource Management Policy in Zimbabwe's Communal Lands

WILLIAM A. MUNRO

Introduction

For well over half a century, perceptions of deepening ecological degradation have shaped agrarian policies in Zimbabwe's small-scale agricultural sector, known as the communal lands, where the bulk of the population lives. The patterns and extent of environmental stress vary widely depending on the histories of settlement, the implementation of land-use planning models, and the ecology itself. But perceptions of degradation have focused particular attention on the destructive effects of bovine stock pressure on the quality of vegetation and the integrity of the soil. Ever since the 1930s, agricultural officers have sought to curb environmental stress by limiting cattle numbers according to the land's scientifically determined carrying capacity. Despite this long history of efforts to sustain a stable environment, however, land degradation and 'overstocking' continue to be dire problems.[1] In this context, both the effectiveness of the policy paradigm and the interpretations of environmental decline that inform it solicit scrutiny. Why have resource-management strategies, implemented by an apparently 'strong' state with effective agricultural agencies, achieved a poor record of success? What can the persistence of this problem tell us about relations between the state, rural resource users, and nature? What precisely are the ecological dynamics of environmental change in the communal lands? What should appropriate resource management policies look like?

This chapter aims to cast light on these questions by revisiting the history of

---

[1] See, for instance, L. Cliffe, *Policy Options for Agrarian Reform in Zimbabwe: A Technical Appraisal* (Rome: FAO, 1986); World Bank, *Zimbabwe: Land Subsector Study* (Washington, DC, 1986); Y. Biot, P. Blaikie, C. Jackson and R. Palmer-Jones, *Rethinking Research on Land Degradation in Developing Countries*, World Bank Discussion Paper 289 (Washington, DC: World Bank, 1995); S. Moyo, P. Robinson, Y. Katerere, S. Stevenson and D. Gumbo, *Zimbabwe's Environmental Dilemma: Balancing Resource Inequities* (Harare, 1991); Government of Zimbabwe (GOZ), *Prospectus for Environmental Assessment Policy in Zimbabwe* (Harare: Ministry of Environment and Tourism, September 1993).

livestock management policies in the communal lands, and reviewing patterns of continuity and change in the policy orthodoxy of stock limitation. In Zimbabwe, policy-makers, technical officers, administrators, and the public in general understand the problem of overstocking socio-politically, as rooted in the colonial history of racial land apportionment, population pressure in the communal lands, and an expansion of arable production into grazing areas. Nevertheless, for the purposes of resource management, policy-makers approach the problem *technically* as an outcome of the relationship between individual stockholders and nature. This technical approach is operationalized through the two key concepts of 'carrying capacity' and 'offtake': once one has established the appropriate carrying capacity of a piece of land, one creates institutions that will constrain stocking levels from exceeding that capacity. Carrying capacity, defined conventionally as the 'maximum animal numbers which can graze each year on a given area of range, for a specific number of days, without inducing a downward trend in forage production, forage quality, or soil', can be determined by a technical assessment of topland soil quality, vegetation and rainfall. Offtake can be determined by the incentive of stockholders to raise and sell their herds profitably over time. Thus, resource management draws on a 'simple arithmetic of the grazing problem',[2] according to which carrying capacity and offtake are correlated to measurable, and relatively stable, 'tipping points' of sustainability.

This 'arithmetic' is not as simple as the policy paradigm suggests. In the first place, the ecological tipping point is never absolute, but is always relative to a desired socio-economic or environmental outcome: the point at which livestock livelihoods can merely be sustained is not the same as the point at which livestock can be raised at marketable values; the point at which resource depletion can be halted is not necessarily the same as the point at which regeneration is still possible. As a result, both carrying capacity and offtake, as analytical categories for guiding policy, are complicated by economic objectives, ideas about development, and calculations of risk that inform resource-use patterns. In the second place, defining tipping points requires both that changes in range quality over time are measurable on a clear set of criteria – such as plant succession and a relative increase in less palatable species – and that such changes are clearly attributable to stock pressure on the range. This may be readily done where vegetation systems are stable and ecologies are at equilibrium. But where the ecology is vulnerable to destabilization by irregular climatic shocks such as intense drought, the degree to which changes in range quality are cyclical, or are caused by climatic factors, or by excessive stocking, or by the patterns of resource use dictated by the range management system itself, is much less clear. Chronic uncertainty makes ecological 'tipping points' immensely difficult to define.[3] Such uncertainty may be overcome by defining carrying capacity conservatively, so that reduced profits in good years are offset by sustained

[2] A.J.B. Hughes, *The Role of Cattle in Tribal Society* (Salisbury: Department of Conservation and Extension [Conex], Ministry of Agriculture, 29 May 1964), p.2.

[3] See R.H. Behnke, I. Scoones and C. Kerven (eds), *Range Ecology at Disequilibrium: New Models of Natural Variability and Pastoral Adaptation in African Savannas* (London: Overseas Development Institute, 1993); I. Scoones, 'The Dynamics of Soil Fertility Change: Historical Perspectives on Environmental Transformation from Zimbabwe', *The Geographical Journal*, 162, 2 (1997).

productivity in drought years.[4] But such a strategy may antagonize stockholders if it does not fit their own understanding of local ecologies, engage their framework for risk management, or address the ways they value livestock.

It is evident, then, that the 'simple arithmetic' of stock management policy in Zimbabwe cannot be taken for granted. Indeed, it rests on a broader policy-formulation framework, which has several essential components:

(i) a technical assessment of the resource base, which generally draws its justification from the epistemological authority of scientific knowledge;

(ii) a social and economic interpretation of the existing farming system (and its faults);

(iii) a social and economic model of the desired system (and its advantages);

(iv) a determination and deployment of appropriate technology to effect a shift from the existing to the preferred model; and

(v) the design and establishment of effective and appropriate institutions, especially with regard to property rights, to achieve the desired outcomes.

These components provide a starting point for evaluating the policy paradigm. Clearly, resource management is inseparable from the exercise of state power. Indeed, a wide-ranging literature documents the emergence of an 'ecological managerialism' in twentieth-century Africa, through which governments deployed conservation and development policies as mechanisms of domination over both people and nature. As Fiona Mackenzie has suggested, applying science to the environment offered colonial governments 'both a discourse of knowledge and a discourse of power' through which to manage the lives of their subjects.[5] In effect, the political authority of the state and the epistemic authority of scientific knowledge wielded by its officers are intimately linked in the formulation of resource management policy, and in sustainable development more generally.[6]

This perspective cautions us to pay close attention to the ways in which policy-makers 'read' and interpret landscapes. But it does not mean that government

---

[4] See N. Abel and P. Blaikie, *Land Degradation, Stocking Rates and Conservation Policies in the Communal Rangelands of Botswana and Zimbabwe*. ODI Pastoral Development Network Paper 29a (London: Overseas Development Institute, May 1990). The argument is that high offtake rates can only be maintained if farmers recognize an improvement in welfare. Given the assumption that overstocking causes degradation, as Abel and Blaikie point out (p.12), 'the assumption is therefore that destocking carries its own reward, and that if individuals could provide "mutual assurance" through an appropriate institution, de-stocking could become individually and collectively rational because the productivity increase would provide the necessary incentive'.

[5] F. Mackenzie, 'Selective silence: a feminist encounter with environmental discourse in colonial Africa', in J. Crush (ed.), *Power in Development* (London: Routledge, 1995), p.101 (Mackenzie is paraphrasing Mudimbe). For theoretical development of similar arguments, see also J. Ferguson, *The Anti-Politics Machine* (Minneapolis: University of Minnesota Press, 1992); N. Peluso, 'Coercing conservation: the politics of state resource control', in R. Lipschutz and K. Conca (eds), *The State and Social Power in Global Environmental Politics* (New York: Columbia University Press, 1993). Also W. Beinart, 'Soil Erosion, Conservation and Ideas about Development: A Southern African Exploration, 1900–1960', *Journal of Southern African Studies*, 11 (1984); D. Anderson and R. Grove, *Conservation in Africa: People, Policies and Practice* (Cambridge: Cambridge University Press, 1987); W. Adams, *Green Development: Environment and Sustainability in the Third World* (London: Routledge, 1990).

[6] It is partly for this reason that some scholars and activists have rejected 'the development project' in its entirety as inappropriately imperialist and modernist; cf. the 'new traditionalism' in India.

science misapprehends the actual state of natural resources. Nor does it mean that the power to define authoritative interpretations of natural resources is seamless, or that such mechanisms of control are always successful. One reason for the ineffectiveness of resource management policies has been the ability of users to resist them, often in quotidian ways, not only because policy prescriptions disrupt their livelihood strategies or reduce their access to crucial resources, but also because they clash with precepts of practical knowledge rooted in long experiences of working the land. Many analysts have noted that indigenous knowledge frequently underwrites more effective interpretations of environmental processes than the science of government agencies, and environmental policy-makers increasingly acknowledge that resource management strategies are unlikely to be effective or sustainable unless user communities are included in policy planning and design as well as implementation.[7] Yet a simple trade-off between peasant practical know-ledge and power-inflected government science cannot provide a stable basis for policy-making. Not all peasants have the same interests and expertise, vis-à-vis either the state or nature, and there may be serious disjunctures between community interests and individual incentives. Nor are all state imperatives undesirable, especially if they pertain to the protection and security of the poor.

In the context of these cross-cutting currents of knowledge, power, and ecological dynamics, one can appreciate the core categories of policy formulation – carrying capacity, overstocking, offtake – as both technical and political categories. They are never simply the expressions of scientifically derived formulas, but are always understood in terms of particular institutional settings – those institutions that govern rights of access, use, alienation, and exchange, as well as those institutions that substantiate public authority – and they always invoke arguments about the public good. How policy-makers understand the problem will shape the kinds of institutions they choose. These institutions will, in turn, shape future policies. Thus the resource management paradigm engages a complex interplay between the managerial imperatives of state agencies, the claims to epistemological authority advanced by technical personnel, and the political ecology of resource-using communities.

This chapter evaluates Zimbabwe's resource management paradigm by examining these dynamics over time. It reveals a politically resilient model, rooted in traditions of ecological managerialism and shaped by inappropriate institutional structures and epistemological claims, that has taken little account of the dynamic interactions between stockholders, their livestock and local ecologies, and may in places have exacerbated soil breakdown by limiting livestock mobility. Thus the conventional categories of range management may be inadequate to underwrite the construction of ecologically sustainable resource management institutions. The discussion proceeds in four sections. The first sketches socio-economic conditions in the communal lands today. The second surveys the development of a destocking paradigm during the colonial period. The third considers the legacy of that history in the post-independence period. The fourth evaluates the current state of

---

[7] One might call this the populist/indigenous model of analysis. See, for instance, R. Chambers, *Rural Development: Putting the Last First* (London: Longman, 1983); P. Richards, *Indigenous Agricultural Revolution: Ecology and Food Production in West Africa* (London: Hutchinson, 1985).

ecological interpretation. The conclusion links the discussion to the policy framework sketched above, and raises challenges for a re-assessment of resource management policy in the communal lands.

The Communal Lands

The communal farming areas in Zimbabwe lie mainly in ecological zones characterized by low, erratic rainfall and soil types with limited capacity to retain moisture (classified in Zimbabwe's agro-ecological survey as regions IV and V). Though these areas are regarded as by and large unsuited for intensive dryland cropping, the preponderance of Zimbabwe's population (with estimates of up to 70 per cent) occupy them on small family holdings, farming maize, millets, and groundnuts, with additional cash-cropping of cotton and tobacco. The communal lands (CLs) are subject to intensifying population pressure. Land is held under traditional or communal tenure, in which arable land is farmed by individual households and grazing land is communal. The state claims the right to allocate land holdings but its power to do so is highly contested by traditional authorities, local notables and lineage elders. Land and livestock holdings are stratified on interlocking indices of region, class, gender, age, and lineage that shape patterns of productivity.

Livestock plays an integral role in communal area farming systems, fulfilling a number of functions including the provision of draft for plowing and transport, of manure for fertilizer, and milk and meat. Stockowners are generally reluctant to sell their cattle, and look instead to build up their herds. Cattle are largely owned and controlled by men, and livestock holdings are markedly skewed.[8] In the CLs today, ownership of sufficient livestock for draft purposes is limited to about 30 per cent of farmers, and a sizeable minority of households – over 40 per cent in almost all parts of the country – have no cattle. Non-stockholders find that their crop husbandry is debilitated by limited access to draft and manure, which they must obtain by borrowing, hiring, or work-sharing.[9] Pressure on poorer farmers to enter draft arrangements is exacerbated where transport shortages force them to haul their

[8] Cliffe, *Policy Options for Agrarian Reform*, pp.25–9; S. Chipika, 'Livestock ownership and inequality with particular reference to cattle: the case of some Communal Lands in Zimbabwe', in B. Cousins (ed.), *People, Land and Livestock* (Proceedings of a workshop on the socio-economic dimensions of livestock production in the communal lands of Zimbabwe) (Harare: Centre for Applied Social Sciences, University of Zimbabwe, 1989), pp.373–86; K. Truscott, 'Women and tillage: strategic issues posed by farmer groups', in Agritex, *Tillage, Past and Future* (Proceedings of Workshop at Institute of Agricultural Engineering, Harare, March 1991), pp.34–6. Estimates of stock-ownership and stock-holding have varied quite widely, and it is likely that different surveys use different measures (and therefore measure different things). Agritex's Wedza base-line study, for instance, places the figure at 58 per cent. However, all studies agree that the pattern of ownership is heavily skewed. Sandford has estimated that there are nowhere near enough cattle in the communal areas as a whole to guarantee each peasant family a draft team. (Stephen Sandford, *Livestock in the Communal Areas of Zimbabwe*, Report prepared for Ministry of Land, Resettlement and Rural Development, Harare, 1982).

[9] See Agritex, 'Wedza Baseline Study: Summary and Analysis' (mimeo, n.d.), pp.3–4; Truscott, 'Women and tillage'. Also E. Jassat and B. Chakaodza, *Socio-Economic Base-Line Study of Rushinga District (Mashonaland Central Province)* (Harare: Zimbabwe Institute for Development Studies, 1986), pp. 53–5.

produce to marketing depots on scotch-carts or sledges. In some areas farmers deal with this problem by means of a group approach to draft exchange. In others informal co-operation and exchange is not the norm, and labor performed by non-family members tends to be paid for, at a local rate.[10] Commercialization of draft has become a lucrative source of income for stockholders, and this places an increasing strain on co-operative relationships, control of labor, and rights to grazing, manure, stover, and draft.[11] It is likely that the relationship between stock-ownership and labor supplies influences crop selection and production patterns.[12]

These agrarian transformations are complex and variable. They are directly related to patterns of incorporation into the capitalist economy, as unequal access to land, livestock, technology and agricultural capital has widened gaps between households with differential capacities to generate agricultural surpluses. They also play out in culturally sedimented relationships of lineage, wealth and gender that engage issues of labor exchange, cattle loaning and sometimes land-sharing. These transformations are directly salient to the formulation of resource management policy, in as much as they demonstrate the significance of relationships between stockholders and non-stockholders in shaping patterns of resource use. They place a premium on cattle ownership, and shape the nature of resource 'user communities'. Where there is severe population pressure on the land, cropped areas encroach on grazing areas with the effect of reducing carrying capacity without increasing marketable surplus. Thus, as Lionel Cliffe has noted, there is a kind of paradox in the national land question, whereby there are too many people and cattle on the land at the same time as there is a shortage of labor, underutilized land, and a lack of enough oxen to till the land.[13]

For policy-makers, these conditions create a powerful conundrum: livestock are essential to communal areas' farming systems, but neither wealthier nor poorer farmers are eager to control stock numbers, partly because of their relationships with each other. For wealthier peasants cattle are a source of power, patronage and accumulation, while for poorer farmers cattle provide some security against drought years (as savings) and against dependence on wealthier neighbours (as a factor of production). It is this conundrum that resource management policy must address.

[10] See, variously, Department of Veterinary Services, *Monthly Reports*, July and August 1988; M. Bratton, 'Farmer Organizations and Food Production in Zimbabwe', *World Development*, 14, 3 (1986); Interview, migrant landholder from Gutu. Zimbabwe Women's Bureau, *We Carry a Heavy Load: Rural Women in Zimbabwe Speak Out* (Report of a Survey, Harare, 1981), p. 24; M. Drinkwater, *The State and Agrarian Change in Zimbabwe's Communal Lands* (New York: St. Martins Press, 1991).

[11] I. Scoones and K. Wilson, 'Households, lineage groups and ecological dynamics: issues for livestock development in Zimbabwe's communal lands', in Cousins, *People, Land and Livestock*, p. 29. M. Steele, 'The Economic Function of African-owned Cattle in Rhodesia', *Zambezia* 9, 2 (1981), p. 30 notes that large stockholders were able to make poorer peasants into virtual clients, and cattle 'became an important agency in social stratification'. Results of a preliminary survey in the Chinamora communal land (October 1988) indicate that the largest stockholders are frequently local businessmen.

[12] During the 1950s and 1960s, colonial extension officers based their extension advice to African farmers on the amount of labor or draft they believed the farmers could command.

[13] Cliffe, *Policy Options*, p. 23; cf. the Deputy Minister of Lands, Agriculture and Rural Resettlement (*Hansard*, vol. 15, no. 29, 14 September 1988): 'Mr. Speaker Sir, the problem of livestock improvement in communal areas is compounded by the fact that people are short of cattle for draught power, manure, milk, beef *et cetera* and yet at the same time there is overgrazing in those areas.'

Ecological 'Crisis' and Stock Control in Colonial Zimbabwe

Rural administrators and agricultural officers first became seriously concerned about environmental destruction in the African reserves (known as Tribal Trust Lands after 1962, and as communal areas after independence in 1980) during the late 1920s, as entrepreneurial African farmers responded to expanding market opportunities by adopting newly available plowing technologies.[14] For Africans, access to cattle was increasingly a requisite for the accumulation of wealth, less as a source of cash income from beef production than as a source of draft, manure, and transport. Colonial records indicate a dramatic increase in African-owned cattle during the 1910s and 1920s, both in numbers and in rate of herd expansion, though these figures may reflect the stabilization of rural administration and improved record-taking rather than actual herd trends.[15] In the eyes of officials, increased ox-plowing not only created the 'threat' of African agricultural competition with white settler farmers, it also raised the spectre of land shortage and ecological degradation in the reserves, due to over-stocking and over-cropping.

As land pressure in the reserves increased, alarm over 'over-crowding,' 'over-stocking,' and 'a vicious and expanding circle of destruction' fueled a conservationist zeal within the administration that found expression in the 'centralization' strategy of the 1930s, compulsory destocking in the 1940s, and ultimately in the development phase of 'progress by compulsion' associated with the 1952 Native Land Husbandry Act. The centralization policy, as well as the Land Husbandry Act, called for clear demarcation of arable and grazing land into separate blocks, with grazing areas to be further divided into paddocks for rotational grazing. By settling individual landholders on 'economic units', the Land Husbandry Act aimed to secure a farming system of stable and productive peasant households with low levels of proletarianization and stratification. The Act provided explicitly for destocking measures, and also allowed for confiscation of cattle that 'strayed' from grazing blocks, a policy that stymied any stockholder's strategy of using resources that lay outside of the designated grazing area.

It is not difficult to trace in these nature-management policies the use of technical arguments to achieve political objectives. As Terence Ranger has noted, cattle destocking was a measure for managing agrarian class formation by reducing differentiation among cattle owners.[16] But efforts to regulate African agriculture are noteworthy not only for their relationship to the larger political economy of colonial capitalism but also for the ideas about resource degradation and modernity

---

[14] T. Ranger, *Peasant Consciousness and Guerrilla War* (Berkeley: University of California Press, 1985), pp. 60–65; L. Bessant and E. Muringai, 'Peasants, Businessmen and Moral Economy in the Chiweshe Reserve, North-Central Zimbabwe 1940–1968' (mimeo, 1990), pp. 5–7, 12–14.

[15] See, for instance, Government of Southern Rhodesia, *Statistical Handbook of Southern Rhodesia*, 1945, p.12; *Report of the Advisory Committee on the Development of the Economic Resources of Southern Rhodesia [Phillips Committee Report]*, 1962, p. 158.

[16] Ranger, *Peasant Consciousness and Guerrilla War*, p. 230. The Land Husbandry Act did allow for some alienation of stocking rights and skewing of cattle holdings; see *What the Native Land Husbandry Act Means to the Rural African and to Southern Rhodesia: A Five Year Plan that will Revolutionise African Agriculture*, (Native Affairs Department, 1955), p. 5. Native commissioners frequently selected stock for reduction on the basis of 'least social disruption' rather than considerations of equity.

that shaped relations between the state, farmers, markets, and nature. Under the rubric of colonial capitalism and racial domination, policies towards nature and its regulation were framed by a bifurcated political vision in which market-oriented commodity production would be undertaken principally by large-scale white commercial farmers, and African husbandry would rest on a compromise between 'traditional' (subsistence) farming and commercial principles. At the same time, the adoption of sound agricultural practices would be a civilizing measure for Africans, for existing farming strategies were understood to be primitive, backward, and based largely on superstition. E.D. Alvord, the Chief Agriculturist to the Native, who devised and directed the centralization program, conceived a direct link between the technical innovations of modern, rational scientific agriculture, exemplified in the program, and human social and cognitive progress.[17]

This ideology had a profound and lasting impact on official interpretations of the landscape and of human/nature interactions. The link that officials perceived between technological innovation and cognitive progress placed an additional premium on adherence to the model. From the outset, therefore, policy-makers inexorably centered the apparatuses of environmental assessment and policy design on ideas of what farmers *should* be doing rather than on why they were doing what they did. In assessing the resource base, officials paid inadequate attention to the complexities of existing land-use systems. While they recognized the constraints imposed by highly variable rainfall and sandy soils, they sought sustainable solutions by adapting the environment to the model and took little account of the flexible resource-use strategies that enabled existing farming systems to adapt to variable and uncertain environments. In this model-driven approach to environmental assessment, the cause of degradation was so clearly perceived – poor farming methods cause soil erosion, soil erosion causes land degradation, land degradation is demonstrated by decreasing productivity – that the actual processes of environmental change, and the role of human agency in them, were not carefully and systematically examined. Policy-makers stringently presented resource management as a technical problem amenable to technical solutions based on scientific assessment.

The scientific foundation for agricultural and ecological thinking was entrenched in the 1950s, when the government carried out a national agro-ecological survey that came to provide the baseline for all subsequent development strategies. Written mainly with an eye to boosting the productivity of white settler farmers, the survey expressly excluded the role of humans in the environment, on the argument that human interventions are impermanent and that 'for a land classification to have any stable value...it must be based on the permanent characteristics, that is the dominant natural conditioning factors'.[18] The survey established an authoritative map of

---

[17] Alvord, who as an agricultural missionary took his civilizing mission seriously, articulated this connection in terms of a 'gospel of the plough' (a kind of 'swards into plough-shares' vision). For intriguing accounts of this vision, see E.D. Alvord, 'Development of Native Agriculture and Land Tenure in Southern Rhodesia' (typescript, Waddilove, 1956 [?]), esp. pp. 4–12; *idem.*, 'The Great Hunger', *NADA*, 6 (1928); L. Gann and M. Gelfand, *Huggins of Rhodesia: The Man and his Country* (London: Allen and Unwin, 1964), pp.129–30.

[18] Federation of Rhodesia and Nyasaland, *An Agricultural Survey of Southern Rhodesia, Part I: Agro-Ecological Survey* (Salisbury: Government of Southern Rhodesia, n.d. [1962?]), p. 2. In the survey this sentence appears in italics.

ecological resources that divided the country into five distinct agro-ecological zones, each matched with specific models of appropriate agriculture. As such, it provided a categorical framework for agricultural and resource-use policy based on an understanding of the ecology as a stable (timeless) system: policy-makers should understand the structure of incentives that inhibit users from adopting sound, rational, and sustainable resource-use methods and design mechanisms that will shift those incentives towards the adoption of appropriate strategies.

As an authoritative knowledge regime, this understanding proposed a relatively simple population-resource relationship that could be managed by defining the carrying capacity of the land established by the agro-ecological zone (measured by prevalent vegetation, predominant soil type, and estimated annual rainfall) and matched to the appropriate farming system. In the reserves, these estimates were inexorably sullied by the unpredictability of the climate and the local variability of soil types as well as by population inflows resulting from colonial land apportion- ment policies and the eviction of African families from white-owned farms. Causally, range degradation arising from reduced grazing area was different from degradation resulting from excessive cattle numbers, though the effects might be the same. Yet it was the effects that were of most interest to colonial administrators. Ecological monitoring tended to be based on visual assessments of soil erosion, as demonstrated by gully formation, which conservation-minded native commis- sioners ascribed rather presumptively to the movement and grazing patterns of livestock.[19] These patterns themselves were the result less of absolute cattle numbers than of mobility constraints imposed by colonial land-use planning initiatives. But those initiatives were beyond the power of local administrators to remedy.

The construction of a coherent policy framework based on 'carrying capacity' and 'offtake' was undermined by both politics and the operating assumptions of the economic model. The Native Affairs Department, which implemented the agro-ecological survey in the reserves, used it to vindicate the Land Husbandry Act – at the time collapsing under sustained rural resistance – rather than to assess the state of available agrarian knowledge. In this agrarian blueprint, the role of cattle was determined by their economic value, measured by two criteria. The first was their ability to produce manure as fertilizer to boost African agricultural returns. The second was their commercial value in beef slaughter markets. Appropriate stocking levels were determined first by evaluating the cropping capability of land, on the basis of 'economic' family units, and then matching this land area with the fertilizing capacity of manure. The rule of thumb was that one acre of land would be kept fertile by one beast, and therefore 'the acreage of arable land should equal the number of cattle that can be carried'.[20] According to this vision, land holdings, cattle numbers and arable production would all be in equilibrium and appropriate carrying capacity could be easily determined.[21] Once

---

[19] See especially, A. Wright, *Valley of the Ironwoods* (Cape Town: Bulpin, 1972); M.A. Stocking, 'Relationship of Agricultural History and Settlement to Severe Soil Erosion in Rhodesia', *Zambezia*, 6,.2 (1978).

[20] D.A. Robinson, 'Soil Conservation and Implications of the Land Husbandry Act', *NADA* (1960), p.33.

[21] B. Floyd notes [in *Changing Patterns of African Land Use in Southern Rhodesia* (Lusaka: Rhodes- Livingstone Institute, 1959), p.126] that the Act drew criticism at the time for being an essentially

equilibrium was achieved, the commercial value of beef became significant: excess cattle would be sold.

Neither of these criteria exercised much purchase on the realities of agrarian life in the reserves. As population pressure increased, arable production expanded into designated grazing areas, and statutory controls on stock mobility prevented farmers from responding by moving their cattle. As a result, by the mid-1950s, many of the reserves were both understocked and overstocked in terms of the objectives of the management system. On the one hand, as the Native Department acknowledged, 'the grazing areas at present are incapable of carrying the number necessary to provide the manure required for the total arable area'.[22] On the other hand, evidence of increasing stock pressure on the designated rangeland was growing. For this reason, the government became preoccupied with increasing livestock offtake in the African reserves.

Notwithstanding the deep intrusions of conservation policies into the livestock management systems of African farmers, government attempts to regulate stocking levels depended heavily throughout the colonial period on their ability to persuade or coerce African stockholders to sell their cattle at state-run sales.[23] As in other parts of colonial Africa, however, the government consistently structured pricing policies against African farmers, setting prices for low-grade stock to subsidize (mainly European-owned) high-grade stock. Such machinations were not lost on African farmers, who participated sporadically and reluctantly in state-run sales. But the politics of markets was more complex than farmer resentment against the state and pricing policies.[24] Frequently, African owners would bring cattle long distances to auctions only to spurn the prices offered and trail them home again. This suggests that African stockowners were playing a quite intricate market game.[25] On the one hand, they were quite willing to participate in cattle markets if the price was right. On the other hand, they used the sales to establish cash values for their cattle so that they could negotiate in separate markets the local relations of

---

[21] (cont.) agricultural act. The resilience of this rule of thumb in determining estimates of carrying capacity was remarkable; see for instance D.L. Barnes, 'Problems and Prospects of Increased Pastoral Production in the Tribal Trust Lands', *Zambezia* 6, 1 (1978).

[22] *An Agricultural Survey of Southern Rhodesia: Part II – Agro-Economic Survey* (Federation of Rhodesia and Nyasaland, n.d [1961?]), p.97. Such 'understocking' was also noted by Steele, 'The Economic Function of African Owned Cattle', and Barnes, 'Problems and Prospects of Increased Pastoral Production', p.50.

[23] Prior to 1943, the Native Affairs Department insisted that all sales were voluntary. But as one Native Commissioner acknowledged, a suggestion from the NC was frequently interpreted by local Africans as an instruction. This particular NC did not regard this as a bad thing. See *Report of Commission on Sale of Native Cattle in Areas Occupied by Natives* (Salisbury: Government of Southern Rhodesia, 1939), paras.147–54, 165–87, and *passim*.

[24] In 1961, the Robinson Commission reported that Africans complained that they received lower prices than Europeans and blamed the NC, who arranged cattle sales. See *Report of the Commission Appointed to Enquire into and Report on the Administrative and Judicial Functions in the Native Affairs and District Courts Departments* (Salisbury: Government of Southern Rhodesia, 1962) p.20. The Extension Research Officer's first interim report on African cattle sales in 1965 pointed out (para. 19) that many African farmers 'fail to see why the extra profits resulting from [free and communal grazing] should not be reaped by themselves; rather than by the Cold Storage Commission or some other purchaser'.

[25] Here I draw mainly on the Ministry of Internal Affairs, *Monthly Summary of Cattle Sales*, 1960–79.

accumulation, trade and patronage with shopkeepers, butchers and kinfolk.[26]

Thus, agricultural officers had adopted a measure of value for cattle that was far too narrow to take full cognizance of the factors influencing farmers' decisions. Since arguments for offtake were based on this measure, they could not properly assess the costs of offtake to farmers, and therefore could not evaluate the impact on the farmers' welfare. Certainly, policy-makers lacked a clear appreciation of how the values of cattle, as agricultural commodities, were mediated both through local social relationships and through the farming system. As a result, the formal cattle market afforded the state little purchase either on stocking levels in the African areas or on the economic decisions of African farmers.

The obvious unwillingness of African farmers to sell cattle fueled increasing frustration among agricultural and administrative officers. Yet, given the importance of stock management to the relationships between the state, African farmers, and nature, it is startling to recognize how little policy-makers knew about African herds or the use of cattle in reserve economies. Since most data on African agricultural activity came from the Native Department, the database for policy decision-making was scattered and unreliable. This was partly because Native Department data did not include cattle exchanges or disappearances within the reserves, and partly because data collection and presentation depended on the energy, interests and ideology of particular native commissioners and their personnel.[27] These were marked throughout the colonial period by the conviction that African stock management practices were primitive and inefficient, and that the relationship between African stockholders and their cattle was essentially cultural rather than economic. Africans were understood to prefer quantity over quality of cattle, because of their role in ritual and religious occasions, and their social value in conferring prestige and status on their owners. This view could explain both why African farmers were not price-responsive in organized markets and why they declined to sell cattle in drought years, allowing them to become debilitated or even to die.[28] But it could not take proper account either of how people used cattle, or how cattle used the environment.

The quality of data on ecological trends was also undermined by shifts in administrative boundaries, long distances from district offices, and institutional rivalries between the Native Department, which exercised administrative and technical jurisdiction over the reserves, and the Ministry of Agriculture, which held most research talent and resources but had little clout in the reserves. The upshot

---

[26] Mtetwa has also suggested that African farmers engaged in 'target selling', ie. selling to raise cash for a specific targeted purchase; decisions made on this basis might well change the way in which the value of a sale was calculated. See R. Mtetwa, 'Myth or Reality: The 'Cattle Complex' in South East Africa, with Special Reference to Rhodesia', *Zambezia*, 6, 1 (1978), p.28.

[27] Central Statistical Office (CSO) surveys occurred only every ten years, starting in 1949. The CSO's 1959/60 *Survey of African Agriculture* delivered estimates of cattle numbers and deaths in the reserves that were startlingly different from those of the Native Department. Indeed, the *Survey* implied that almost all agricultural data issuing from the Native Department were suspect.

[28] For a pithy expression of this perspective, see P. Le Roux, A. Stubbs and P. Donnelly, 'Problems and Prospects of Increasing Beef Production in the Tribal Trust Lands', *Zambezia*, 6, 1 (1978). The authors note that, since most cattle are owned by older people, a certain resistance to change is to be expected.

was that the standards of ecological assessment applied to white-owned commercial farms were never replicated in the reserves, though the methodological assumptions of such assessment were. By the 1960s, at least some technical officers were recognizing that the assumptions of the development model, and the paucity of knowledge about African cattle management, provided inadequate bases for formulating livestock management policy. This became clear in two reports written by Conex's extension research officer, one (a 1964 internal report) on the role of cattle in tribal society, and the other in 1966 (for the Standing Committee on African Beef Production) on 'the reason why (African) farmers were not selling cattle'.[29] The 1964 report highlighted the paucity of useful knowledge about African stockholding, concluding that the government had 'very little quantitative information' on either non-economic uses of African cattle or economic uses other than cash sales, although it was clear that 'these uses exert a major influence on cash sales'. Against this background, these reports illuminate several important aspects of the relationship between stock management and environmental degradation as interpreted by policy-makers.

First, they showed the shortcomings of policy-makers' preoccupation with official cattle sales to regulate stock numbers. In his 1966 report, the extension officer could not fathom the point of asking why African farmers were not selling cattle when in fact they *were* selling cattle at organized sales, though not at the same rate as white commercial ranchers and at widely fluctuating annual rates. Moreover, there were no accurate data on the numbers of cattle being slaughtered or sold locally outside of formal channels, though the monetization of the rural economy meant that this was probably increasing. Official notions of offtake were incoherent.[30] More broadly, the reports demonstrated that official thinking about African stock and range management was not based on any clear and consistent principle that could provide the basis for an appropriate stock/vegetation mix. It was not just distorted markets but distorted notions of appropriate market behaviour that made policy ineffective.

Second, the reports showed the shortcomings of 'cattle-complex' conceptions of African livestock management. If one of the least important economic considerations for African stockholders was the sale of slaughter stock, one of the most important was the supply of draft. Placing the demand for draft at the center of stockholders' interest in livestock might alter the valuation of cattle for policy-making. As the extension officer noted, individual herds tended to be very small, and there was a paucity of breeding cows to allow draft oxen to be replaced before they were about 9 years old. From this perspective, the cultural and social importance of cattle could not be taken as overriding, and an argument could be made for increasing productivity not simply by encouraging African stockholders to increase their offtake but by *replacing* older draft oxen with younger stock through appropriate (probably subsidized) marketing arrangements. In effect, *destocking* might be mitigated by *restocking*, i.e. by actively manipulating the age-sex structure of herds rather than simply encouraging cattle sales or imposing stocking rates set by

[29] Hughes, *The Role of Cattle in Tribal Society*; idem., *First Interim Report by Extension Research Officer to Standing Committee on African Beef Production* (Salisbury: Conex, Ministry of Agriculture, 28 February 1966).

[30] *Ibid.*, p.3.

the state. This suggestion, which had already been floated in 1960 by the Director of Native Agriculture, offered an alternative policy paradigm based on a different conception of the value of cattle, as well as a different conception of the market as policy instrument. Though it appeared again as a policy recommendation in the late 1970s, it would have been costly and was never taken seriously.[31]

Third, the reports highlighted the paucity of policy-makers' knowledge about African herd size, age-sex structure, and even the ways in which African farmers defined their herds. For the purposes of policy-making, the composition of herds was indeed very complicated. The number of cattle on an individual's dip-card did not necessarily correspond to the number of cattle he was running. Large stockholders were known to 'herd out' portions of their herds among kin and clients. As the extension officer noted,

> Either the 'dip-card' or 'single byre' herd may contain animals belonging to persons other than the card holder, while he in turn may have beasts of his out on 'loan' elsewhere; which beasts will appear on someone else's dip-card. So, if one asks an individual how many cattle he has he may truthfully answer that he has 31. Yet only 8 of these may appear on his own dip-card, on which he may also have 4 animals belonging to other people.

Thus, 'in practice several individuals may have accepted rights over a single animal... *In this context, the main thing to remember is that the registered owner of a beast may be very restricted in regard to the decisions that he can make about its disposal.*'[32] Under these conditions, stockholding patterns could not be disaggregated from social and power relations in African communities, and the selection of stock for reduction was likely to spark social conflict. For this reason, native commissioners resisted compulsory destocking against the insistence of the Natural Resources Board, and often bent the rules to maintain the peace.[33]

The apparent flexibility of rights in stock certainly provided a kind of 'shell game' strategy through which stockholders could contest officials' demands to destock by 'hiding' their cattle among kin and neighbours. But, in as much as it indicates that patterns of stock deployment were both socially and physically mobile, it highlights an additional point about the relationship between stock-holding and range management. As discussed in further detail below, farmers moved their cattle in dry seasons and drought years to take advantage of specific key ecological resources, such as sponges, drainage lines, or patches of desirable grass species, that lay outside the boundaries of designated grazing areas, and sometimes beyond the village or reserve boundaries. For this reason, farmers resented

---

[31] Robinson, 'Soil Conservation and Implications of the Land Husbandry Act', p.33; Le Roux, *et al.*, 'Problems and Prospects of Increasing Beef Production', p.44.

[32] The *First Interim Report* distinguished various types of 'herd', including the cattle-right herd, the dip-card herd, the single-byre herd, the 'owned' herd, and the 'family' herd. On cattle clientelism, see Steele, 'The Economic Function of African Owned Cattle', p.30. I am grateful to Pius Nyambara and Guy Thompson for discussions on this point.

[33] The Natural Resources Board, which recommended compulsory destocking in 1943, expressed both the degree of panic at land degradation and thorough-going disdain for such squeamishness, largely on the dubious grounds that destocking had been successful – and indeed appreciated by African farmers – in South Africa. See Natural Resources Board, *Memorandum on the Conservation of Natural Resources on the Land Occupied by Natives*, 25 May 1943. On NAD resistance, see Ranger, *Peasant Consciousness*, p.88.

veterinary controls that restricted the movement of cattle.[34] Indeed, native commissioners, concerned more with social peace than with the effectiveness of African range management systems, recommended to the Natural Resources Board in the 1940s that such strategies be permitted once again. Though the Board endorsed this recommendation, it was never implemented. Throughout the colonial period, range management specialists were inclined to interpret strategic patterns of resource use as slovenly (and even immoral) farming practices and a lack of control over livestock.[35]

Taken together, these points indicate two important features of resource management policy in the reserves. First, the tenuous ability of the colonial authorities to exercise effective management over reserve-area cattle populations demanded flexible policy implementation on the ground. Second, the development model itself rested on tendentious assumptions and remarkably weak information on the role of cattle in the developing relations of power, patronage and accumulation in the communal areas, on the role of cattle in African farming systems, and on the relationship between African livestock and the environment. Systematic research, or even records, on these issues were hard to find and became harder in the years following the unilateral declaration of independence, when the government's commitment to research virtually disappeared, funding and logistical support were in short supply, and control of the reserves became increasingly precarious.[36] From the early 1960s, stock control programmes disintegrated in the face of popular resistance. The 1960s and 1970s were a period of *ad hoc* and often panicky conservation measures, as the state tried to re-assert control over an increasingly defiant rural population. Increasingly, the government recognized the need to secure local co-operation through participation – or at least co-optation – and devised a variety of local institutions, such as Farmers' Clubs, Tribal Land Authorities and Tribal Development Groups, through which to pursue its rural management strategies.

In the spirit of its 'new approach' to social management, the government moved to develop a livestock management system based on multi-paddock grazing projects managed by locally constituted committees (though planned by agricultural officers). The aims of this initiative were twofold. First, it would establish a livestock management system based on constant consultation and participation of the community. Second, it would establish new boundaries of inclusion and exclusion in resource-using communities, and thereby inculcate new forms of responsibility among users for the resources under their management. In general, the initiative was understood as '[changing] the system of tenure from a communal one to a community system whereby the tribesmen have responsibility for a certain defined area of grazing and arable, giving the agricultural staff an ideal development

[34] *Report of Commission on Sale of Native Cattle*, paras.12–20, 87, 88. This was particularly irksome for Africans who wanted to move from one reserve to another, and would be forced to sell their holdings before they left, often at a lower price than they would have to pay to re-stock in their new area.

[35] For a moralizing tone, see especially NRB, *Memorandum on the Conservation of Natural Resources*, pp.7–8. On the lack of livestock control, see Le Roux et al., 'Problems and Prospects.'

[36] Notably, in the *First Interim Report* the extension research officer called for more research and also counselled against deploying a new investigative team until the government knew more precisely what it was that it wanted to know.

area in which to start the drive for production.'[37] In 1968, the administration began to develop grazing schemes in Victoria Province, where long-term state intervention in agriculture was well established. Expansion of this project was inhibited by the escalation of the liberation war, and few grazing schemes became securely established. Nevertheless, these initiatives provided an institutional model for post-independence agrarian technocrats to build on.

Livestock Management and the Environment in Post-Independence Zimbabwe

In the 1960s, the possibility of re-assessing the bases of resource management policies, and devising a new paradigm that looked beyond a market-oriented concept of offtake, foundered on the rocks of UDI politics. Agricultural specialists, while repeatedly noting that the most trenchant cause of environmental pressure in the reserves was human overcrowding resulting from land policies, proceeded with research and development projects as if it could be held constant. After independence in 1980, the politics of development shifted, new research opportunities opened up, and the question of how to interpret ecological processes in relation to existing farming systems arose with new urgency.

The post-independence government was committed to improving the quality of life for rural citizens but was inhibited by the retention of colonial-era land apportionment structures. In addition, policy-makers were deeply perturbed by increasing levels of soil erosion, dwindling vegetation cover, and the fact that stocking numbers in some communal lands were at three times the recommended level.[38] Given increasing land scarcity and population density in the CLs, conservation-oriented policies of stock control and veld management took on added political significance in as much as they affected the government's aims to promote economic productivity and rural welfare. Economically and environmentally sustainable development, as the Chavunduka Commission on the agricultural industry noted, was a complex problem that would require 'both a sense of history and of community'.[39] But solutions to the dilemmas of resource management in the CLs were also necessarily political: how were stocking rates to be enforced without evoking resistance or exacerbating poverty and economic differentiation?

Many commentators have noted that the government, notwithstanding its populist rhetoric of social transformation, devised development policies that resonated markedly with the development ideas of the 1950s and 1960s.[40] Though this

---

[37] M. Froude, 'Veld Management in the Victoria Province Tribal Areas', *Rhodesia Agricultural Journal*, 71, 2 (1974), p.29. For a more general discussion, see also B. Cousins, *A Survey of Current Grazing Schemes in the Communal Lands of Zimbabwe* (Centre for Applied Social Sciences, University of Zimbabwe, 1987).

[38] See especially World Bank, *Zimbabwe: Land Subsector Study* (Washington, DC: World Bank, 1986), pp.15–17; Ministry of Local Government, Urban and Rural Development, *Manicaland Provincial Development Plan*, 1986, p.122.

[39] *Report of the Commission of Inquiry into the Agricultural Industry*, [Chavunduka Commission], 1982, p.7.

[40] On this resonance, see R. Bush and L. Cliffe, 'Agrarian Policy in Migrant Labour Societies: Reform or Transformation in Zimbabwe?', *Review of African Political Economy*, 29 (1984); M. Drinkwater, *The State and Agrarian Change in Zimbabwe's Communal Areas* (New York: St. Martins Press, 1991); J. Alexander, 'State, Peasantry and Resettlement in Zimbabwe', *Review of African Political Economy*, 21

continuity is generally explained in terms of structural economic forces, state weakness, or the class interests of the emerging elites, it is worth noting that for technical policy implementation the government needed policies quickly but had to draw on socio-economic and technical data about rural conditions that were minimal, scattered, and disorganized. Though the agricultural experimental stations had done some cattle production research in the reserve areas, much of it had involved trials conducted under controlled conditions based on the assumptions of commercial commodity production under equilibrium ecological conditions. Grassland and soils research, though good, was scattered. In 1983, a government agronomist wrote that, although peasant farmers exercised very poor soil management practices, 'the long-term effects of continuous cropping of these soils under poor crop and soil management practices on their production potential have not been investigated'.[41] Two years later, animal nutritionist B. Mombeshora noted that, 'The important livestock population in the communal areas have been almost completely ignored and, with the exception of some dipping and vaccination data, almost nothing is known about them.' Although CL livestock fed almost entirely from natural pastures, residues on arable lands, and possibly from browse, 'the relative importance of these different sources of fodder, both in total and in different seasons, is not known'.[42] In short, when it came to environmental protection and development in the CLs, the government faced urgent imperatives to act, weak institutions to act through, and limited agrarian information on which to base its actions.

In light of these conditions, it is noteworthy that government agencies continued to blame ecological degradation on the poor land management techniques of African farmers. Provincial and district development plans consistently described rural agrarian conditions in terms of 'poor husbandry methods and poor utilisation of farmland', 'unplanned settlement', and 'poor land management' especially by the encroachment of settlement and cropping into grazing areas. This perception of peasant farmers as the culprits of environmental damage was accompanied by a heightened sense of urgency about ecological decline in the CLs. Taking over the existing apparatuses of rural development, agricultural officers were seized by an apprehension that resource-use controls had largely collapsed in the wake of the liberation war, not only because of resource-use methods, but also because of population mobility and the spread of 'squatting'. They desperately feared a loss of ability to regulate rural society, a fear heightened by the state's inability to regulate land transmission and tenure at the local level. In policy-making circles, this fear fueled a sense of impending ecological crisis similar to that expressed by colonial officers when they felt their grasp on the countryside slipping – though the extent, nature and impact of land degradation have remained a matter of serious debate.[43]

[40] (cont.) (1994); see also Beinart, 'Soil Erosion, Conservation and Ideas about Development'.

[41] N.A. Mashiringwani, 'The Present Nutrient Status of the Soils in the Communal Farming Areas of Zimbabwe', *Zimbabwe Agricultural Journal*, 80, 2 (1983), p. 73.

[42] B. Mombeshora, 'Livestock production research', in M. Avila (ed.), *Crop and Livestock Production Research in Communal Areas*. ZAJ Special Report No.1. (Harare: Department of Research and Specialist Services, 1985), pp. 84, 87. For a provocative perspective on colonial-era research, see J. Oliver, 'Nature and Nurture in Animal Production: A Heretical View.' *Zambezia*, 7, 2 (1979).

[43] For contending views, see World Bank, *Zimbabwe: Land Subsector Study*, p.16; Cliffe, *Policy Options for Agrarian Reform*, appendix IV; Cousins, *A Survey of Grazing Schemes*, pp. 69–76.

Determined to assert its rural hegemony, the government sought a resource management framework that would meet the criteria of rational land-use planning mandated by colonial development models, and also the objectives of sustainable development advanced by the World Conservation Strategy, which the government adopted in 1980. In a startling echo of centralization policy, the state devised an agrarian policy that called for a systematic and controlled separation of grazing from arable land, the promotion of 'proper' land-use methods (safeguarding slope areas, containing and reclaiming gullies to curb erosion and siltation, protecting dam catchments and preventing streambank cultivation), 'modern' methods of farming (crop rotation and cultivation of appropriate crops), and the management of grazing areas by paddocking, fencing and applying appropriate stocking rates. Within this framework, livestock management presented a special challenge because it involved rights of access to and use of common resources: how could the state recognize farmers as having a 'most fundamental' right to graze cattle, yet also curb that right?

Initially, agricultural economists, influenced by Hardin's paradigm of the 'tragedy of the commons,' argued that control could be exercised by opening up the land market for freehold tenure, or by individualising grazing rights, so that stocking rates would be controlled by the farmers themselves through a localized market.[44] In 1984, senior officials in Agritex proposed a National Land Use Programme under which each member of a community would be allocated an equal grazing 'share'. Those with fewer cattle than represented in their share could either purchase animals to fill their quota or sell their grazing rights, on an annual basis, to other members of the community. The plan, however, raised political objections at the highest levels of policy-making, and was immediately rejected.[45] Not only would it have been likely to encourage a fairly rapid process of economic stratification, it would also have disrupted a wide-ranging and overlapping panoply of rights and powers already entrenched in rural communities. The government was well aware that measures to regulate public resources from above could rekindle the tension between central and local control of land use, and between forced and voluntary change, which characterized the history of state-peasant relations in colonial Zimbabwe. It seems likely that the dumping of the National Land Use Programme reflected an appreciation within the government of the political dangers of tampering with people's stock rights.

Unwilling to abandon the technocratic model of resource management, the government nevertheless recognized the need to make resource management less managerial and coercive. Consequently, it set out to embed the technocratic model in a populist strategy of developing participatory resource management institutions in rural communities through which it might encourage a self-policing sense of local responsibility to the ecology. The key institution of this policy framework was to be the village development committee (VIDCO), which was conceived as a democratic, participatory institution that would not only bring popular will to bear on policy-makers, but would also provide the foundation for a populist one-party

---

[44] World Bank, *Zimbabwe: Land Subsector Study*, p. 30; Cousins, *A Survey of Current Grazing Schemes*, pp.76–84; Interview.
[45] Cousins, *Survey of Current Grazing Schemes*, pp. 23–4.

democracy.[46] This approach allowed the government to bring political and technical thinking together in a land-use policy which called for stock control through community-based grazing schemes co-ordinated by VIDCOs. The grazing schemes were to be operated by means of multi-rotational fenced paddocks grazed on short duration rotations. In effect, the state resurrected the concept developed in the late colonial period, facilitated now by ready funding from NGOs and aid agencies eager to provide expertise and funding. For NGOs, as for many state officials, its main advantage was that it rested on local participation and community accountability. Resource-using communities would elect their own grazing committees to draw up and implement local bylaws for regulating stock use of grazing resources and enforcing stocking rates.

These institutional innovations were not accompanied, however, by any serious technical reconsideration of the relationship between stocking rates and ecological decline, despite a surge of new research into the role of cattle in CL farming systems.[47] First, the strategy deploys a conservative conception of carrying capacity. Since the CLs are subject to highly variable rainfall patterns, the challenge is to determine an appropriate stocking rate that does not result in overgrazing in a dry year and underutilization in a good year. As a result, government policy (as well as NGOs engaged in grazing scheme development) have tended to follow the conventional wisdom of range science and set recommended stocking densities at fairly low levels in the interests of maintaining range stability.[48] Second, the strategy deploys a liberal conception of offtake. Conservative stocking rates can only be maintained if high offtake rates are maintained, and this depends on farmers recognizing an improvement in welfare rooted in a direct relationship between destocking and range improvement.[49] Third, the strategy rests on an understanding that productivity is the measure of value for livestock: if farmers can be assured of the survival and improved quality of their livestock, they have an incentive to keep fewer cattle. This understanding rests on the belief that productivity increases per beast will more than compensate for losses caused by reduced density, a notion that implicitly links development to

---

[46] See GOZ, *National Conservation Strategy*, p. 23; W. Munro, 'Building the Post-Colonial State: Villagization and Resource Management in Zimbabwe', *Politics and Society*, 23, 1 (March 1995).

[47] See especially Gesellschaft für Agrarprojekte in Ubersee M.B.H. (GFA), *Study on the Economic and Social Determinants of Livestock Production in the Communal Areas – Zimbabwe: Final Report* (Hamburg, March 1987); J.C. Barratt, 'An Economic Model of Animal Draught Power in Agropastoral Farming Systems in Zimbabwe', paper presented to a Workshop on a Systems Approach to the Analysis and Planning of Communal Resource Management, University of Zimbabwe, 8 August 1989; *idem.*, 'Valuing Animal Draught in Agropastoral Farming Systems in Zimbabwe', paper presented at Workshop on Tillage: Present and Future. CIMMYT FSR Networkshop Report No.2, March 1991. Much of this new research focused on productivity-enhancing *economically* sustainable growth rather than environmentally sustainable development.

[48] Cf. CARD, *Pfumai Livestock Development Project: Livestock Management Recommendations* (Masvingo, August 1988), pp.10–12; Abel and Blaikie, *Land Degradation, Stocking Rates and Conservation Policies*, p. 12.

[49] *Ibid.* Abel and Blaikie propose the development of a mutual assurance institution that differs markedly from the market mechanism recommended by the extension research officer in 1964, not least in that it depends on an entirely endogenous dynamic, whereas the colonial market mechanism would demand external structuring of the market.

the market.[50] Finally, the model assumes a stable ecology whose condition can be measured by the state of topland vegetation.

Ironically, what is absent in these assumptions is the 'sense of history and of community' that the Chavunduka Commission called for. There is a powerful continuity in the technical model, a market-oriented rationality, and a sense of environmental urgency that drives resource management policy. But there is a sharp tension between this continuity and the state's commitment to participatory resource management. In particular, despite a dramatic increase in socio-economic information about the role of cattle in CL farming systems, the model for *environmental* protection takes little account of the role of production tasks (including the relationship between livestock ownership and access to labor) in the ways that farmers value their livestock. Nor does it take into account relations between owners and non-owners within 'user communities'. Nor does it examine the relationship between farmers' resource-use strategies and their responsiveness to a conservative stocking rate. In short, all the dilemmas of policy assessment raised by the extension research officer in 1964 and 1966 arise again in the post-independence era: to what extent does policy address actual conditions of resource use in the CLs? What is the relationship between the policy model and the strategies and ideas of farmers? What is the ultimate objective of the policy, and whose interests should it actually protect?

Re-interpreting Ecology?

In a 1949 publication in which he discussed the concept of land carrying capacity in Northern Rhodesia, William Allan noted that such a concept could not be meaningfully defined separately from the system of land use upon which it was based. In recent years, range ecologists concerned with semi-arid regions of Africa have taken up this point again. As Behnke and Scoones argue, 'it makes little sense to speak about overgrazing or understocking unless managers also specify the kind of management system they wish to institute and frame their assessment in terms of the appropriate stocking density for that system.'[51] In Zimbabwe, as we have seen, resource management policies have not been conceived in strong relational or interactive terms, either among community members or between stockholders and natural resources. They have taken little account either of local power relations or of the ways that stockholders and their cattle actually use the range. Thus they have

---

50 L. Mhlanga, Address to National Conservation Congress, printed in *Zimbabwe Science News*, 16, 12 (December 1982); G.L. Chavunduka, 'Social Aspects of Conservation', *Zimbabwe Science News*, 16, 12 (December 1982); *Report of the Chavunduka Commission*, pp. 49–50. This assumption raises the vexed question of peasants' apparent non-responsiveness to beef prices in determining production policy. For relevant arguments, see *Report of the Commission of Inquiry into the Zimbabwean Beef Industry*, 1981; M. Blackie, 'A Time to Listen: A Perspective on Agricultural Policy in Zimbabwe', *Zimbabwe Agricultural Journal* 79, 5 (1982), p. 54; K. Gobbins and H. Prankerd, 'Communal Agriculture: A Study from Mashonaland West', *Zimbabwe Agricultural Journal*, 80, 4 (1983), p. 175.

51 W. Allan, *Studies in African Land Usage in Northern Rhodesia*. Rhodes-Livingstone Papers No.15. (Cape Town: Oxford University Press, 1949), pp.1–2; R. Behnke and I. Scoones, 'Rethinking range ecology: implications for rangeland management in Africa', in Behnke et al., *Range Ecology at Disequilibrium*, p. 6.

drawn weak links between existing farming systems and the preferred (sustainable) environmental outcomes.[52] A re-evaluation of ecological trends in the light of these relationships may suggest a reconsideration of the operating assumptions of resource management policy.

An initial question is why CL farmers are often not responsive to a conservative stocking rate, which promotes herd stability by maintaining lower stock densities in good years in order to reduce mortality in drought years. The difficulty is that such a strategy may not make most sense where periodic but unpredictable climatic shocks, such as intense drought, will degrade the range regardless of the stock management system. In such conditions, as Westoby and others have argued, stock-holders may be best served by *basing* their resource-use strategies on uncertainty and reckoning their relationship to the ecology interactively: rather than trying to prevent range degradation, they should respond flexibly to it when it occurs by trying to 'seize opportunities and to evade hazards, as far as possible' through strategies of 'opportunistic management'.[53]

There is considerable evidence that CL farmers adopt two types of opportunistic strategy to take account of an uncertain environment: calculated risk-based herd attrition, and strategic use of key ecological resources. First, they allow animal numbers to build up in good years in the knowledge that they will decline dramatically in drought years, and they rebuild their herds from the survivors.[54] As we have noted, this strategy frustrated colonial agricultural officers, some of whom bemoaned 'the cruelty suffered not only by the animals which succumb to poverty but also by the others who manage to pull through as bags of skin and bone...', and claimed that critics of destocking could not 'appreciate the greatly reduced value of the survivors or the rapid deterioration of the land'.[55] Yet, as we have seen, stockowners have real economic and political reasons not to sell their stock. In addition, they may have quite dependable ideas about how readily their herds will rebuild after a dry spell, based on knowledge of available resources that are not part of official stocking calculations (though how long a dry spell will last, which they do not know, remains a major constraint on their calculations).

Second, farmers use 'key resources' that lie outside of designated grazing areas to sustain herds through the dry season and through the cropping season in drought years. This strategy, too, had long been noted, though its significance had perhaps been misunderstood. In 1943, the Natural Resources Board addressed the paradox of the apparently low mortality of stock in apparently depleted reserves (in the

[52] As recently as 1993, the government's *Prospectus for Environmental Assessment Policy in Zimbabwe: Public Background and Discussion Paper* (p.3) noted that 'Zimbabweans have recently considered the social, economic and biophysical features of sustainable development and attached greatest importance to the biophysical aspects of sustainability'.

[53] M. Westoby, B. Walker, and I. Noy-Meir, 'Opportunistic Management for Rangelands not at Equilibrium', *Journal of Range Management*, 42 (1989). See also Behnke *et al.*, *Range Ecology at Disequilibrium*; S. Sandford, *Management of Pastoral Development in the Third World* (Chichester: Wiley and Sons, 1983).

[54] According to the Department of Veterinary Services Annual Report for Masvingo Province (p. 3), over 3,000 poverty deaths were recorded in the province's communal lands in the first three months of the 1987/8 agricultural year, and it was estimated that 10 per cent of the total herd was lost after January 1988. See also monthly reports from Masvingo and Matabeleland South provinces.

[55] NRB, *Memorandum on the Conservation of Natural Resources*, pp. 7–8.

terms quoted above), noting that cattle feed 'not on the grass, which has disappeared, but on the foliage of trees and shrubs'.[56] The Board assumed that these browsing resources were food of last resort. But this did not necessarily mean that the range was being depleted by excess stock numbers. It might mean that cattle had moved away from sour grasses as they became unpalatable with seasonal change. Or it might mean that bush encroachment was under way as part of a long-term ecological cycle in which both livestock and climatic variations played a role though neither was determinant, and that litter from the bush would eventually replenish soil nutrients to reinvigorate grasses.[57] To make a causal determination, however, required a finer analysis of the relationship between cattle grazing habits, dynamic patterns of vegetation growth and depletion within local micro-ecologies, as well as the ability of stockholders to make opportunistic use of those patterns across time and space, than the prevailing population-resource paradigm allowed.

In the mid-1980s, a group of social scientists conducting new research on the socio-economic and ecological dynamics of Zimbabwe's fragile rangelands began to advance this perspective and to suggest that in fact CL farmers deploy a kind of environmental knowledge that challenges government notions of carrying capacity and may upset established conceptions of environmental degradation.[58] Several agricultural ecologists adopted the position that African savanna ecosystems are inherently unstable, with rainfall variabilities so high that livestock and grass populations rarely reach a stable level. However, they suggested, such systems are resilient in as much as spatial and temporal variations in perturbation allow them the ability to recover.[59] If the dynamics of ecological resilience are recognized and incorporated into resource management institutions, it is possible that stocking rates can be raised without producing irreversible degradation. But this requires a nuanced appreciation of the relationship between resource use and resource availability – fodder, grass, browsing, water – in CL farming systems.

This perspective has been explored most extensively by Ian Scoones, who argues that farmers make sophisticated ecological distinctions between nutrient-poor sandy soils, which are relatively stable in dry years, and nutrient-rich clay soils, which are more volatile. For grazing purposes, farmers like to move their livestock between these two zones to take account of climate shifts. Thus, their grazing strategies are sustainable because they are both temporally and spatially mobile (a strategy, as we have seen, that the Natural Resources Board remarked approvingly in 1943, but that frequently clashes with veterinary controls, with neighbors' use of land or with the property rights of commercial farmers).[60] On this basis, Scoones calls for 'a

---

[56] NRB, *Memorandum on the Conservation of Natural Resources*, p. 8. In 1974 field trials, Dankwerts also discovered that apparent range depletion had not seriously affected calving rates.

[57] On the importance of livestock grazing habits in evaluating range use, see J. J. Jackson, 'Some Observations on the Comparative Effects of Short Duration Grazing Systems and Continuous Grazing Systems on the Reproductive Performance of Ranch Cows', *Rhodesia Agricultural Journal*, 69, 5 (1972). On the role of woody biomass in the grazing systems of the reserves, see Barnes, 'Problems and Prospects.'

[58] See especially Cousins, *People, Land and Livestock*; GFA, *Study of the Economic and Social Determinants*.

[59] Abel and Blaikie, *Land Degradation, Stocking Rates and Conservation Policies*, p. 14.

[60] My discussion here is based on I. Scoones, 'Patch use by cattle in dryland Zimbabwe: farmer knowledge and ecological theory', in Cousins (ed.). *People, Land and Livestock*; idem. 'Coping with Drought: Responses of Herders and Livestock in Contrasting Savanna Environments in Southern

different way of thinking about "carrying capacity"' that recognizes different sources of nutrition for livestock at different times of the year so that one can design a livestock development programme strategically around what particular resources are being reduced, when, and why. Key resources should be the central focus for designing any management scheme, for the danger of environmental degradation lies not in the destruction of topland grazing but in the degradation of key resources that create the possibility of opportunistic, adaptive use of a hetero-geneous environment. He proposes that the notion of 'economic carrying capacity' should be replaced with a notion of 'ecological carrying capacity' that uses drought years as a baseline because 'it is resource availability at the close of the dry season in a drought year that ultimately determines the lower limit of "ecological carrying capacity"'.

These arguments, drawn from new developments in non-equilibrium range ecology, suggest that policy-makers need to pay closer attention to opportunistic patterns of resource use in areas where local ecologies are highly variable, complex, and marked by uncertain climates.[61] Two issues are particularly important. The first is the status of wetlands. If technical appraisals that focus on toplands have indeed missed the point, ecological degradation may be assessed by considering whether wetlands are being systematically degraded. To date, however, wetland research has been sporadic and scattered, and the answer is unclear.[62] Again, the history of environmental policy has muddied, rather than clarified, the issue. Control of wetlands was established under the Water Act of 1927 and the Natural Resources Act of 1941 (amended in 1976 and 1981). These Acts, written to address the environmentally destructive farming techniques of white commercial farmers, effectively excluded wetlands from cultivation, though there is evidence that they had been regularly and effectively cultivated by African farmers for generations.[63] This legislation reflected and fueled an assumption that cultivation of dambos and streambanks not only degraded the soil and dried out watercourses but also denied livestock access to dry-season grazing. Yet the prohibition of wetland cropping disrupted the relationship between livestock and cropping in the farming

---

[60] (cont.) Zimbabwe', *Human Ecology*, 20, 3 (1992). For related arguments, see also Drinkwater, *The State and Agrarian Change*.

[61] This approach raises significant methodological questions about the analysis of human/nature relations, for neither human actions/decisions nor 'natural' ecological trends can be assigned causal predominance over time. As agents changing the landscape (range quality), human efforts and natural processes may be inextricable: it is the *relationship* between them that is the causal agent. For useful ruminations on this point, see Scoones, 'The Dynamics of Soil Fertility Change'; D. Good-man, 'Agro-Food Studies in the "Age of Ecology": Nature, Corporeality, Bio-Politics', *Sociologia Ruralis* 39, 1 (1999). This may have important ramifications for the actual construction of a policy paradigm on the basis of this approach: there is very little room in a state's institutional system of reckoning for recognizing nature as a social agent.

[62] T. Matiza and S. Crafter (eds), *Wetlands Ecology and Priorities for Conservation in Zimbabwe* (Gland, Switzerland: IUCN, 1994).

[63] See J. Rattray, R. Cormack and R. Staples, 'The Vlei Areas of S. Rhodesia and their Uses', *Rhodesia Agricultural Journal*, 50, 6 (November–December 1953); Robinson, 'Soil Conservation and Implications of the Land Husbandry Act'; R. Whitlow, 'Vlei Cultivation in Zimbabwe: Reflections on the Past', *Zimbabwe Agricultural Journal*, 80, 3 (May–June 1983); *idem*. 'Research on Dambos in Zimbabwe', *Zimbabwe Agricultural Journal*, 82, 2 (March–April 1985).

system, and placed additional strains on local relations of agrarian production.[64]

Furthermore, the state's continued reliance on this legislation has complicated the transfer of practical knowledge between CL farmers and government extension agents, especially where implementation of the legislation has been sporadic. In the first place, the existence of the legislation has inhibited fruitful interactions between farmers and extension officers on the sustainable use of wetland resources. Either extension officers have avoided the issue because not to do so would require them to enforce the legislation; or they have enforced the legislation. In either case, close research on the actual land-use practices of farmers has been inhibited. In the second place, where there is degradation it has been difficult to determine its rate, its degree of irreversibility, or indeed its actual causes (such as the comparative levels of damage done by continued cropping or increased livestock use). In short, it is unclear today – paradoxically in part *because of* rather than *despite* the legislative framework of resource management – how the environmental degradation of the communal lands is to be assessed, either in terms of the wetlands, or in terms of the toplands, or in terms of the relationship between these resource areas.

The second issue is the appropriateness of grazing schemes based on rotational paddocking for sustainable resource management. Rotational paddocking schemes do not respond well to the demands of flexibility created by uncertain climates and highly variable local ecologies. In these contexts, the grazing habits of cattle play a very important role in determining patterns of pressure on forage resources. For instance, cattle will graze in 'patches', selecting sour grasses early in the growing season while they are still palatable and nutritious, and later moving to sweeter grasses. Their preferences for different parts of different grasses also varies seasonally. As J.J. Jackson has shown, where there are variations in grass type and the moisture status of soil within a paddock, resting land on a rotational basis may in fact lead to overgrazing of patches, especially of sweeter grasses, thereby changing patterns of palatable forage by allowing sour grasses to become more dominant.[65] In effect, rotational paddocking may change the dynamics of the local ecology in ways that do not promote its sustainability.

New ecological research, then, suggests that there are serious reasons to question the technical paradigm of carrying capacity/offtake and the policy paradigm of rotation-based grazing schemes, and to base policy provisions on close local research. It does not resolve the question of whether irreversible environmental degradation is taking place in the communal lands. But it does suggest that degradation might have been misread: livestock numbers and movement patterns may not be the principal cause of degradation and techniques of range evaluation appropriate to more stable ecologies may not be appropriate to the highly varied local ecologies of the semi-

---

[64] As one chief, cited in Joy MacLean's memoir, complained: '[F]irst of all we are told it is not right to plant at the bottom of a vlei or on the streambanks. This is ridiculous for it is damper there and it is easier for the women to water the crops if need be. These are the best places to plant. Then we are told to manure our lands with cattle manure in order to get bigger crops – but who is to spread the manure on the land?... The lands round our huts belong to many different people. How can we decide who will put on the manure?' See J. MacLean, *The Guardians* (Bulawayo: Books of Rhodesia, 1974), p. 205.

[65] Jackson, 'Observations on the Comparative Effects of Short Duration Grazing Schemes', especially pp.100–1.

arid areas. If this perspective is right, a more appropriate policy paradigm may be one that protects key resources such as wetlands, does not constrain livestock movement between them, and deploys a more variable concept of carrying capacity.

At the same time, innovative ecological research alone cannot provide the solutions for resource management policy-making. Even if a serious reconceptualization of livestock/range relations is in order, it must be accompanied by a reconsideration of the social institutions governing the commons, since resources and access rights other than grazing – including water and wetland cropping – are also at stake. Grazing schemes, for instance, often do not fit isomorphically with village communities made up of households with significantly different livestock holdings and levels of production. Paddocking frequently cuts stockholders off from vlei areas, or excludes non-stockholders from cultivation opportunities. Poorer households that lack cattle or labor resources (often headed by women) are likely to want to retain access to wetlands for cropping, though this may clash with the interest of large stockholders in wetland grazing. As Ben Cousins has noted, patch use by cattle 'reveals a mismatch between the recomended technology (fenced paddocks and Short Duration Grazing) and local ecological dynamics (livestock rely heavily on the key resource of vlei grazing, most of which lies outside of the paddocks)', which can '[undermine] embryonic management institutions and [fuel] an internal power struggle within the community in respect of issues of access to and control over common property resources'.[66] Indeed, adoption of grazing schemes has often been driven by wealthier farmers' attempts to establish exclusive rights over land resources.

Questions of political will, definitions of property rights, and incipient community conflicts are thus central to the development of resource management policy. These issues are not easily resolved. Authority and control remain a major source of tension between agricultural officers and community members. The scientism of bureaucratic ideology persists, often sustaining the disdain of agricultural officers for farmers' knowledge. The impetus for most grazing schemes has originated outside of the community, in Agritex or in development-oriented NGOs. Because farmers want selected benefits, and different farmers sometimes want different benefits, the by-laws which govern and enforce their operation are difficult to agree on – especially if they assign stocking rates – and also tend to originate outside of the community, either in the District Council or in Agritex.[67] Within some grazing schemes – and communities more broadly – ambiguity persists among allocating authorities, and neither government officials nor community agents have an overriding voice.

These conditions exacerbate pressures for government agents to exert control. But the institutional imperatives of government agencies, which stress upwardly accountable management, do not facilitate locally flexible policy design and implementation. On the other hand, the dominant planning role played by Agritex frequently raises suspicion or anger where individuals' farming capabilities are compromised by conservation policies. In this situation, NGOs can play a

---

[66] B. Cousins (ed.), *Institutional Dynamics in Communal Grazing Regimes in Southern Africa* (Centre for Applied Social Sciences, University of Zimbabwe, 1992), pp.7–8.

[67] Statement by the Minister of Lands, Agriculture and Rural Resettlement, 22 February 1989 [61/89/SM/SK/EMM]; see also *The Herald*, Harare, 16 August 1988.

significant implementing role, because they can pay attention to local specificities without having to meet the larger structural demands of the state or the monitoring requirements of aid agencies. Several international NGOs currently participate in grazing scheme development. But they share with government agencies a predilection for technical and managerial regulation: strong leadership and management, appropriate training, and the need for by-laws that sanction access to resources. Operating within the government's policy and epistemological framework, they tend to regard paddocking, and the exclusive rights it entails, as the keystone of resource protection. Despite a broad rhetorical commitment to participatory resource management and local flexibility, therefore, policy-makers and agrarian technocrats by and large cling to model-driven development, to national administrative and planning structures, and to broader managerial objectives.[68] When it comes to the incorporation of community members as active participants in designing resource management strategies, theory and practice tend to diverge.

Government agents do not bear the entire blame. Within communities, conflicts between various power brokers – traditional authorities, district councillors, wealthy farmers or businessmen, VIDCOs – can undermine the institutional integrity of grazing schemes. Community-level conflicts sometimes erupt when schemes try to exclude non-members from traditional grazing areas, erect fences, or are dominated by the interests of wealthier farmers. Non-owners sometimes join schemes only because they are engaged in reciprocal arrangements.[69] Conflicts consequently arise between farmers, groups of farmers, or communities *within* a VIDCO area, thereby complicating the definition of the 'community' that should draw the benefits.

Where grazing schemes reflect lines of local socio-economic stratification, they run the risk of *straining* local community relations. Yet it is clear that the success of grazing schemes depends heavily on community acceptance. Of greatest concern is the situation of the rural poor – those lacking livestock or occupying poor land – who are most vulnerable to climatic freaks. Given the role of cattle in relationships of reciprocity, power and obligation, any management policy that seeks to respond to the spatial and temporal flexibility of a non-equilibrium ecosystem cannot proceed effectively without taking account of socio-economic institutions such as tenure, labor co-operation systems and livestock loaning arrangements that facilitate or inhibit the movement of cattle in time and space. In effect, the resource 'user community' must be defined in terms of relationships between stockholders and

[68] As the CARD *Livestock Management Recommendations* for the Pfumai livestock development project noted bluntly: 'Some farmers in other Communal Lands have been very reluctant to erect fencing as this restricts the movement of cattle and the use of "key forage resources" – this has led to the suggestion that only these key resources should be fenced (Scoones 1986) and that the stock should be allowed to forage over large areas. However, in essence this is contrary to Government's ideas of VIDCO development and the assumption of responsibility of the VIDCO for its village area resources' (p. 15); see also Cousins, *A Survey of Current Grazing Schemes*, p. 68.

[69] See U. Otzen et al., *Development Management from Below: The Potential Contribution of Co-operatives and Village Development Committees to Self-Management and Decentralized Development in Zimbabwe* (Berlin: German Development Institute, 1988), p.106; B. Cousins, *Room for Dancing On: Grazing Schemes in the Communal Lands of Zimbabwe* (University of Zimbabwe, Centre for Applied Social Sciences Occasional Paper, NRM Series, 4/1992, 1992); World Bank, *Zimbabwe: Land Subsector Study*, p. 31.

non-stockholders as much as of interactive relationships between stockholders and nature if policy is not to promote social conflict or the impoverishment of the powerless (perhaps leading to more destructive resource-use practices).

Conclusion

There is evidence of severe soil loss and land pressure in many parts of Zimbabwe's communal lands today. Over time, policy-makers have designed strategies to relieve this stress based on a model of livestock management that now seems institutionally and epistemologically inadequate in terms of all the components of a policy paradigm sketched at the beginning of this chapter. (i) Technical appraisals of the resource base have been based not only on narrow conceptions of authoritative knowledge and inflexible interpretations of ecology, but on weak, scattered and uneven data. These conditions, in conjunction with state imperatives of regulation, have resulted in a technicist model-driven policy framework rather than one sensitive to local knowledge and local conditions. (ii) Partly because it conceives resource degradation narrowly as an effect of the impact of individual users on natural resources, the policy framework ignores the fact that farmers' relations with the ecology are both interactive human/nature relationships and social relation-ships. (iii) The failure of the model to take account of local agrarian power relations, and the role of livestock in those relations, has fueled tensions between the productivity interests of large stockowners and the risk management interests of the rural poor. (iv) Such tensions, which are frequently exacerbated by the preferred technology of paddocking, might actually exacerbate ecological stress and fuel social conflict where the poor are denied access to necessary resources or are used for labor to fence and re-seed paddocks that do not benefit them. (v) Both property institutions and participatory management institutions remain ambiguous, drawing their regulatory authority from outside, and their political authority from local power structures (VIDCOs, chiefs, party). Notwithstanding a rhetorical commit-ment to local participation, government agencies have not incorporated farmers' ideas into policy design, and have maintained a firm managerial and regulatory hold on resource management institutions.

Any re-assessment of ecological degradation and its solutions must take account of all these factors. The discussion here suggests that the challenge is not simply to replace managerial state agencies with sympathetic NGOs or participatory local institutions, but to reconsider the systems of practical knowledge that current policy marginalizes, and to incorporate local resource users more actively into policy design. Yet this requires a careful evaluation of local knowledge and a broad interpretation of categories of 'users'. The knowledge regimes of farmers are as likely to be shaped by interests, power relations and institutional considerations as those of government agents.[70] People use the same resources for different purposes.

---

[70] For warnings along these lines, see also J. McGregor, 'Conservation, Control and Ecological Change: The Politics and Ecology of Colonial Conservation in Shurugwi, Zimbabwe', *Environment and History*, 1, 3 (1995); K. Wilson, '"Water Used to be Scattered in the Landscape": Local Under-standings of Soil Erosion and Land Use Planning in Southern Zimbabwe', *Environment and History*, 1, 3 (1995).

Thus, it may be necessary to reconceive access rights as multiple and integrated across time, space, agrarian usage, and social structure. Given the structure of knowledge and the imperatives of regulation inherent in the state, the challenge is to design institutions that can patrol such rights without generating local conflict or becoming overly mechanistic. This will require careful and sustained political negotiation at both national and local levels if management institutions are to protect the welfare interests of the poor and weak. Indeed, if participatory institutions are to be the mechanism for sustainable resource use, those institutions must make a distinction between 'users' and 'community,' and take into account the interests of non-users as well as users.

PART FIVE

Policy, Producers & Resources

9

Littering the Landscape
Environmental Policy in Northeast Ethiopia

DESSALEGN RAHMATO

Anyone travelling in the early 1990s along the main road to Wällo and the north-east would have seen the landscape littered with the debris of the environmental rehabilitation program of the Derg, the military government which collapsed in 1991. They would have seen ruined stone terraces and checkdams, with the stones strewn all over the slopes and now posing a great erosion hazard; disused bunds, many of which were partially demolished; and heavily disturbed woodlots and forests. As is shown in Table 9.1 below, the cost of this program, which was implemented mainly in the second half of the 1970s and the 1980s, may well exceed US$900 million, all of it provided by Western donor agencies. About a third of this cost was borne by the World Food Programme (WFP) in the form of food aid totaling one million metric tons. Other actors whose contribution to this massive investment was significant were the European Union, the World Bank, SIDA, Canada, and dozens of Western NGOs. While some of the afforestation schemes, area closures, and terraces still remain intact, the bulk of the environmental assets created in these years has either been demolished or fallen into disuse.

This chapter will attempt to examine why the gigantic conservation program of the Derg achieved so little, and why the massive resources invested in it were largely wasted. I shall argue that the explanation for the program's failure should be sought in the nature of the dominant environmental discourse in the country, and the conflict between what I call the environmentalism of the state and the environmentalism of the peasant. The period covered extends between the mid-1960s, when the first set of environmental legislation was initiated, and the beginning of the 1990s when the Derg regime was overthrown.

The Environmental Discourse

Since the early 1950s, environmentalists in Ethiopia, most of them expatriates, have raised the specter of large-scale environmental disaster leading to widespread economic ruin. The picture they have painted is a grim one: the country's forests, its top soil, wild life and biodiversity are being lost on a massive scale, and if this continues unabated, the very survival of the nation will soon be threatened. Many of these expatriates were employed as technical advisors to the Ethiopian government by donor agencies such as FAO, UNESCO, and the World Bank. Very few of them stayed long enough to acquire an in-depth knowledge of the country. Only a small number had the opportunity to travel through the countryside and acquaint themselves with the agricultural and environmental practices of the peasants. Hardly any of them undertook serious research on land and farm management among the peasantry.

Writing in the early 1950s, E.H.F Swain, an Australian conservationist and a FAO technical advisor to the Ethiopian government, underlined the urgency of a forest conservation policy with the argument that the 'very life and future of Ethiopia' was in the balance. The traditional practice in the country, he argued, was the uninhibited denudation of virgin forest for cultivation. The country's primeval forest, he maintained, had been in retreat from north to south by a thousand years of remorseless destruction.[1] Swain was one of the earliest naturalists to argue that originally the greater part of the country was covered with forest, but through years of wanton destruction only a fraction of the forest cover was now remaining. Logan, who had visited the country earlier, cautiously cited oral tradition as evidence for views similar to Swain's,[2] but it was Swain's claims which became the basis of conventional wisdom in the decades thereafter. Swain's estimates of the loss of forest cover, as well as those of subsequent expatriates, were not based on reliable measurements but rather on guesswork or on methodologies of dubious validity. Another conservationist, W.C. Bosshard, also a FAO advisor to the government, was similarly emphatic. Forest destruction in the highlands, he argued, 'has started a process which, sooner or later, will come to an end, but this end, however, will be a tragic one. It will result in the total loss of soils and the drying up of the High-lands which will thereby become absolutely uninhabitable.' Formerly, Bosshard contended, 50–55 per cent of the country's surface area was forested, but thanks to wanton deforestation only 6 per cent remained under forest cover. Anticipating the Derg by nearly two decades, he recommended a massive program of afforesta-tion, involving 'the co-operation of the nation as a whole', though even with the 'help of all the people' it would take 100 years to reforest 25 per cent of the highlands.[3]

Leslie Brown, a British naturalist actively involved in the wildlife conservation schemes in the country, wrote a monograph pointedly titled *Conservation for Survival: Ethiopia's Choice*. In it, he expressed pessimism about the prospects for the country's environment. 'I see this beautiful and potentially productive country,' he sadly noted, 'sliding … rapidly towards early and complete environmental and

[1] E. H. F. Swain, 'Forestry Possibilities in Ethiopia', *Unasylva*, VI, 1 (1952), pp. 15-20; Final Draft Forestry Proclamation for Ethiopia, unpublished mimeo (Addis Ababa, 1953); FAO, *Report to the Government of Ethiopia on Forestry Policy and Forest Development*, FAO Report No. *321* (Rome, 1954).

[2] W. E. M. Logan, *An Introduction to the Forests of Central and Southern Ethiopia*. Institute Paper No. 24, (Oxford: Imperial Forestry Institute, 1946).

[3] W. C. Bosshard, *Report to the Government of Ethiopia on Forestry Development*. FAO Report No. *1143* (Rome, 1959), pp. 2, 10.

consequently economic ruin'.[4] After a brief visit to the famine-ravaged provinces of Wällo and Tigray in 1975, Brown declared that the highlands of the two provinces, which he found denuded, 'were originally covered in dense forest of woodland with glades of verdant grass'.[5]

Similar stories of sylvan catastrophe were told by Breitenbach, a German advisor to the Ministry of Agriculture (MoA) in the 1960s, and by Swedish foresters in the 1970s. Breitenbach estimated the rate of deforestation to be 20,000 hectares per year while the Swedes put the figure at 200,000 ha. In the 1980s, Wood and Ståhl described the progress of deforestation as follows (they cite no evidence for their statement): before settlement, the country's high forests covered 35 per cent of the country; this was reduced to 16 per cent in the 1950s, 3.6 per cent in the early 1980s, and 2.7 per cent in 1989. The Ethiopian Forestry Action Programme (EFAP) endorsed this argument without question.[6]

In the 1980s, the danger posed by soil erosion, which was said to be taking place on a large scale in most parts of the country, became another element of the expatriate environmental discourse. Hans Hurni, a Swiss soil scientist, who served as advisor on conservation to the MoA, claimed that the rate of soil erosion in the highlands was 1.5 billion tons per year. He felt quite certain that at this rate of land degradation, all the top soil from the country's farm land would be washed away in 100 to 150 years, resulting in complete and irreversible agricultural destruction.[7] The Ethiopian Highlands Reclamation Study, a project initiated and financed by FAO, was even more pessimistic. It estimated that nearly 2 billion tons of soil was being removed from the highlands annually, and that in 25 years' time about 18 per cent of the country's farm land would go out of production, affecting the livelihoods of some 10 million peasants.[8]

The writings of the expatriates have been influential in shaping the environmental thinking of Ethiopian officials, and have left a legacy of environmentalism rooted in ecological calamity. Ethiopian environmentalists and policy planners came to believe, especially from the latter part of the 1970s, that the danger of large-scale environmental collapse was indeed imminent, and that urgent measures were needed to avoid such a catastrophe. This legacy promoted a policy framework that can only be described as 'unilateral' and 'state centered', with a strong tendency to exclude pluralist approaches. Informed as it was by a high sense of urgency, the 'apocalyptic' outlook tended to be impatient, and intolerant of other experiences,

[4] Leslie Brown, *Conservation for Survival: Ethiopia's Choice* (Addis Ababa, 1973), p. 2.

[5] Relief and Rehabilitation Commission, *Drought Rehabilitation in Wällo and Tigre. Report of a Survey and Project Preparation Mission* [prepared] by Leslie Brown (Addis Ababa, 1975), p. 8.

[6] F. von Breitenbach, 'National Forestry Development Planning', *Ethiopian Forestry Review*, 3/4 (1962), pp. 41–68; National Forestry Programme for Ethiopia, 'National Forestry Programme for Ethiopia, Phase One: A Three Year Development Project', unpublished mimeo (Addis Ababa, 1997); IUCN, *Ethiopia: National Conservation Strategy. Phase I Report*, based on the work of A. Wood and M. Ståhl (Gland, Switzerland: IUCN, 1990); EFAP, *Ethiopian Forestry Action Program, Draft Final Report. Vol. II, The Challenge for Development* (Addis Ababa, 1993).

[7] Hans Hurni, 'Ecological Issues in the Creation of Famines in Ethiopia', paper presented to the National Conference on Disaster Prevention and Preparedness Strategy for Ethiopia (Addis Ababa, 1988).

[8] EHRS, *The Degradation of Resources and an Evaluation of Actions to Combat it.* Working paper 19, prepared by M. Constable (Addis Ababa, 1984).

particularly those of the smallholder farmer. Its policy prescriptions laid strong emphasis on the practical and technical side of conservation; the underlying assumption being that land degradation was caused by natural processes and could be brought to an end only if people understood these processes and learned to employ the right conservation measures. The failure to go beyond this one-sided outlook, to recognize that in fact environmental problems are a product of the complex interplay of economic, political, social and natural processes, led Ethiopian environmental planners and decision-makers up a blind alley, and in consequence, conservation policy ended up becoming inflexible, authoritarian and bureaucratic. Indeed, during the latter half of the 1980s, environmental policy contained repressive overtones and was employed to extend state domination over the rural population.

The story of the environmental depletion of the country has been variously criticized by Ethiopian researchers,[9] but it is worth looking at again for the light it sheds on the ideological assumptions of the expatriates. Implicit in this 'discourse of catastrophe' are a number of ideological preconceptions about the country's ecological, agricultural and demographic history. Hoben has emphasized that hidden behind this 'environmental narrative' is an overriding neo-Malthusian argument, which postulates an unremitting population growth as the primary cause of environmental depletion. While the demographic paradigm is certainly significant, my reading of the narrative suggests a different interpretation. I submit that at the core of the narrative is a set of assumptions about the land user and his agricultural and environmental practices. Lurking beneath these assumptions is the view that the peasant farmer does not comprehend the forces behind land degradation and has been employing agricultural and land-use practices that are responsible for accelerating the degradation process. The accent here is on human mismanagement, born of ignorance or disregard for the consequences, rather than on human population pressure as the neo-Malthusian argument stresses.

Varieties of Environmentalism

We may distinguish two varieties of environmentalism in this period: the environmentalism of the state and the environmentalism of the peasant farmer. The underlying assumptions of state environmentalism were the following: the state was considered the guardian of the country's natural resources, and the primary emphasis was on the *protection* of these resources rather than their *utilization*. Environmental protection involves bringing natural resources under state ownership or control. It was the responsibility of the state to choose the appropriate conservation technology and 'transfer' such technology to the population concerned. These assumptions could only lead to a 'unilateral' course of action, and an

---

[9] Melaku Bekele, 'Forest History of Ethiopia from Early Times to 1974', MA Thesis, University College of North Wales, Bangor, 1992; J. P. Sutcliffe, *Economic Assessment of Land Degradation in the Ethiopian Highlands: A Case Study*, National Conservation Strategy Secretariat, Ministry of Planning and Economic Development (Addis Ababa, 1993); Allan Hoben, 'Paradigms and Politics: The Cultural Construction of Environmental Policy in Ethiopia', *World Development*, 23, 6 (1995), pp. 1007–21.

undemocratic approach. Thus, many of the major conservation initiatives invariably deprived rural communities of control over and access to natural resources. To many a peasant, 'conservation' came to be synonymous with the appropriation of local resources by the state. Conservation was frequently restrictive and exclusionist. Measures for benefit-sharing were rarely incorporated in policies. Conservation schemes did not take into account the interests of rural land users, and the benefits of any such schemes were not shared by them. Conservation policy tended to be inflexible and narrowly conceived. The objective of soil conservation, for instance, was merely to control erosion and not to promote the sustainable utilization of the land. Finally, conservation programs were always imposed from above, and rural land users – the people directly affected by environmental change – were rarely consulted.

Moreover, the worst enemy of state environmentalism under the Derg was the state itself, many of whose rural policies worked against environmental objectives. Collectivization, villagization and resettlement, which were carried out on a large scale in the 1980s, were accompanied by extensive deforestation and soil erosion. The radical land reform of the post-Revolution period and the periodic redistribution of farm plots undermined security of holdings and discouraged peasants from investing in the land or employing conservation measures. Peasants interviewed in Wällo in 1995 all pointed to the Derg's land policy as the main cause of the deterioration in traditional land management practices, which, they said, were effective in the past in controlling soil erosion. Traditional grass bunds known locally as *weber*, drainage ditches constructed along the contour, terraces, and a variety of other agronomic techniques were cited as the main indigenous land management measures.[10]

The environmentalism of the peasant contrasts sharply with this, though it is hard to find in clearly articulated form. Broadly speaking, peasant farmers are not keen on the mere protection of natural resources such as forests, pasture and water sources, but rather prefer their sustainable use. Central to peasant environmentalism is the issue of access to and control over basic environmental goods. Conservation in this context places strong emphasis on *management through use.*Traditionally such management was undertaken in a variety of ways: a) In the past, local resources were managed through community authority structures which were responsible for enforcing rules and regulations; such forms of indigenous local governance have now all but disappeared in most parts of the country. b) Management by religious sanctions or community convention is also an old tradition; sacred groves, church forestry, woody burial grounds, holy springs, etc. are some good examples of this.[11] In the past, religious institutions played a much greater role in the protection of forests and water sources, but this role has been undermined by the political changes under way since the 1950s. c) Management by the use of conservation-based farming practices is very widespread.[12] However,

[10] Interview with peasant in Ambassel *wäräda*, Wällo, May 1995.
[11] Dessalegn Rahmato, 'Environmentalism and Conservation in Wällo before the Revolution', Paper prepared for Conference on Environment and Development in Ethiopia (Debre Zeit, 1997).
[12] Tahal Consulting Engineers, 'Study of Traditional Conservation Practices', Report prepared for the Ministry of Agriculture (Addis Ababa, 1988); Dessalegn Rahmato, *Famine and Survival Strategies: A Case Study of Northeast Ethiopia* (Uppsala, 1991).

successive agricultural and tenure reforms have undermined traditional farming practices, and this has weakened the conservation element of land husbandry.

These two forms of environmentalism – those of the state and the peasant – constituted two different perceptions and valuations of nature, and two different approaches to the utilization of environmental resources. They confronted each other against a backdrop of increasing natural resource loss and scarcity. The state believed that the peasant was responsible for land degradation and hence it was its responsibility to protect the environment. In contrast, the peasants harbored a deep suspicion of the motives of the state, and frequently wished nothing more than to keep the latter at arm's length. Over the years, the state's claim to stewardship over forests, pasture and similar resources was contested by the peasants, often through the use of covert action and 'everyday' forms of resistance. These included the illegal felling of trees, the grazing of livestock in closed areas, the uprooting of tree seedlings planted in forest schemes, and the invasion of national parks.[13]

Thus, I take environmental policy to be a contested terrain, with the state attempting to impose its environmental values and programs on an unwilling and frequently resistant peasantry. In what follows I shall take a closer look at the interplay between state and peasant environmentalism by examining the evolution of conservation policy in the period under discussion. While the study deals with national policy initiatives undertaken at the time, I shall make use of my research findings from Wällo to illustrate the impact of these initiatives.

Environmental Policy: Ideas and Justifications

Until the second half of the 1960s, conservation was understood in a different sense from what it came to be in the 1980s. First, it was taken to mean the preservation of wildlife and the protection of their natural habitat. Policy-makers were keenly interested in using the country's wildlife resources to stimulate the tourist industry as part of the development initiatives under way at the time. The success of the national park schemes of British East Africa was taken as a model to emulate. Secondly, conservation also meant the preservation of the country's historical heritage. The protection of antiquities, as it was then called, involved the preservation of ancient monuments, historical documents, and relics of archaeological interest. The objective here was not only historical and cultural, but economic as well.[14] It was thought that investing in natural as well as historical-cultural resources would enable Ethiopia to compete with the East African countries for the tourist trade. Thirdly, conservation gave impetus to the protection of the country's forests. However, while a few steps were taken to encourage tree-planting, protection took the form mainly of bringing the forests under state ownership, and legislation to this effect was issued in the mid-1960s. In part the objective of the legislation was to enlarge the sources of state revenue: rights over timber harvests and other forest products, taxes on timber exports, payment for timber permits, and royalties, all of

---

[13] Dessalegn Rahmato, 'The unquiet countryside: the collapse of socialism and rural agitations 1990–1991', in Abebe Zegeye and S. Pausewang (eds), *Ethiopia in Change* (London, 1994).

[14] J. H. Blower, 'Ethiopia: Wildlife Conservation and National Parks', unpublished report to UNESCO (Paris, 1971).

which the new forest laws made possible, provided additional revenue to the government.[15] However, in the second half of the 1960s, the imperial government began to show greater concern about environmental problems, and legislation setting up government bodies responsible for environmental protection was issued at the beginning of the 1970s.

In the post-Revolution period, state environmentalism became imbued with a new sense of purpose and a new urgency. The slow and tentative initiatives of the imperial regime now gave way to operations whose scale and tempo were unprecedented. Environmentalists and decision-makers appeared to be inspired by a clearer vision, and to be armed with more definite objectives. The initial impetus behind this accelerated and large-scale conservation program was *famine* and international *food aid*, the first providing the rationale and the second the resources for conservation on a national scale. It is ironic that the Derg, which was shunned like a pariah by many in the donor community, relied on massive Western food aid for its conservation program for almost a decade and half.

The famine of 1973–4, which was one of the worst environmental disasters to befall the rural population of Wällo and the northeast, is a landmark in terms of changes in environmental and agricultural policy. This disaster undermined the legitimacy of the imperial regime and provided powerful ammunition to its opponents on the eve of the Revolution. It also later served the Derg as justification for changes in agricultural and environmental policy. In the past, drought and famine were considered natural calamities and acts of God, but from the late 1970s government sources began to attribute such disasters to what was called 'uncontrolled human activity'. Such activity, which was thought to have been going on for centuries, was said to be responsible for the loss of valuable natural resources on a large scale. Here was a 'radical' environmental discourse, befitting a 'radical' government, a discourse which appeared to be based on historical and scientific evidence. Land degradation, it was argued, was caused not by divine providence but by backward agricultural practices, a 'primitive' system of land use, and high population pressure. The combined effect of these factors was the continuous and massive denudation of the land and the impoverishment of the rural population. 'Soil erosion has reached exceedingly dangerous levels', the government believed, and the rate of deforestation was estimated to be 200,000 hectares per year – a figure borrowed from the expatriate discourse.[16] The 'uncontrolled' activity referred not just to the peasant's 'backward' farm techniques, but also to the expansion of agriculture and the deforestation of the land. In the government's view, the peasant was thus both the victim and the cause of environmental disaster.

Similar arguments were also provided by expatriate technical advisors to explain the recurrent famine and natural disaster. The EHRS document noted above, which was prepared at the onset of the famine of the 1980s, reflected the thinking

---

[15] Imperial Ethiopian Government, 'State Forest Proclamation, Proclamation No. 225, 1965'; 'Private Forest Conservation Proclamation, Proclamation No. 226, 1965'; 'Protective Forest Proclamation, Proclamation No.227, 1965'; 'Awash National Park, Order No. 59, 1969'; 'Simien National Park, Order No. 60, 1969'; 'Wildlife Conservation Organization, Order No. 65, 1970'; 'State Forest Development Agency, Order No. 74, 1971', *Consolidated Laws of Ethiopia* (Addis Ababa, 1972).

[16] Relief and Rehabilitation Commission, *Combatting the Effects of Cyclical Drought in Ethiopia* (Addis Ababa, 1985), pp. 9–17.

of many of the technical experts in the donor community; the dramatic revelations which were presented as important 'findings' were used to underscore the urgency of large-scale conservation. These 'findings' were not, however, the product of proper scientific investigation but estimates based essentially on subjective criteria. Only 20 per cent of the total land area of the highlands, the document argued, was relatively free from erosion. About half of the highlands, some 27 million hectares, was significantly or severely eroded. Over 80 per cent of the erosion regularly occurred on cropland. Average net soil loss from cropland was said to be 100 tons per hectare per year. The average yield decline due to erosion was 2 per cent per year, which was equivalent to 120,000 tons of cereals per year – enough food to feed up to 4 million people. The Wällo highlands were said to lose 0.5 per cent of soil annually, with yearly crop and grass yield declines of 3 and 1 per cent respectively.[17]

The causes of this massive erosion were said to be high population pressure, and poor farming and livestock management practices. The farming population, it was argued, far exceeds the carrying capacity of the land, particularly in Wällo and the northeast where erosion is most severe; population pressure has led to the cultivation of more fragile ecosystems and more marginal lands, thereby accelerating the erosion process. In addition, the document noted, farmers do not practise sustainable agriculture: plowing and harvesting methods expose the soil to erosion, and soil and water protection techniques are virtually unknown. Due to prior deforestation in the areas of high population density, and the consequent shortage of woodfuel, the document went on to argue, dung and crop residues are removed from farmland, thus contributing to soil erosion and loss of fertility. All through this analysis, no mention is made of the role of the state and the impact of its economic, social and political policies on resource degradation. It is as if the state was invisible.

While the arguments of both the government and the technical specialists are broadly similar, there is nevertheless a difference of emphasis, which is worth noting here. The Derg placed particular stress on the underdevelopment of peasant agriculture and the backwardness of farm technology, while, in contrast, the technical experts were strongly neo-Malthusian and saw the demographic factor as the most significant cause of land degradation. Behind the Derg's 'backwardness' argument lay its agricultural collectivization program.

Let us now turn briefly to food aid and its justifications. At the beginning of the 1980s, when the volume of food aid delivered to the country began to increase appreciably, government officials showed a lack of clear understanding of the politics and economics of international food aid. It was naively believed that disposing of their surplus food would not only serve the interests of donor countries, but would also be in the best interests of starving countries like Ethiopia. At this time too, Food for Work (FFW) schemes were seen by officials as a great innovation: FFW would contribute to food security and at the same time enable important development work to be undertaken. From the very beginning the government set ambitious objectives for FFW: it was meant to support environmental protection and rehabilitation activities, to provide employment opportunities to the rural population in the food-deficit areas, to improve the nutritional status of

[17] EHRS, *The Degradation*, Chapters 6 and 7.

the poor, and to help reduce poverty.[18] Officials at the Ministry of Agriculture, which was the executing agency for a large part of the FFW schemes at the time, believed that FFW would be a great success because it offered the right incentives for the right purposes.[19] As we shall see, the optimism of the early years was short-lived; officials soon came to realize that food aid and conservation were unlikely bedfellows, and that in the end neither food security nor land rehabilitation would be achieved.

Conservation: Strategies and Measures

It is worth noting for comparative purposes that the imperial regime issued far more conservation legislation than the Derg but the latter's environmental program greatly overshadowed the earlier efforts. In fact, despite the legislative framework and the implementing bodies set up in the 1960s and early 1970s, there was limited conservation activity in the pre-Revolution period. The Derg, on the other hand, operated without legislative backing, relying instead on mass mobilization and forced labor campaigns.

The earliest endeavor at land rehabilitation in the northeast of the country was undertaken at the initiative of provincial governors and local self-help groups in the latter part of the 1960s. In Wällo, the chief governor, *Däjazmach* Mammo Seyum, who was committed to conservation, spent considerable time and effort to reforest denuded hillsides in the central part of the province.[20] Similar attempts were made in Tigray to plant trees on some of the bare lands in the province with the support of the provincial administration.[21] The Ethiopian Forestry Association, which was set up in 1960, had earlier launched a 'farm woodland' (or agro-forestry) campaign to encourage peasants to plant trees on their plots as an economic and conservation measure.[22]

However, these ventures had limited impact and were soon superseded by donor-funded Food for Work conservation schemes. US surplus food supplied under PL 480 was used to carry out reforestation and the construction of low-cost rural roads and small water projects, first in Eritrea in 1970, and then in Tigray and Wällo in 1971 and 1972 respectively; this was the first FFW undertaking in the country. The

---

[18] Kebede Tato, 'Soil Conservation and Forestry Development through Food for Work: the Ethiopian Case', mimeo (Addis Ababa: Soil Conservation Research Project, 1992); Berhe Wolde-Aregay, 'Twenty Years of Soil Conservation in Ethiopia: A Personal Overview', Regional Soil Conservation Unit, SIDA (Nairobi, 1996).

[19] Kebede Tato, personal interview, 5 February 1993. MoA's Soil and Water Conservation Department, established in 1970 and rebaptized as the Community Forestry and Soil Conservation Development Department (CFSCDD) in 1979, played a major role in the government's environmental rehabilitation program in the 1980s. Kebede Tato was the head of the Department until the end of the 1980s when he became the co-manager of the SCRP; he was at the time a dedicated conservationist. He was succeeded at CFSCDD by Berhe Wolde-Aregay.

[20] Dessalegn, 'Environmentalism'.

[21] T. W. Crawford, 'Law and Development in Ethiopia: An Application of the Seidman Model to a Reforestation Program', MA thesis, Fletcher School of Law and Diplomacy, Medford, MA, 1973.

[22] For the activities of the Association and the ideas debated by its members see *Ethiopian Forestry Review*, 1961–2.

program lasted until 1974 when it was replaced by several FFW projects funded by the World Food Programme (WFP) whose intervention was largely a response to the drought and famine of 1973–4. The main activities undertaken were reforestation and soil and water conservation in the drought-prone areas of the country. However, at the end of the 1970s, the government, with the eager collaboration of WFP, launched an ambitious program of environmental rehabilitation covering the greater part of the highlands. WFP's relatively small-scale, individual FFW activities were consolidated under one super-project called Rehabilitation of Forest, Grazing and Agricultural Lands Project (or Project 2488), the 'watershed' or 'catchment approach' of which became the cornerstone of the government's conservation strategy. Project 2488, which was managed by the Ministry of Agriculture (MOA), grew to be the largest FFW program in Africa and the second largest in the world in the 1980s, but it was not the only environmental program in the country. Most of the NGOs working in the rural areas were engaged in similar activities, as were the European Commission, SIDA, the World Bank and others, but these were all dwarfed in scale and resource commitment by WFP/MoA's massive program.

The main implementing agency of the program, and the manager of the food resources supplied by WFP, was the MoA, in particular its Department responsible for community forestry and soil conservation, CFSCDD. In the early 1980s, WFP provided food aid to both this Department and state forestry, another unit attached to the MoA, but over the years CFSCDD became the dominant partner. The main components of the conservation program consisted of afforestation, on-farm and hillside terracing, area (or hillside) closures, and gully control. Initially, the work was carried out in some 20 catchments, but by the end of the decade the number had grown to 117 scattered throughout the highlands, with a total area of 3.5 million ha. In Wällo, conservation work was undertaken in 11 catchments each on average measuring 44,000 ha.[23]

State conservationism, especially from the 1980s onwards, came to be equated with the construction of physical structures on land considered to be at risk, through the incentive of FFW. Large numbers of peasants were mobilized by the MoA as well as local-government officials to construct bunds, terraces, micro-basins and checkdams in return for which they were paid in food. Tree seedlings were similarly planted on terraced hillsides. The scale of the operations and the haste with which they were carried out meant that the activities were poorly planned, and the measures adopted did not take into account the diversity of farming systems and agro-ecologies in the country. The selection of sites and the choice of conservation technology were determined by MoA agents without consultation with the surrounding peasantry. It was clear that the MoA was driven by the food delivery: it felt compelled to expand its 'absorptive capacity' and to utilize as much of the food aid delivered by the WFP as possible. Maximizing FFW opportunities was considered important, and extending them to as many localities as possible was regarded as fair.[24] The primary motive was the distribution of food to the needy, and the benefits of conservation were considered of secondary importance. By the

[23] L. G. Sewel, *Technical Report of the International Consultant on Soil and Water Conservation*, EFAP (Addis Ababa, 1992); Berhe, 'Twenty Years'.

[24] Kebede, Interview.

third quarter of the 1980s, FFW programs managed by the MoA, NGOs and others employed between 1.5 and 2.0 million peasants annually. Over the years, FFW became so inextricably linked with conservation that it inhibited all voluntary and participatory activities, and ruled out alternative conservation strategies.[25]

The flow of cereal food aid to the country as a whole, for both emergency and non-emergency purposes, in the period from the mid-1970s to the end of the 1980s shows an interesting trend. A gradual increase is discernible all through the 1970s, but deliveries rose sharply in the 1980s. In the period 1975–80, WFP's commitment for environmental rehabilitation amounted to 105,000 metric tons (mt) of food aid; for the years 1980-2, it reached 145,000 mt, an increase of 38 per cent. In the three following years, i.e. the famine years of 1982–5, WFP delivered 250,000 mt, but commitments were comparatively low, 100,500 mt, in the following two years, 1985–7, again reaching a high of 378,272 mt in 1987–93.[26] If we look at the total flow of food aid supplied by all donors in the period in question, the increase in delivery in the 1980s is even more dramatic. From a relatively low figure of 76,000 mt in 1977/8 the supply of cereal food aid rose to 356,300 mt in 1982/83 and in the famine years of 1984/5 reached 868,900 mt, a 400 per cent surge. Interestingly enough, food aid deliveries never returned to their pre-famine levels, even though the environmental crisis had abated. In 1989/90, the country received 537,500 mt of cereal food aid, in 1989/90 the figure had risen to 893,900 mt, and in 1991/92, it reached a massive 1.1 million mt.[27]

The evidence compiled by Aylieff indicates that from 1977/8 to 1990/1, a total of 6.7 million mt of food aid was provided to the country by Western donors. On the average 65 per cent of the food aid delivered was utilized for emergency relief, the rest was payment for FFW programs of which environmental rehabilitation, supported by the WFP, NGOs, and bilateral and multilateral donor agencies, was by far the most important component. It is thus quite evident that considerable resources were invested in land rehabilitation by a diverse group of donors and NGOs in the period from the mid-1970s to the early 1990s. While the major portion of this investment was in the form of food commodities, there were a number of donors that provided assistance in other forms. The extent of this investment will probably never be known accurately, but I estimate that the total resources committed may have amounted to more than US$900 million. Table 9.1 shows my estimate of the breakdown of the cost by principal donor agencies.

[25] There is considerable source material on conservation in the Derg period, but much of it consists of 'grey literature', ie. unpublished works by consultants, civil servants, academics, donor agencies and NGOs. I have made use of some of this literature in this paper. For a recent study, see Yeraswork Admassie, 'Twenty Years to Nowhere: Property Rights, Land Management and Conservation in Ethiopia', Ph.D. dissertation, Uppsala University (1995). For Wällo, see Alemneh Dejene, *Environment, Famine and Politics in Ethiopia* (Boulder, CO: 1990). For a study of the farm economy of Wällo, see Mesfin Wolde-Mariam, *Suffering Under God's Environment: A Vertical Study of the Predicament of Peasants in North-Central Ethiopia* (Berne, 1991).

[26] K. Adly, 'New Approaches on Food Aid Activities', Paper prepared for WFP's Food-for-Work Workshop (Addis Ababa, 1992).

[27] John Aylieff, *Statistical Summary of Food Aid Deliveries to Ethiopia 1977–1992* (Addis Ababa: WFP Food Aid Information Unit, 1993); WFP, *World Food Progamme and Food Aid in Ethiopia* (Addis Ababa, 1995) gives even higher figures.

Table 9.1 Estimated cost of environmental protection, 1974–93 (US$m.)

Agency	Resources invested
WFP[a]	275.00
EC[b]	150.00
UN Agencies[c]	50.00
Bilateral donors[d]	100.00
NGOs[e]	25.00
Ethiopian Government[f]	300.00
TOTAL	900.00

Note: a) Figure has been rounded. b) Includes support for PADEP, N and S Shoa projects, etc; also support to MoA environmental program, but not aid to WFP. c) FAO, UNDP, UNICEF, UNSO, etc. support to individual projects and also to MoA's forestry and environmental program. d) Germany, Canada, Scandinavian countries, World Bank, etc. support to MoA's environmental program and to individual projects. e) Individual environmental activity, and donor support to NGOs. f) MoA budgetary outlays. Conn 1990 and Adly give different figures, the former much higher than the latter.
Sources: Adly, 'New Approaches'; Aylieff, *Statistical Summary*; Aytenew Birhanu and J. Aylieff, *Inventory, Map, and Analytical Review of Food and Cash for Work Projects in Ethiopia* (Addis Ababa, 1992); Luther Banga, *Reducing People's Vulnerability to Famine: An Evaluation of BAND AID and LIVE AID-Financed Projects in Ethiopia* (Douala, 1991); G. Conn, 'Forest Resource Base, Identification, Conservation, and Rational Use in Ethiopia', Paper presented to the Rotary Club (Addis Ababa, 1990); 'Reforestation in Ethiopia', GTZ Forestry Project (Addis Ababa, 1992); Institute of Development Studies, Sussex University and Institute of Development Research, Addis Ababa University, 'Evaluation of EU/EC-Ethiopia Cooperation', Draft Main Report (Brighton, 1995); SIDA, *SIDA Support to Welo Region*, SIDA Welo Mission (Addis Ababa, 1986); WFP, *World Food Programme and Food Aid*.

It is worth looking briefly at a few of the main conservation measures undertaken by the government and the responses of peasants to them; the discussion does not include conservation work by NGOs and others. We shall begin with afforestation. Tree-planting was the earliest conservation measure promoted by both the imperial and Derg regimes. However, there was (and still is) no clear policy on the ownership of the trees. Under both regimes, there was a strong tendency to undertake reforestation on land owned or claimed by the state. However, the attempt to encourage individual tree-planting was frustrated by the lack of tree policy as well as the uncertainties of the land tenure system. The Derg considered that a system of state forestry, on the one hand, and community forestry on the other would address some of the difficulties, but in the end this aggravated the problem instead of ameliorating it. State forestry under the Derg posed a threat to peasant livelihoods: it encroached on farm land, evicted households living on or near it, and took away land that was customarily used for grazing. Many of the forests in question were enlarged by expropriating farm land and pasture.[28] The resources of these forests were closed to the surrounding peasantry who were not allowed to graze their animals in them, nor cut grass or wood; forestry personnel provided no material or technical support to peasants who wished to establish their own woodlots.

Community forests were woodlots owned or planted by members of the community on land belonging to a Peasants' Association, village, co-operative, etc.; the purpose was to provide forest-based products for the benefit of the members

[28] Ministry of Agriculture, *The Last Forests of Ethiopia* (Addis Ababa: Forestry and Wildlife Conservation and Development Department, 1989).

involved. Community forests were also envisaged as providing environmental benefits: they were expected to reduce erosion, rehabilitate degraded land and support agricultural production.[29] These schemes were relatively small in size, not more than 80 ha., and were intended to complement state forestry; they included eroded hillsides which were planted by the local population through FFW programs. Occasionally, community woodlots were established on pasture land and land of cultural significance to the communities concerned, and were resented for these reasons by the peasants. The local peasantry had no say in any of the decisions affecting the planning and management of the forests; government agents decided what species to plant, where to plant them and how the schemes were to be managed. Many peasants were not convinced of the benefits of community forests, and some suspected that the schemes were a means of promoting collectivization. Indeed, the peasants carried on covert resistance against the schemes: tree seedlings planted had a low survival rate because the peasants deliberately planted them badly, and there was increased livestock grazing on newly planted areas. In addition, poor site selection and inadequate management of woodlots contributed to the low survival rate which in many parts of Wällo was between 33 and 48 per cent;[30] due in part to frequent droughts, survival rates of below 20 per cent were not uncommon.[31] Perhaps the thorniest issues, which eventually led to the failure of community forestry, were the ownership and utilization of the forests as well as the land on which they were established. While in principle the forests belonged to and were intended to benefit the people in the community, in practice the community had no control over the resources and not infrequently it was the government which felled the mature trees to sell elsewhere. Once the peasants realized that the forest assets were not really theirs, they began to harvest the trees illicitly, often under cover of darkness.

The Derg's conservation program laid heavy emphasis on the construction of bunds and terraces, which were believed to be effective measures against soil erosion. These structures were built on denuded hillsides as well as on cropland. It was strongly argued that the highest rate of erosion occurred on peasant farm-land, and that the correct physical measures would not only control erosion but also lead to improvements in crop yield for the land user. Spearheaded by the Soil Conservation Research Project (SCRP), the research arm and technical advisor to the MoA, the drive to construct bunds, terraces and *fanya-juu* continued through-out the 1980s and cost an inordinate amount of aid resources.[32] They were uniformly constructed, with stone or soil, at vertical intervals of one meter, and were recommended for all agro-ecological zones and farming systems. The decision

[29] H. Sjoholm, *Guidelines for Development Agents on Community Forestry in Ethiopia* (Addis Ababa: CFSCDD, 1989).

[30] Ministry of Agriculture, 'Assessment Report of Field Activities of CFSCDD' [in Amharic] (Addis Ababa, 1989).

[31] M. Bendz and P. A. Molin, *Trees Grow in Wällo, Mission Report* (Vaxjo, 1988); Shawel Consult, *Evaluation of the Community Forestry Program in Ethiopia, Vol. I Main Report* (Addis Ababa, 1989).

[32] See Hans Hurni, *Guidelines for Development Agents on Soil Conservation in Ethiopia* (Addis Ababa, 1986) for the variety of physical measures recommended for use. *Fanya-juu* is a Kenyan term referring to a ditch with the dug soil piled on the upper side of the slope to form a barrier against runoff.

Table 9.2 Estimates of conservation work, 1976–90 ('000ha)

Measures	Berhe's Estimate	Sewell's Estimate	CFSCDD Estimate
On-farm bunds	771	1000	445
Hillside terraces	233	280	175
Area closure	390	375	135
Afforestation	448	150	181[a]

Note: a) Includes state forests, peri-urban fuelwood plantations and CFSCDD projects.
Source: Berhe, 'FFW and the Community' (Berhe's later work gives different figures); Sewell, *Technical Report*; CFSCDD, 'Some Facts on Soil and Water Conservation Program in Ethiopia', in *Proceedings of the 20th National Crop Management Conference, 28–30 May 1988*, Institute of Agricultural Research (Addis Ababa, 1989).

to adopt a uniform technical design was justified on the grounds that the peasant could not follow complex directions and that conservation technology had to be simple and easy to implement. Over a period of a decade, large areas of the country were treated with physical structures both on-farm and on hillsides. Owing to inadequate planning and management, and also the food aid factor, far too many bunds and terraces were built. Not infrequently, bunds were built on farmland without the consent, participation or even knowledge of the owners of the land. Moreover, many of the structures were poorly built and soon became an erosion hazard. Maintenance work received very little attention, confirming the view that, for MoA agents in the field, conservation was a once-only exercise. Many peasants were unhappy with physical conservation, but they participated in the program for the employment opportunities it provided. On many occasions, peasants destroyed the bunds and terraces that they had been paid to build in order to be paid again to rebuild them.[33]

In the early 1990s, both WFP and MoA officials declared that the environmental program undertaken in the previous 12 to 15 years, i.e. Project 2488 and its predecessors, was successful and had been of great benefit to the peasantry.[34] It was claimed that 1.8 million hectares of land had received one form of treatment or another, and that this amounted to 1.5 per cent of the total land mass of the country, or about 7 per cent of the highlands.[35] But these were figures provided by the MoA and should be treated with caution. Estimates of some of the work accomplished in the country as a whole are given in Table 9.2. The discrepancies in the figures shown in Table 9.2 indicate that there is no reliable statistical evidence of the magnitude of the results achieved.

Of the 1 million mt of WFP-supplied food grain used for environmental rehabilitation in the country since the 1970s, nearly one-third was spent in Wällo; however, the record of achievement here is far from outstanding. MoA archives indicate that by the end of the 1980s some 30,000 ha of land was reforested, and 40,000 ha of bunds and about 45,000 ha of terraces were constructed on farm plots and hillsides respectively. The records show that 10 million tree seedlings a year were produced

[33] Kebede, Interview.
[34] WFP, *WFP Food for Work Workshop*, Annex 3 (Addis Ababa, 1992).
[35] Berhe Wolde-Aregay, 'FFW and the Community Forests and Soil Conservation Program in Ethiopia, Abstract', in WFP, *WFP Food for Work Workshop*; Adly, 'New Approaches'.

in MoA nurseries in the province; the survival rate of planted seedlings was atrociously low, however, and the greater part of the nurseries' product wasted.

The Failure of State Environmentalism

Most of the assets created by the WFP/MoA conservation program were demolished during the disturbances following the collapse of the Derg, or have fallen into disuse. In the end, the result was a colossal wastage of resources and a stinging rebuff to state environmentalism. The story of the destruction of the environmental assets created under the Derg is too complicated to recount here, and a few words on the subject will have to suffice.[36] The collapse of the Derg in 1991 unleashed an outburst of violent activity in the countryside, which included, among other things, an attack on conservation measures. In Wällo, the assault on state forestry was triggered by the Derg army, which was deployed there in large numbers. Soldiers indiscriminately cut down trees, without permission from the authorities, to use as fuelwood in their camps or to sell to urban consumers. Deforestation by peasants followed, and by the end the damage was considerable. MoA records indicate, for example, that up to two-thirds of the forests planted in North Wällo under Project 2488 may have been 'harvested' illegally in 1990 and 1991; in South Wällo, an inventory carried out by the MoA in the latter part of 1991 showed that the area's forest cover had been reduced by 62 per cent. Peasants also demolished on-farm bunds as well as hillside terraces; areas enclosed for regeneration were invaded and grazed. At a workshop organized by the WFP in 1992, MoA officials declared that 65 per cent of the conservation measures in the country still remained intact;[37] however, while it is impossible to determine accurately, the true figure was probably in the neighborhood of 35–40 per cent. As I have argued in the work noted above, peasants' involvement in the destruction of environmental assets must be seen as an act of protest against a conservation program which was imposed on them, and which ignored their needs and flouted their experience and practical knowledge.[38]

The conservation strategy of the Derg was flawed from the outset. The catchment approach adopted since the early 1980s, in which the focus of conservation activity was the catchment and not the household farm, placed the peasant in an insignificant position. Under this approach, the individual farm had to submit to the larger interest of the catchment, and what was good for the catchment was believed to be good for the peasant's plot. Moreover, catchment conservation had a limited range of options, and was mainly concerned with physical structures and reforestation. Secondly, food aid, and the MoA's drive to expand FFW opportunities, became in the end self-defeating. Very few durable assets were created with FFW, since the peasants were more interested in the food than in the benefit of the works constructed. Besides, FFW discouraged proper and timely maintenance of conservation measures. While FFW extended the MoA's leverage and outreach in the countryside, the Ministry became dependent on food aid, without which it was

---

[36] Dessalegn, 'Unquiet Countryside'.
[37] WFP, *WFP FFW Workshop*.
[38] Dessalaegn, 'Unquiet Countryside'.

unable to undertake many of its other programs. Moreover, the MoA was deaf to alternative conservation approaches (such as biological or agronomic approaches, for example) because they required less labor and hence less food aid. The choice of conservation technology was based not on its effectiveness but rather on how many employment opportunities it would open up.

Both the WFP and the MoA were frequently criticized by expert opinion for their heavy reliance on physical conservation. But it was peasant criticism which highlighted the deleterious effects of the new conservation technology. The peasants rejected the narrowly spaced bunds imposed on them, on the practical grounds that they made plowing difficult and time-consuming. Moreover, these structures took away valuable farm land – estimates range between 10 and 15 per cent – and their benefits, in terms of increased production, failed to compensate the land user. Level structures, in particular *fanya-juu*, often caused waterlogging in high rainfall areas, while graded ones were frequently impractical because of lack of suitable water courses to channel excess runoff. The peasants were unhappy about stone bunds because they said they harbored rodents that destroy crops. They complained about area closures because these involved the loss of access to grazing land; closures were not accompanied by alternative sources of pasture. Underlying these criticisms was the view that the traditional conservation technology was more effective and less costly.

In May 1995 we conducted several interviews with peasants in central Wällo on the subject of land degradation and environmental protection. All our informants were keenly aware of the problem of soil erosion and the environmental impact of forestry and deforestation. Several of the elders recounted the story of recurrent deforestation in their localities in the previous six decades in response to ill-advised state policies, following political unrest or in times of civil war; they noted that the degradation of resources was aggravated as a consequence. Many highlighted the differences between traditional conservation practices and those introduced by the Derg, and cautiously suggested that the former were better and easier to employ. They stressed that traditional practices employed physical and vegetative/biological measures in a dynamic combination. Common resources, such as pasture land and forests, the peasants pointed out, were managed in the past by informal community institutions which regulated, and resolved conflicts arising from competition over their use, and protected them against encroachment by outsiders.

Peasants have a more dynamic view of environmental protection and attempt to adapt conservation to their immediate agricultural needs and to changes in the environment. They prefer temporary physical structures to permanent ones: a bund or grass *weber* is 'moved' periodically from one part of the field to another in response to shifts in soil movements and changes in the landscape and in cropping needs. Conservation measures are incorporated into the local farming system and are not separate from it; good environmental management is part of good farming practice. Unless the fields are located on rocky slopes, soil and plants are used to construct physical structures, as these are easier to employ and to remove when the need arises. All this contrasts sharply with state conservationism with its heavy emphasis on physical engineering, durable structures, and the separation of conservation and farming practices. The government program paid scant attention to local farming systems or the economic needs of farmers.

The effectiveness of peasant practice *in the past* has been attested to by the few travelers who made it their business to observe and comment upon local farming techniques. A. B. Wylde, who traveled through the northeast of the country and carefully observed the economy and ecology of the area in 1896, is full of praise for the farming and land management skills of the local peasants. He cites terracing, contour bunding, drainage ditches and crop rotation as some of the main techniques of land management. He was impressed by the extensive use of irrigation, and the skill with which water was managed to grow a wide variety of crops. While he saw extensive areas of the northeast which were treeless, he observed or passed through a good number of large forests in both north and south Wållo. A conscientious observer, he makes no mention of land degradation, which suggests that it was not a significant problem at the time. Major F. Joyce, an agricultural expert attached to the British expeditionary force in Ethiopia, also traveled extensively in the country, including the northeast, in the latter part of 1941. The Ethiopian farmer, he notes, 'knows his job and does it extremely well'. Although a good deal of the northeast he traveled through was treeless, he observed no erosion *on cultivated land*, and he attributed this to the wide variety of indigenous conservation measures peasants employed. He describes bench terracing, contour bunding covered with grass or other vegetation, contour hedges, strip cultivation, bench cropping, diagonal drainage ditches, and contour plowing as the main techniques used by farmers in the northeast and elsewhere in the country to control erosion. He compares peasant crop fields with those of large-scale farms managed by Italian concessionnaires and states that erosion and gullying were common on the latter and largely absent from the former.[39]

It was only in the last years of the Derg that senior government officials began to be seriously concerned about the deficiencies of the conservation program and the damaging effects of the FFW approach. Conservationists within the Ministry of Agriculture now started to re-examine the work that had been undertaken for nearly a decade and found much to criticize. The work of CFSCDD, which was responsible for the management of Project 2488, was said to have been carried out poorly, without proper planning and in an unco-ordinated manner. It was realized that FFW had undermined the objectives of environmental rehabilitation: the peasants, it was noted, were more concerned with the food offered than with creating durable environmental assets.[40] At the end of the 1980s, a new conservation strategy was drawn up, once again at the initiative of Western donors and with the involvement of expatriate advisors, and without sufficiently examining the experiences of the past. The new document advocated the 'sustainable utilization of natural resources', and made a plea for 'participatory approaches'; however, it was never implemented as the Derg regime collapsed shortly afterwards.[41]

---

[39] A. B. Wylde, *Modern Abyssinia* (Westport, CT, reprinted 1970); F. de V. Joyce, 'Notes on Agriculture in Ethiopia', *East African Agricultural Journal*, VIII, 3 & 4, IX, 1 (1943).

[40] Ministry of Agriculture, Assessment Report, *Proceedings of the First Departmental Workshop of the Community Forests and Soil Conservation Development Department held at Awassa 5–12 October 1987* (Addis Ababa, 1989).

[41] Office of the National Committee for Central Planning, *Towards a National Conservation Strategy for Ethiopia. Report of the Proceedings of the Conference held in Addis Ababa 22–25 May 1990* (Addis Ababa, 1991); IUCN, *Ethiopia*.

This chapter started out with a review of the environmental 'narrative' of the expatriates and its influence on government policy. I shall conclude the discussion with a brief look at the Soil Conservation Research Project, another expatriate voice which was closely connected with the conservation program of the Derg. The SCRP, jointly funded by the Ethiopian government and the Swiss development agency, was established in 1981 at the initiative of Hans Hurni who became its manager for its first five years and continued to exercise influence long after that. The objective of the Project was to undertake research on soil erosion, and to monitor and evaluate the effectiveness of the conservation measures implemented by the MoA. Its research consisted in the main of data on soil loss collected from erosion plots located in several parts of the highlands; Hurni and his assistants believed the data were accurate and reliable because they were said to be based on scientific research.[42] The SCRP did not itself implement conservation measures, but its influence on MoA officials regarding the choice of conservation technology was considerable. From the very beginning it was exclusively concerned with physical structures, and, despite the criticism of expert opinion and even the MoA itself,[43] it did not consider other conservation options until the early 1990s, that is, until the collapse of the Derg and its conservation program.[44]

Behind the SCRP's 'scientific' research and recommendations were a set of assumptions about the peasant and peasant attitudes. First, Hurni and his team regarded the peasant as either a passive object of conservation policy, or altogether irrelevant. What was important for them was the technical soundness of the chosen conservation technology and its proper implementation. Karl Herweg, a student of Hurni, and co-manager of the SCRP in the late 1980s, observed at a briefing in 1993, at which this writer was present, that the SCRP began to respond to peasant reaction only with the fall of the Derg, and that before then 'the farmer as such did not exist for us'. Secondly, it was assumed, and frequently expressed in writing, that the peasant neither understood the dynamics of land degradation nor was receptive to conservation. The uniform technical guidelines adopted in the early 1980s noted above, a recommendation of the SCRP, were justified on the grounds that the peasant was too ignorant to grasp anything other than simple guidelines. In a book edited by Kebede Tato and Hans Hurni, the two editors give the following reasons why they are not enthusiastic about the participatory approach to environmental rehabilitation: 'In many circumstances real participation by farmers is not feasible because of their obvious lack of knowledge about the processes of degradation and about the means used by outsiders to intervene positively'.[45] Similar sentiments are expressed in several of SCRPs progress reports.[46] Thirdly, it was believed that, since land rehabilitation would immediately result in improvements in crop yield, it was not necessary to obtain the consent of the peasant, as he would see for himself the benefits of conservation soon enough. However, research work carried out by the

---

[42] See SCRP, *Progress Report* for the years 1982 to 1989 (1982–91).

[43] Ministry of Agriculture, *Proceedings*.

[44] The SCRP was closed down a few years ago and its work was assigned to the regional agricultural bureaux. SCRP issued a series of *Progress Reports* from 1982 to 1991; Hurni was the main author of these *Reports* for the years 1982–6.

[45] Kebede Tato and Hans Hurni (eds), *Soil Conservation for Survival* (Ankeny, 1992), pp. 6–7.

[46] SCRP, *Third Progress Report* (Year 1983), 4 (1984), pp. 48-54.

SCRP itself at the end of the 1980s revealed that bunds and similar physical measures did not contribute to improvement in yield, thus confirming what the peasant had suspected all along.[47]

One of the SCRP's important contributions was said to be the collection of soil loss and runoff data from test plots established in various parts of the highlands. These data were used to provide information about the rate of soil loss on farms and the extent of land degradation in the highlands. Hurni's claim that farm lands were losing as much as 300 tons of soil per hectare per year, and his prediction, noted earlier, that the country's top soil would be washed away in 100 to 150 years were based on data collected from these test plots. However, expert opinion has shown that the SCRP's research is methodologically flawed, and its findings are unreliable. Sutcliffe has argued that the Ethiopian Highlands Reclamation Studies' and Hurni's annual calculations of soil loss are based on questionable assumptions and their figures are higher by many orders of magnitude.[48] The distinguished tropical soil scientist, Michael Stocking, has demonstrated in a recent work that erosion rates derived from experimental test plots suffer from scale effects, are liable to gross errors and therefore of 'dubious validity'. He shows that the attempt to extrapolate from small experimental erosion plots to catchment, regional and national scale, which is what SCRP research had been doing for nearly a decade, may exaggerate 'actual sediment loss by a factor of 100 or more'. Real slopes, he argues, 'have a balance of erosion and deposition throughout the length. Consequently, on a per hectare basis, small [test] plots suffer higher soil loss than large plots'.[49] Erosion plots, he goes on to note, provide a classic case of experimental interference: the measurement itself intrudes on the process being measured, and hence the results from field plots are 'extremely unlikely to resemble those on a real field'. Moreover, it should be noted that not all the soil washed down from the Ethiopian slopes is lost forever; some of it is deposited in valley bottoms where most of the country's farmland is located. Thus what is a loss for the steep lands becomes a gain for the valley floor. The erosion plot only catches sediment that passes into the sludge container below, i.e. the sediment that is 'lost'; it does not 'catch' the sediment that is deposited elsewhere. We may thus safely conclude that the ecological calamity predicted by Hans Hurni as well as the soil loss data collected by SCRP were based on flawed research and unreliable evidence.

I believe Hurni and the SCRP should be held partly responsible for the disastrous record of the Derg's conservation program. Hurni had a mind-set similar to that of the colonial conservationist in Africa, who in the 1940s and 1950s imposed unpopular conservation measures on an unwilling African farmer without understanding the dynamics of his farm and livestock management

---

[47] Karl Herweg, 'Major Constraints to Effective Soil Conservation. Experiences in Ethiopia', Paper presented to the 7th International Soil Conservation Conference, Sydney, Australia (Addis Ababa, 1992).

[48] Sutcliffe, 'Economic Assessment'.

[49] Michael Stocking, 'Soil erosion: breaking new ground', in M. Leach and R. Mearns (eds), *The Lie of the Land. Challenging Received Wisdom on the African Environment* (Oxford: James Currey, 1996), p. 150.

practices.[50] Hurni's writings reveal that he had only superficial knowledge about Ethiopian agriculture and little regard for the laboring peasant. He marched through the countryside like a feudal warlord armed with *fanya-juus* and graded bunds with hardly any concern for the views and reactions of the farming population.

In conclusion, I submit that the conservation program of the Derg ultimately failed not simply because of its misguided policies and undemocratic ways but also because it was resisted, now passively, now actively, by the peasant population. From the start the environmentalism of the state and that of the peasant were on a collision course because each had different perceptions and different objectives. While all peasants eagerly participated in the FFW activities made possible by Project 2488, they had neither trust in the value nor confidence in the quality of many of the environmental assets they helped create. Most peasants rejected both the rhetoric and the technology of the conservation program in a variety of ways, the most dramatic of which was the violent anti-conservation agitation that greeted the fall of the Derg.

[50] Colonial conservation policy is similar in many respects to that of the Derg. In both instances, the state undermined smallholder agriculture by a series of anti-peasant land tenure legislation, and imposed unpopular conservation measures, which in most cases consisted of physical structures. Peasants supported the independence movement in eastern and southern Africa because of their opposition to colonial conservation. Just like in Ethiopia, African farmers demolished the conservation measures immediately following the end of colonialism. See Michael Stocking, 'Soil Conservation Policy in Colonial Africa', *Agricultural History*, 59, 2 (1985), pp. 148–61; Adrian Wood, 'Zambia's Soil Conservation Heritage: A Review of Policies and Attitudes towards Soil Conservation from Colonial Times to the Present', in Kebede and Hurni, *Soil Conservation*, pp. 156–71.

10

Social Differentiation, Farming Practices & Environmental Change in Mozambique

MERLE L. BOWEN, ARLINDO CHILUNDO
& CESAR A. TIQUE

Land issues are probably the most contentious topic in Mozambique since the Frelimo government and the Renamo opposition movement signed a peace accord in 1992, ending a 16-year-old war. The government's land reforms (for example. the selling of state farms and the granting of land concessions) combined with privatization – key aspects of Mozambique's transition to capitalism in the late 1980s – have incited intense competition for resources, especially land. Smallholders – a euphemism for family sector farmers or peasants – have contested land grabbing by government members, Frelimo party supporters, Renamo backers and foreign investors in the countryside and the peri–urban areas.

This chapter examines the struggles of small-scale farmers to gain access to resources, especially land, in Nampula province and how recent government policy changes have affected their local management systems.[1] Its thesis is straightforward. Since the mid-1980s, environmental degradation has taken place because of political and economic changes in Mozambique that have incited competition for land and other resources.[2] Privatization has led to struggles for land. Small-scale farmers have been squeezed by government policies that favor parastatal enterprises,

---

[1] *Small-scale farmers* refer to agricultural producers in both the family and the private sectors. Typically, the term *family* or *smallholder* sector refers to agricultural production that is primarily engaged in by family members, though additional manual labor may be used during peak periods, such as for weeding and harvesting. It is distinct from the private sector which uses hired manual labor or mechanization to engage in agricultural production. Although the terms are not entirely satisfactory, as Myers points out, and there is some overlap, they are still widely used in Mozambican policy and legislation: G. Myers, 'Competitive Rights, Competitive Claims: Land Access in Mozambique', *Journal of Southern African Studies*, 20, 4 (1994), pp. 603–33. In this chapter, we are concerned with small private farmers, who cultivate at least 10 hectares of cotton, use machinery, hire labor, and receive inputs from a concessionary or private company, as well as smallholders who cultivate less than 5 hectares and may or may not grow cotton.

[2] Environmental degradation is defined here as 'any change or disturbance to the environment perceived to be deleterious or undesirable', Donald Johnson *et al.*, 'Meanings of Environmental Terms', *Journal of Environmental Quality*, 26, 3 (1997), pp. 581–9.

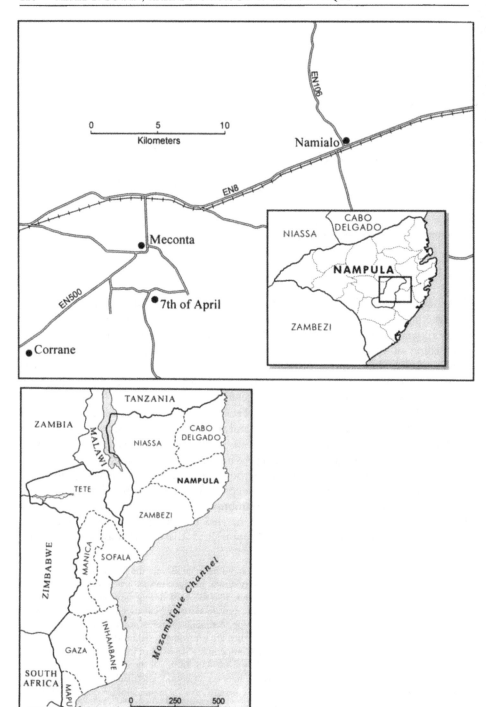

Figure 10.1 Map of administrative posts, Meconta District, Nampula

foreign multinational corporations and large-scale private farmers (domestic and foreign). Given that small farmers consist of socially differentiated producers with varying degrees of secure access to land and complementary inputs, their response to these changes has not been uniform. Small private cotton contract farmers who rented tractors and received other inputs from joint venture companies (JVCs) complained of soil compaction. In contrast, land-poor family farmers who did not cultivate cotton experienced chronic problems with pest resistance on their food crops. Unlike cotton contract producers, they did not have access to insecticides and the traditional method of intercropping was increasingly unsuccessful in combating pests.

This research complements earlier studies by examining small-scale production of cotton and food in northern Mozambique.[3] Through an analysis of production in Namialo, an administrative post in Nampula province, we assess the impact of post-colonial agricultural policies on farming practices.[4] We examine socio-economic and environmental changes that have arisen with government attempts to promote cotton production. To conduct the study, we administered an in-depth questionnaire to 33 households in rural Nampula province in 1995 and return interviews in 1996. We also held discussions with local officials, traditional authorities, and company representatives in Namialo, an area with a long history of cotton production.[5] Namialo is located within the zone of influence of the Sociedade de Desenvolvimento Algodoeiro de Namialo (SODAN), a revived cotton concessionary company that has benefited from Mozambique's privatization policies. Within its zone of influence, SODAN buys and processes raw cotton produced by family and small private farmers.

Though the area of Namialo is small, its present and past experience with cotton production is exemplary. Namialo was in the zone of influence of the former colonial company, the Companhia dos Algodões de Moçambique (CAM); the former state farm, the Empresa Estatal de Algodão de Nampula (EEAN); and now SODAN. All three companies have had significant influence throughout the province. Thus Namialo's experience with cash cropping from the colonial period up to the present can be taken as illustrative of company policy on production throughout Nampula. The chapter begins with an overview of Frelimo's policies from nationalization to privatization in Nampula province. It then examines small farmers' perceptions and management of environmental changes in cotton-producing regions.

[3] M. Anne Pitcher, 'Conflict and Cooperation: Gendered Roles and Responsibilities within Cotton Households in Northern Mozambique', *African Studies Review*, 39, 3 (1996), pp. 81–112. MoA/ Michigan State University Research Team, *A Socio-Economic Survey in the Province of Nampula: Agricultural Marketing in the Smallholder Sector*, Working Paper No. 4E (Maputo: Ministry of Agriculture, January 1992); *A Socio-Economic Survey in the Province of Nampula: Cotton in the Smallholder Economy*, Working Paper No. 5E (Maputo: MoA, January, 1992); *Evolution of the Rural Economy in Post-War Mozambique: Insights from a Rapid Appraisal in Monapo District of Nampula Province*. Working Paper No. 16 (Maputo: MoA, January 1994).

[4] Namialo, Meconta Center, 7th of April, and Corrane are the four administrative posts that form Meconta District. Namialo is located about 100 km from the city of Nampula, the provincial capital.

[5] Merle Bowen, Arlindo Chilundo, and César Tique conducted the interviews both individually and collectively over two years.

Frelimo Policies: From Nationalization to Privatization

Following independence in 1975, the Frelimo government created agro-industrial state farms to produce cotton on former cotton concessions in each one of the country's provinces. The Empresa Estatal de Algodão de Nampula (EEAN) was created in 1980 and took over the facilities formerly belonging to CAM, which had been nationalized in 1977. It consisted of 10 cotton gins, and related equipment and infrastructure. Besides producing cotton and sisal directly on its land, the state farm supervised, bought and ginned cotton produced by the family sector in Nampula province as the old cotton concessionary companies had done previously.

Farmers experienced major changes during the socialist period (1975–87) when agricultural production systems were changed to agro-industrial state farms and co-operatives, and local people were concentrated into communal villages. By the late 1970s, cotton production was in decline on family plots, state farms and collective fields. The EEAN was unable to increase or even maintain output in the family sector or to manage its own affairs. Even before the war began to disrupt production, cotton output was dropping. By 1977, family sector production had decreased to 32 tons, down from 49 tons in 1974 in Meconta District.[6] Output was negligible in many regions by the mid-1980s. The reasons were many. State institutions submitted conflicting plans on how to organize cotton, so there was confusion at every stage of the cultivation cycle. The state-run organizations responsible for supplying services to the family sector delivered cotton seeds, insecticides and sacks for harvesting late or not at all. The state officials responsible for buying cotton from producers offered low prices and ran badly organized markets that did not start on time. They regularly classified cotton as low-grade and often delayed payments to producers. Even when producers did receive cash for their cotton, by 1980 there were few consumer goods available for them to purchase.[7]

The government also attempted to foster co-operative cotton production on the abandoned settler farms. In Namialo smallholders who had become laborers on the settler farms took over the land of the departed owners, established a communal village called the 25th of September, and engaged in collective production of cotton in 1976. The Ministry of Agriculture donated two tractors to the communal village to be used in agricultural production, mainly cotton, as well as seeds and insecticides. Yet each year the members cultivated less land; the collective fields shrank from 125 hectares in 1976/77 to only 30 ha. in 1978/9.[8] The quantity of land cultivated by the members was extremely low: 0.23 ha per member. Some of the reasons to account for the poor results included lack of skilled labor, low wages, and production plans designed by the District Agricultural Directorate. For most members, co-operatives represented a new form of production where the economic results were low and insecure. They only worked the minimum amount of time, unprepared to divert their labor to a form of production that offered dubious results. Although there are no data on the extent of soil erosion, co-

---

[6] Centro de Estudos Africanos (CEA), *Cotton Production in Mozambique: A Survey, 1936-1979* (Maputo, 1980).

[7] *Ibid.*, pp. 43–4, 51–4.

[8] *Ibid.*, p. 73.

operative members recalled that soil fertility declined, which resulted in lower yields over time.[9] Three years after the collective fields were established, the members of this pilot communal village were heavily dependent on state subsidies.

Even with regard to mechanized production and processing in its own area, the state farm was both inefficient and expensive. A government report found that 42 per cent of the equipment at the EEAN was broken and that costs outstripped receipts nearly 5 times in 1986/7.[10] Part of the poor performance was attributed to the interruption of production and the destruction of equipment by the war. In addition, the progressive devaluation of the metical forced up the cost of primary materials. Finally, poor management practices and over-borrowing had led to debt, while the lack of skilled labor and the use of inappropriate technology had decreased productivity.

Compounding these economic difficulties were the political changes introduced by Frelimo. After independence, the government removed *régulos* as local political authorities, denouncing them as collaborators of the old colonial regime. Local and district administrators assumed many of their responsibilities, allocating land, co-ordinating production programs and supplying materials to farmers. *Grupos dinamizadores* (dynamizing groups or local party activists) and party officials were also involved in reorganizing production. However, they had neither the skills nor the experience to co-ordinate the annual cycle of cotton production. The abolition of *régulos* probably contributed to the sharp decreases in family sector production.

The second problem was the organization of communal villages. Following the adoption of Marxism-Leninism in 1977, Frelimo ordered rural dwellers to establish communal villages. According to the official propaganda, the objectives of communal villages were twofold: the concentration of the rural population to provide basic services (schools, stores, and health posts) and political facilities (party cells and state apparatus), while simultaneously laying the foundation for new forms of collective production. Villagization was also seen as the most efficient method for bringing rural producers under direct state control. Initially villagization was voluntary but, over time, forced villagization occurred in many areas in northern Mozambique, particularly in areas affected by the war. For example, the government forced all dispersed farmers to move to communal villages in Nampula and Zambezi, provinces hard-hit by the war in the mid-1980s. It is estimated that between 1981 and 1985 villagization increased in Nampula province by around 400 per cent.[11]

In Namialo, the village 25th of September was a pilot communal village and received assistance from both the provincial and district administrations. It had a primary school, a health post, a judicial post, a consumer cooperative, a bakery, a carpenter's shop, and other businesses.[12] Many people moved to the communal village because it provided social services and other incentives. By 1980, about 800 families lived in the village. Given that cotton was the dominant crop grown on the collective fields, the village leaders decided that each family should be allocated between 0.5 and 4 hectares to grow their own food crops. Soon after this

[9] Interview, João Muarapaz, 1995.
[10] Pitcher, 'Conflict and Cooperation', p. 91.
[11] Alex Vines, *Renamo: Terrorism in Mozambique* (London: James Currey, 1991).
[12] CEA, *Cotton Production in Mozambique*, p. 72.

regulation, land became scarce around the village. There was not sufficient land to grow food crops, collect firewood and hunt. Many people gradually abandoned the communal village, until finally, with the expansion of the war in Meconta District after 1984, the communal village system collapsed.

Though Namialo never experienced heavy fighting nor occupation by Renamo as did areas to its north, insurgents entered the town on at least two occasions and destroyed buildings.[13] Nevertheless, it was considered a relatively safe haven and *deslocados* (displaced people) from other parts of Meconta District and areas in the province resettled there. Although there are no data available to show the number of people who moved into Namialo, the administrator recalled that the population increased substantially from 1986 to 1992.[14] Resources became scarce, as people were unable to travel outside the main village. As the war cut off access to markets, store shelves emptied and residents had nothing to buy. Food shortages emerged in the area, forcing local people to farm outside the main village during the day under insecure conditions and to spend the night in Namialo.

Given these war-related problems, Namialo inhabitants made intensive use of natural resources in accessible areas. Farmers no longer left their field fallow or rotated their crops. Many farmers grew only drought-resistant crops – such as cassava, groundnuts and millet, for their subsistence needs. In many cases, however, the soil characteristics were not appropriate for the type of crops grown, which resulted in the loss of soil fertility and soil erosion. Manuel Celestino, a family farmer, described the situation:

> I opened up a new field in 1986 in the area under the *regulado* Mugila in Namialo. I farmed cassava, groundnuts and beans together. However, I got very low yields because the soil was not suitable. It was *N'tapo* (a sandy soil). I didn't have an alternative since there was no land available during that period mainly because of the war. All the areas accessible and close to the village had been taken. I had no choice but to plant in those soils that I had no previous experience of. (Interview 1995)

Facing continually declining volumes of cotton and cashew for export, the governor of Nampula threatened African farmers with sanctions. In 1986 Governor Gaspar Dzimba announced that producing cotton and cashew was not a favor but an order of the state.[15] All farmers in cotton-designated zones were to produce cotton. Moreover, those farmers located in communal villages and those displaced in the province were to be given 4 hectares to cultivate various crops (cotton, cashew, maize, cassava, groundnuts, beans, and millet) and 3 hectares for crop rotation. In addition to these orders, local residents were not allowed to travel to areas outside of their settlements during the crop season, without special authorization from the district authorities. Although cotton production increased in the short term, these measures – reminiscent of pre-1961 colonial policies – led to further popular dissatisfaction with the Frelimo government.

As a result of war, a declining economy and political difficulties, the Frelimo government was forced to change its development strategy. After 1987, as part of a

---

[13] Interview, Maria de Graça Tomo, 1996.
[14] Interview with administrator, 1995.
[15] Salomão Moyana, 'Produzir algodão e castanha de cajú não é favor é ordem do Estado', *Tempo*, No. 836 (1986), pp. 12–15.

World Bank-mandated structural adjustment program, the state began to sell off the state farms by forming huge joint venture companies with private capital. The EEAN was divided into two joint ventures, with private capital represented by Grupo Entreposto and João Ferreira dos Santos (JFS), respectively.

Namialo is located in the zone of influence of SODAN, a joint venture between the Portuguese private firm, João Ferreira dos Santos, and the Mozambican government for the production and marketing of cotton. It was founded in 1991 (though JFS had been running part of the state farm since 1986 and the rest since 1990).[16] According to the contract – which is good for 25 years and renewable for another 10 – SODAN's land concession totaled 20,000 hectares. The main office is in Namialo and its zone of influence includes the districts of Eráti, Namapa, Muecate, Meconta, and the administrative post of Netia in Nampula province, as well as Chiúre, Cabo Delgado province.[17]

The methods of production that SODAN used in the mid-1990s resembled the pre-1961 production schemes, prior to the abolition of the forced cotton cultivation system.[18] The company relied on low-risk, low-investment methods; limited mechanized production; the privileging of some (progressive) farmers through the use of a block scheme similar to the old *concentração* scheme; and the bulk of production by the family sector.[19] In 1995, it had around 1000 hectares devoted to direct, mechanized production of cotton as well as sisal and maize. Given the high costs of investment and labor, it did not consider expanding this form of production. As in the 1960s, SODAN targeted a small group of large family farmers and encouraged them to join a block scheme. They each cultivated between 3 and 20 hectares of cotton and benefited from company assistance with insecticides, labor and credit. In return, they sold their cotton to the company.

It was the family sector that was responsible for the bulk of production in the SODAN area. In 1996, around 80,000 smallholders were contracted with SODAN. The company provided credit in the form of seeds and insecticide applications – deducting these expenses from the farmer's total sale at the cotton market – in exchange for the right to purchase cotton from these families after the harvest. The company organized market days when it bought the cotton.[20] Typically families produced cotton on 0.5 to 1 hectare (on average), in addition to food crops for their own consumption. Couples performed most of the agricultural tasks on their cotton fields, using only hoes and machetes to clear the land and to plant. Mechanized clearing or harvesting in this sector was rare. Children and additional family members were called on at the peak labor periods of weeding and harvesting to work in return for food or drink.

---

[16] For detailed information on the formation and operation of joint ventures in cotton, including SODAN, see M. Anne Pitcher, 'Recreating Colonialism or Reconstructing the State: Privatisation and Politics in Mozambique', *Journal of Southern African Studies*, 22 (1996), pp. 49–74.

[17] *Ibid.*, p. 93.

[18] Allen Isaacman and Arlindo Chilundo, 'Peasants at work: forced-cotton cultivation in northern Mozambique, 1938-1961', in Allen Isaacman and Richard Roberts (eds), *Cotton, Colonialism and Social History in Sub-Saharan Africa* (Oxford: James Currey and Portsmouth, NH: Heinemann, 1995), pp. 147–79.

[19] *Ibid.*, pp. 94–5.

[20] Interviews, 1995, 1996.

In Nampula, the privatization of cotton production has also hastened the reintegration of *régulos* into the local political economy. At about the time the government began the privatization process, it also began to recognize the traditional authorities again.[21] To encourage and increase cotton production, SODAN (like other JVCs) has recruited *régulos*, giving them a prominent economic role in the countryside. Along with its own company agents, it has relied on *régulos* to supervise production, transmit instructions from the company to the family sector and distribute land – much as they did in the past. For example, during the agricultural season, SODAN officials in Meconta work through *régulos* to exhort people to weed their cotton and to apply insecticides. In return for performing these tasks, SODAN has provided *régulos* with salaries, bonuses, cars, bicycles, and pesticides for their own farms.[22]

The new concessionary policy in Nampula has had a paradoxical impact. On the one hand, it has led to increased cotton production. For example, SODAN's total production rose from 7,000 tons of seed cotton in 1990 to 11,543 tons in 1994.[23] On the other hand, it also has led to land shortages that have hastened environmental degradation. Mr Raimundo Abaina, the President of the Small and Medium Farmers' Association in the District of Meconta, described how the various JVCs had encircled the local community:

> Of course there is not sufficient land. Some of the problems happen because our area is like an island. We have SAMO, SODAN and SEMOC. All of these companies want to have land. That is why we are like an island. For this reason, I had to cut down the cashew trees from my grandparents to clear a new area to farm cotton. (Interview 1996)

In the mid-1990s, the main categories of farmers in Namialo were smallholders, small private farmers, and commercial farmers. Smallholders were the numerical majority and cultivated more land than the other two groups. Yet both commercial and small private farms had steadily increased their hectarage from 1992 to 1997, as illustrated in Table 10.1. With the privatization of cotton production, a notable change has been the growth in the number of small private farms in the area. SODAN has privileged small private farmers, providing them with bank credit, inputs, and extension services, just as the Mozambique Cotton Institute (IAM) did in the late 1960s. According to a report by the National Cotton Institute, 234 private farmers cultivated a land area that amounted, in total, to 8000 hectares.[24]

---

[21] Since the signing of the peace accord in 1992, government officials, company agents, and nongovernmental organizations have contacted and consulted with *régulos*, chiefs, and other traditional authorities to transmit information to local communities, encourage agricultural production, and set up training schemes. Chiefs have also mobilized the local population to repair streets and plant crops, allocated land, and settled conflicts within their communities. According to Pitcher, Frelimo began to recognize the traditional authorities again as part of an attempt to regain legitimacy among people alienated by the policies of the 1970s and 1980s. Pitcher, 'Recreating Colonialism', pp. 70–2.

[22] For example, in the neighboring district, Monapo, Pitcher found that the fields of the chiefs had clearly received more pesticides than those of ordinary people. Their cotton plants were taller and fuller and the bolls were larger: M. Anne Pitcher, 'Disruption Without Transformation: Agrarian Relations and Livelihoods in Nampula Province, Mozambique, 1975–1995', *Journal of Southern African Studies*, 24, 1 (1998), pp. 115–40.

[23] From 1986, when the state began privatizing, to 1994, total cotton exports increased by 50 per cent: Pitcher, 'Recreating Colonialism', p. 64.

[24] Instituto Nacional do Algodão (INA), *Estatística Annual da Produção do Algodão* (Maputo: 1997).

Table 10.1 Land-use changes in Namialo, Nampula Province, 1964–97

Land use type	1964 (ha)	1992 (ha)	1997 (ha)	% Change (1992–97)
Commercial farms	18,665	13,440	21,034	+64.0
Small private farms	–	750	8,340	+1,112.0
Smallholder farms	231,773	205,680	190,320	-9.0
Forest	111,647	125,070	118,130	-9.5
Residential	11,199	28,370	35,476	+12.5
Total	373,284	373,310	373,300	

Sources: DINAGECA topographic map, 1:50,000 (1964); aerial photographs 1:40,000 (1965 and 1992); and César Augusto Tique, 'Local Management Systems as a Response to Land Degradation in Cotton Producing Areas: The Case of Namialo, Northern Mozambique', unpublished paper presented at the Symposium, 'African Savannas: New Perspectives on Environmental and Social Change', University of Illinois, 1998, p. 22.

This figure represented more than an eleven-fold increase in the area cultivated by small private farmers between 1992 and 1997. In the same period, the area cultivated by smallholders decreased by about 9 per cent.

Land Use and Environmental Change

In Mozambique, land degradation is officially attributed to the management practices of small farmers in the family and private sectors. Although the Ministry of Agriculture and Fisheries (MAP)[25] considers factors such as topography, soil type, and frequency and intensity of rainfall as contributing to degradation, it argues that smallholder land management practices greatly exacerbate the problem.[26] For example, it is widely believed that soil erosion occurs mainly on family farms, which constitute approximately 85 per cent of all farmland. During land preparation family farmers clear the land leaving it completely open to rainfall, thus enhancing the detachment of the soil constituents and their transport and deposition by runoff. Yet government policies have shaped the ways local farmers deal with their land – an issue largely ignored at the ministerial level. Farmers have altered their production systems to respond to government policy and shortages of resources.

The information below is based on a survey of 33 small farmer households (family and private), unless otherwise stated. The objective was to gather data on farmers' perceptions of, and strategies to cope with, environmental change – including land degradation and conservation – over the last five decades. Of the 33 households, 17 were smallholder farmers consisting of 10 cotton growers and 7

[25] In January 2000 the ministry changed its name from the Ministry of Agriculture and Fisheries to the Ministry of Agriculture and Rural Development.

[26] S. Reddy and E. Mussage, 'The Rainfall Erosive Capacity Over Mozambique', *Série Terra e Água*, Comunicação no. 25 (Maputo: INIA, 1985); and Ministry of Agricultural and Fisheries, Proagri, *The National Agricultural Program: Land Component Investment Plan*, 1997.

non-cotton growers. Of the 16 smallholder private farmers, 8 held contracts with SODAN and 8 with Ibramugy, an Indian-owned private company.

In the mid-1990s, Namialo was experiencing land degradation, reflected by declining soil fertility, substantial soil erosion, increasing pests and plant diseases in both cotton and food crops as well as falling productivity. When asked what major changes they had observed over time, 30 per cent of the farmers replied that it was declining yields, while 17 per cent said that they had better yields in 1995 than in the recent past. Twenty per cent of the total agreed that there was a reduction in fallow periods, 10 per cent agreed that there was a decrease in the number of termite mounds, and 7 per cent agreed that there was an increase of pests in their fields. Furthermore, all the cotton farmers interviewed reported a greater decline in fertility in cotton areas as compared with areas growing food crops. In the food-crop fields, they reported that the fertility varied with the type of crop and how often it was cultivated.

Smallholder cotton growers

The majority of small cotton growers (family and private) said that they had seen the soil in their farms being washed away, particularly by water. According to these farmers, this process usually took place during the rainy season when their fields were being prepared for cotton. Approximately 67 per cent of the small private farmers described the gullies that appeared in their fields, mainly after heavy rains. Approximately 43 per cent of small farmers (family and private) said that the soils were washed away in their food-crop areas only when there was heavy rainfall. They considered cotton and groundnuts to be the main crops that caused erosion. Previous studies done in the area showed that erosion was prevalent in smallholders' fields, mainly those planted with cotton.[27] The Serno study found high levels of sheet and rill erosion in *Nipati* and *Kotokwa* soils that reduced the surface horizons by 30cm annually.

Small private farmers were cognizant of the effects of erosion in their fields. They reported that water was removing the nutrients from their soils, when small particles were washed downwards leaving the large particles (locally referred to as 'like sand') on the surface. One important factor that led to erosion was the reintroduction of cotton under monoculture practices. In general, cotton was planted in rows with spaces between crops of 20cm and between lines of 80cm, on average, oriented parallel to the sun. Another factor that led to erosion was the use of rented tractors, that arrived after the optimal period for preparing the fields. Given the shortage of private- and company-owned tractors, fields were plowed in mid-December, when the rains had already started. The exposure of plowed soils during the rainy season increased the threat of erosion. Mono-cropping and large spaces between crops and/or rows reduced the ground cover of the soil and, consequently, led to its exposure to raindrops. Sheet erosion was common during the rainy season.

---

[27] FAO/UNDP, *State Farm of Namialo, Province of Nampula, Mozambique: An Evaluation of the Seed Production Farm* (Maputo: Ministry of Agriculture and UNDP, 1981); and G. Serno, 'The Soils of Ribaue, Lalaua, Mecuburi, Muecate, Monapo, Nacaroa, Namapa and Memba Districts, Nampula Province', *Série Terra e Água*, Comunicação no. 77 (Maputo: INIA, 1995).

According to the small private farmers, the nutrient content in the soil surface was very low in their cotton fields after successive cultivation. They had to dig 30cm deep to reach the nutrients, which was difficult to do using hoes. However, deep plowing with machinery increased their production costs. José Pacheleque, a small private farmer, explained that for a tractor 'to plow 20-30cm deep, it took 20 liters of diesel per hectare while to plow only 15cm deep, it took 16 liters of diesel per hectare'.[28]

These farmers also complained about the soil compaction caused by tractors – an issue raised in an INIA study on soils.[29] It was widely known that soil compaction led to runoff, consequently reducing the ability of the soil to hold water after the rains. Furthermore, SODAN recommended that the soil should be broken only for the first 15cm to reduce diesel costs, leaving the soil underneath compacted. This practice reduced water storage capacity, which in turn shortened the growing season. A SODAN tractor driver, who worked on small private farms, described the correct procedure to avoid soil compaction. 'We initially cross the field with the three plow discs and whenever a tractor returns from one side of the bed to the other, its left wheel should pass over the channel created by the disc located on the right side.'[30] But to save diesel, he explained that most tractor drivers pass with the left wheel at the same place as the impression created by the right wheel (over the previous line). He confirmed that mechanical plowing on small farmers' fields often led to gully erosion.

In addition, the majority of small private farmers described how they repeatedly applied the same insecticides, in the same amount, on their cotton plants over the years, which has led to pest resistance. They identified the American bolloworm as the worst pest followed by spodoptera and red bolloworm. Mr Jacinto Massua, a technician at the INIA Experimental Station at Namialo, confirmed that some insects (for example, the aphids [*Aphids gossypii*] and the leaf lizard [*Spodoptera littoralis*]) showed evidence of resisting insecticides.[31] The former showed resistance to toxic organo-phosphate and the latter tended to resist pyrethroids. Nonetheless SODAN still promoted their application. Although the company recommended that cotton farmers apply insecticides five times during the season, many of them applied it only three times because of pest resistance, toxicity and the high price. A growing concern among farmers was that, with the extensive application of these insecticides for cotton, the insects would expand to food crops as well. Furthermore, many farmers reported that they suffered from mild symptoms of insecticide poisoning (itching of various body parts, vomiting, and headaches) due to insecticide application to cotton.[32] These farmers, like most cotton growers in Mozambique, did not wear special protective clothing when spraying insecticides

---

[28] Interview, 1995.

[29] Serno, 'The soils of Ribaue'.

[30] Instituto Nacional de Investigação Agronómica (INIA), *Estudo de Solos do Posto Adminstrativo de Namialo: Entrevistas* (Maputo: INIA, 1995).

[31] Interview, 1995.

[32] This finding substantiated an earlier study which reported farmers suffering from mild insecticide poisoning with the same symptoms: I. Javaid, R. Uaiene and J. Massua, *Insect Pest Survey on Family Sector Cotton Production System in Northern Mozambique* (Nampula: Center for Cotton Research and Seed Multiplication of Namialo, INIA, 1994).

Photo 10.1 Sheet and gully erosion in a small private farm on the outskirts of Namialo.

(Photographs 10.1-10.3 by César Tique)

Photo 10.2 Gully erosion on a non-cotton-producing family farmer's field.

Photo 10.3 Trenches leading to soil erosion and pools of stagnant water following heavy rains in a small private farmer's field prepared by a rented tractor.

Photo 10.4 Pest destruction on the bean crop of a non–cotton-producing family farmer without access to pesticides.

(Photographs 10.4–10.7 by Arlindo Chilundo)

Photo 10.5 Mulch used by small private farmers to provide vegetation cover and protect the soil against erosion.

Photo 10.6 Sugar-cane residue used as fertilizer in family gardens in Namialo.

Photo 10.7 Wood chips used by family farmers to fertilize the soil for vegetables.

Photo 10.8 Family farmers value the fertility of the soil around termite mounds for cultivating maize and other food crops.
(Photographs 10.8–10.10 by Merle Bowen)

Photo 10.9 Multi-cropping of cassava, groundnuts, maize and beans practised by family farmers to protect the soil and manage pests.

Photo 10.10 A rented tractor prepares the cotton fields of a small private farmer.

Table 10.2 Category of farmers and crops cultivated

	Category of farmers	Food crops (ha)	Cotton (ha)
Smallholder family farms	Cotton growers	0.5–2.5	0.5–5
	Non-cotton growers	0.5–2.0	–
Small private farms	Contract cotton growers		
	SODAN	1–3.0	10–20
	Ibramugy	1–3.5	>20

Source: Interviews 1995 and 1996

on their cotton crops. It was assumed that the local farmers absorbed insecticides not only through spraying but also through the drinking water, as most of the insecticide applied was concentrated in the soil and leached into the water courses during the rainy season. This assumption could not be substantiated, however, as analyses of drinking water were not available to show the concentration of these insecticides.

The responses of small farmers to questions about land degradation and their conservation strategies varied considerably, depending on their socio-economic conditions and the resources available. The individual interviews have been grouped into three main categories representing smallholder cotton growers, smallholder non-cotton growers, and small private cotton-growers. Each group of farmers and the size of their fields allotted to food crops and cotton are shown in Table 10.2.

The smallholder cotton growers each farmed at least 0.5 hectares and received seeds and insecticides from SODAN, but they did not qualify for bank credit. In return for these inputs, they sold their raw cotton directly to SODAN. The average area cultivated by these farmers was 2.4 ha. The average area planted per farmer with cotton was 0.80 ha., although there were wide variations among the farmers, ranging from 0.5 to 5 ha. Seasonal workers weeded and harvested the fields of farmers who cultivated between 3 and 5 ha. of cotton, in exchange for maize flour or local beer. In addition, these farmers grew cassava, groundnuts, beans, millet, maize and sorghum.

In 1995 many smallholder cotton growers indicated that in the following season they would switch from growing cotton to food crops for sale. Their main complaint was that SODAN offered a low producer price for cotton that did not cover the cost of production. Labor demands, the time required cultivating cotton, and the high prices of SODAN-supplied inputs like seeds and insecticides were some of the reasons given by smallholder farmers. They planned to grow ground-nuts or beans (*nyemba*), which were both edible and more profitable than cotton. The Director of SODAN confirmed that smallholders were increasingly dissatisfied with growing cotton. He estimated that the number of smallholders cultivating cotton during the 1994/95 season was 10–20 per cent lower than the previous season.[33]

[33] Interview, 1995. According to the Director, food shortages, increased food production (groundnuts and beans), the physical demands of cotton farming, and a general lack of interest accounted for this decline.

Table 10.3 Natural soil fertility matrix

Soils		Average years per group of respondents				Average years	%
		Nc	Sc	SpS	SpI		
Outhako	Cultivate	3	2	2	2	2.25	43
	Fallow	6	4	6	5	5.25	
Ikani (Outhako/*Kotokwa*)	Cultivate	3	3	3	3	3.0	71
	Fallow	4	4	5	4	4.25	
Nipati	Cultivate	5	4	4	4	4.25	100
	Fallow	5	4	4	4	4.25	
Holoku (Outhako/*Nipati*)	Cultivate	3	2	2	2	2.25	60
	Fallow	5	4	3	4	3.75	
M'hiro	Cultivate	4	3	3	3	3.25	81
	Fallow	5	4	4	4	4.25	
Kotokwa	Cultivate	4	4	4	4	4.00	94
	Fallow	5	4	5	4	4.50	
N'tapo	Cultivate	2	3	2	NR	2.33	44
	Fallow	5	5	6	NR	5.33	

Notes: Nc = Non-cotton smallholder farmers; Sc = Smallholder cotton farmers; SpS = Small private farmers financed by SODAN; SpI = Small private farmers financed by Ibramugy; NR = not ranked
Source: INIA, *Estudo de Solos* and interviews 1995 and 1996

For smallholder cotton growers, the most serious land degradation problems they faced were declining soil fertility and pest resistance on their cotton crops. To combat these encroaching problems, they used an intercropping system. Most smallholders knew that intercropping and timely sowing reduced losses from pests. In Meconta District, intercropping was commonly practised for food crops, although smallholder cotton growers intercropped cotton with millet and beans (*boer*). The most important food crops in the area – cassava, groundnuts, millet and beans (*nyemba*) – were intercropped with secondary crops, such as sorghum, maize, sesame, beans and rice. When cultivated using an intercropping system, cassava was combined with beans, sorghum and millet, especially in *Kotokwa* and *Nipati* soils. Generally, farmers who cultivated less than 3 or 4 ha. of cotton practised inter-cropping to provide ground cover against the direct raindrops on the exposed soil.

Smallholders had reduced their practice of shifting cultivation as a conservation method. The total area cultivated in the district had increased as a result of the influx of commercial farmers who had taken up large tracts of land. These farmers, combined with the growth of small private farmers, had increased pressure on land resources, making it difficult for smallholders to practice shifting cultivation.

With the disappearance of shifting cultivation and increased pressure on land resources in Namialo, smallholders reduced fallow periods, particularly on cotton farms. According to the local farmers, fallow periods were determined by the level of natural soil fertility, given the length of time a piece of land was under cultivation, and the type of crops grown. The decline in yields and the appearance of a certain type of shrub locally called *Mazope* (*Tridax procumbens*) and *murupane*

Table 10.4 The matrix for crop rotation variation in the Namialo area

N'tapo	Outhako	Ikani (Outhako/ Kotokwa)	Holoku (Outhako/ Nipati)	M'hiro	Nipati	Kotokwa	
Cassava Groundnuts	Cassava Groundnuts Beans nyemba	Cassava Groundnuts Beans fava	Groundnuts Cassava Beans fava	Cotton Millet Sorghum Beans boer	Cotton Millet Sorghum Beans boer	Cotton Millet Sorghum Beans boer	Year 1
	Millet Maize Cassava Beans jugo	Millet Maize Cassava Beans boer	Millet Maize Cassava Beans boer	Cotton Beans nyemba Beans boer Sesame	Cotton Beans nyemba Millet	Cotton Millet Beans nyemba	Year 2
		Beans nyemba Beans jugo Beans cute	Beans nyemba Beans jugo Beans cute Sesame	Maize Sorghum Sesame Beans boer	Maize Sorghum Sesame Beans boer	Maize Sorghum Sesame Beans boer	Year 3
				Groundnuts Millet Beans nyemba	Groundnuts Millet Beans nyemba	Groundnuts Millet Beans nyemba	Year 4
				Beans jugo Beans cute Beans boer Sesame	Beans jugo Beans cute Beans boer Sesame		Year 5

Source: INIA, *Estudos de Solos*, and interviews 1995 and 1996.

signaled that the land needed to rest.[34] These grass types appeared in low fertility areas. But given the increased competition for land, farmers cultivated land until yields began to drop and shortened the fallow periods. Table 10.3 shows the average number of years necessary for each type of soil to be cultivated and left in fallow in the cotton and non-cotton areas. Based on the ratio between the period in which a certain soil is cultivated and the time left for fallow, *Nipati* and *Kotokwa* are the soils that can be used for at least four years consecutively, without experiencing a significant reduction in yields. *M'hiro* is the soil that recovers fast in the area after fallow periods.

In addition, smallholder cotton farmers who cultivated less than 4 ha. of cotton practised crop rotation both as a soil conservation technique and a method of controlling pests and diseases. The most common rotation occurred in *Outhako*, *Holoku* and *Ikani* soils between cassava, groundnuts and beans (*nyemba*), the initial crops planted. The last two crops were harvested after one year, leaving cassava in the field for one more year, but now planted with millet and maize. In the third year, mainly in *Holoku* and *Ikani* soils, different types of beans were planted; then the land was left to fallow. For example, Sabonete Antonio, a smallholder, cultivated cassava and groundnuts for two years, then rotated with millet and maize.

[34] INIA, *Estudo de Solos*, p. 31.

Maize was planted first, and after it reached 50cm millet was planted. After these crops, the land was left to fallow.[35] Millet, maize and beans (*jugo*) were not only rotated with cassava and groundnuts but also with cotton, particularly in *Nipati* and *Kotokwa* soils. Cotton was planted with millet, sorghum and beans (*boer*) in the same year but with variation in the time of sowing. In the second year, cotton was planted with beans (*nyemba* and *boer*) and sesame in *M'hiro* and with beans (*nyemba*) and millet in *Nipati* and *Kotokwa* soils. In the third year, maize and sorghum were often grown and other small crops were interplanted. In the fourth year, groundnuts, millet and beans (*nyemba*) were usually the crops grown. In the fifth year, only different varieties of beans and sesame were planted. Afterwards the land went into fallow. Table 10.4 illustrates the matrix estimated for crop rotation in different soils used by smallholder farmers in Namialo.

According to the Namialo Agronomic Experimental Station, which has done on-farm experiments, the specific rotation of crops relative to the soil type can improve yields without the need of additional periods of fallow. Rotating crops with different root habits proved to be useful since the crops that made varied demands on the nutrients at different depths (for example, cassava, groundnuts and beans *fava*) were rotated and all of them gave higher yields than the continuous monocropping of each on the same plot. However, in the next crop season, all crops decreased apart from cassava, thus indicating a need for future fallow or other rotation practices.

Smallholder non-cotton growers

The second category, the smallholder non–cotton farmers, cultivated an area of 1.5 ha on average. Many of these individuals settled in Namialo during the war and worked on the cotton farms of small private farmers, commercial farmers, and JVCs in addition to their own plots. Some of these farmers grew groundnuts and beans as alternatives to cotton because of their price, labor demands and markets.[36] Many of them were poor and barely managed to grow enough food crops for family consumption. For example, Afonso Muacamisa, a non–cotton producer, who farmed 1.5 ha, explained why he had a poor harvest in 1996: 'I did not have insecticides. If I had that possibility, like my neighbor, I would have harvested a lot of beans too. But since I did not grow cotton, I could not have insecticides. People who grew cotton had plenty of beans because they had insecticides to fight the pests.'[37] In the Namialo area it was common for farmers to use the same insecticides on both cotton and food crops. Muacamisa did not grow cotton because he had insufficient land and was not well enough. He was too poor to buy insecticides and preferred to spend his money on basic necessities for the family. He complained that he had 'immense difficulties' earning money, like a growing number of the rural poor. Given that chemical inputs were made available through JVCs only for cotton producers, the smallholder non-cotton growers depended on intercropping to control pests. They also used intercropping and minimal tillage to minimize the loss of soil by water as well.

---

[35] Interview, 1995.
[36] The prices for groundnuts and beans were 2500 mt ($0.25) and 1500 mt ($0.18) per kg, respectively, in 1995.
[37] Interview, 1996.

Small private farmers

The third category, the small private farmers, consisted of two groups: those contracted by SODAN and those contracted by Ibramugy, an Indian trading firm.[38] SODAN provided small private farmers, who cultivated at least 10 ha of cotton, with inputs (seeds, insecticides, pesticides, and spray pumps), tractors to prepare their fields, and maize flour to pay their workers. This group benefited from the company's extension services and bank credits directly from SODAN. In return, the farmers had to sell their cotton to the company. In 1995, Ibramugy began to purchase small private farmer-produced cotton on concessions that were contracted out in monopolistic arrangements to SODAN and other companies. SODAN, like other cotton concessionary companies, faced severe limitations in serving all its clientele in its respective area of influence.[39] The company did not have the adequate infrastructure or the organization to service all cotton producers. However, to be eligible for the Ibramugy package (seeds, insecticides, and tractor rental), a farmer had to cultivate 20 ha of cotton.[40] Many farmers explained that they switched from SODAN to Ibramugy because the latter offered a higher producer price for cotton and provided the services on time.[41] Ibramugy introduced competition where no real market existed and in so doing, challenged the dominance of the agro-export companies. Initially, this type of intervention benefited the producers.[42] The private farmers also grew food crops – such as cassava, groundnuts and millet – both for market and family consumption.

While many small private farmers preferred contract farming with Ibramugy, they found it difficult to cultivate at least 20 ha of cotton. Indeed, during the interviews we discovered that some of these 'small private farmers' were in fact smallholder farmers who pooled their resources (for example, land) and signed up under the name of one individual in order to obtain the subsidized inputs. Equally important, they were able to gin the cotton before selling it, and thus received a better price for the product. This practice of smallholder cotton-producers combining to obtain access to resources and inputs had become widespread by 1996. While the strategy made sense for these farmers, it raised questions about the reliability of the National Cotton Institute data on the number of smallholder and small private farmers in Nampula province.

---

[38] Ibramugy was a company that made its money initially in the import-export business, selling second-hand clothes in rural and urban markets. In 1995 the company expanded its economic interests to cotton production, ginning and marketing.

[39] MoA/MSU, 'Socio-Economic Survey'.

[40] In 1995, these farmers occupied a total of 373 ha.

[41] In 1995, SODAN paid 1500 mt per kg and Ibramugy 2500 mt per kg for first-quality cotton: César Augusto Tique, 'Peasant Perceptions of Land Degradation and Conservation in Mozambique: The Case of Namialo, Nampula Province', unpublished M.A. Thesis, University of Illinois, Urbana, 1997, p. 130.

[42] By mid-1996 the relationship between many small farmers and Ibramugy had deteriorated. Ibramugy had changed the terms of credit from a combination of cash and goods (e.g., maize, maize flour and insecticides) to only credit in kind. The company also determined the goods that the farmers had to accept as credit (e.g., imported cloth from India that was not popular among Mozambican women). Interview with Mr Raimundo Abaina, President of the Association of Small and Medium Cotton Farmers, 1996.

Small private farmers who grew a minimum of 10 ha of cotton generally cultivated cotton in a monoculture system, which led to soil infertility. Under pressure to meet the required hectarage criterion set by the companies, they sometimes made local arrangements to leave part of their land fallow.[43] For example, two small private farmers might arrange that each would cultivate half of his/her land and apply for the inputs as one farm. Each would then be able to leave half of his land fallow for three consecutive years, without the knowledge of the company officials. After this period, the fallow fields were cultivated and the part of the farm that was under cultivation was then left in fallow. This practice of small private farmers teaming up to obtain access to resources and inputs had become widespread by the mid-1990s. The strategy made sense for small private farmers, given the limited labor and resources in the area for cultivating all 10 or 20 ha. Nevertheless, long fallow periods were not possible because the farmers had to comply with the requirements of the company to farm a minimum amount of land with cotton every year. This has resulted in a limited time for the land to regenerate its fertility through a natural process. Not only was the ratio of years of cultivation to years in fallow critical for the maintenance of soil fertility in the area, but also the sequence of crops and rotation was necessary to maximize yields.

Small private farmers also experience soil erosion on their farms. They used three methods to protect soils against water erosion: open drainage channels, grass strips, and mulches. The farmers opened drainage channels at the top of the farm and in the areas where they noticed major downward circulation of water. These channels, which vary from 10–30cm deep, were covered with grass, pieces of wood or piles of stone concentrated on the main course to limit the effects of water abrasion at the bottom of the channel. Grass strips were used with local vegetation called *murrequele* (*Pennisetum purpureum*) intercropped with cotton, which provided excellent soil protection against erosion. However, according to Armando Mavala, a small private farmer, they were abandoning its use because it grew rapidly and competed with cotton.[44]

Mulching was another technique used in the area between crops to provide vegetation cover between cotton lines to limit the effects of raindrops on the surface soil, while the mulch also decomposed and increased organic matter in the soil. Mulching also provided weed control and aeration, which helped to reduce moisture evaporation, and improved water absorption by the soil. In general, the operation broke the surface crust formed after rain. Local extension services also recommended that farmers use terraces. However, according to August Araújo, a small private farmer, they did not follow this advice because the extension services did not provide them with technical assistance.[45]

Conclusion

How have the Frelimo government's privatization policies affected rural households and farming practices in Nampula? What has been the impact of farmers'

---

[43] As noted earlier, small private farmers were required to cultivate at least 10 and 20 ha of cotton to be eligible for the credit inputs supplied by SODAN and Ibramugy, respectively.

[44] Interview, 1995.

[45] Interview, 1995.

changing land-use patterns on the environment? For whom is land degradation a problem? Privatization has exacerbated unequal access to land and other resources. Our research indicates a growing socio-economic disparity between non-cotton producers and cotton producers. Farmers who grew 0.5 ha of cotton under contract farming arrangements with concessionary companies benefited from the variety of services that these firms offered. Furthermore, there were strong variations among cotton producers, with advantages for the small private farmers in the district of Meconta. The joint venture companies, working through their agents – company overseers and *régulos* – provided different input packages and market networks for smallholders and small private farmers. In return for subsidized inputs and guaranteed markets, farmers changed their land-use patterns and practices to meet the company-required number of hectares under cotton cultivation and production targets. Consequently, environmental degradation has taken different forms on smallholder and small private farms. Moreover, given the structure of the local rural political economy, certain forms of environmental degradation were more serious for some categories of farmers than others.

In Namialo, the non-cotton growers, who cultivated between 0.5 and 2.0 hectares, were the most disadvantaged group.[46] They sold few agricultural products and relied on off-farm income. The majority depended on weak or non-existent markets for agricultural inputs (seeds, fertilizers, and insecticides) and farm implements. As pointed out by the Food Security Project (FSP), which conducted studies of family sector households in the 1990s, 'smallholder use of chemical inputs in Mozambique is limited almost exclusively to the cotton zone of Nampula and Cabo Delgado provinces'.[47] And most of these inputs were made available through JVCs only for cotton growers and especially for cotton itself. Soil infertility was a chronic problem for this category of resource-poor farmers.

The JVCs and their local rural agents have influenced small farmers' land-use patterns and farming practices. Smallholders with SODAN contracts had to cultivate a minimum of 0.5 ha in order to get access to inputs and a guaranteed market. To meet their cotton quotas, smallholders reduced their crop rotation practices. They also shortened fallow periods, partly because of fears of land-grabbing by state officials and private farmers. While mechanically prepared fields promised better cotton production results, companies did not have sufficient tractors to meet demand and they therefore privileged private farms. While company agents encouraged smallholders to apply insecticides on their cotton crop, poorer households penny-pinched on insecticides partly because of the cost. But there were other reasons why smallholders did not apply the optimal number of treatments to their cotton. Although the companies were supposed to provide insecticides, they were not always made available to smallholders because individual company agents used the association with the state to their own private advantage.

[46] A 1991 study of 343 producers in three districts of Nampula Province found that at least in two of the districts studied, both of which were located in the 'zones of influence' of two cotton joint ventures, those who grew 0.5 ha of cotton had a better standard of living and greater food security than those who did not grow cotton. Those who grew 0.5 ha or less of cotton were no worse off than non-cotton growers: MoA/MSU, 'Socio-Economic Survey', pp. 13 and 15.

[47] MoA/MSU Research Team, *Smallholder Cash Cropping, Food Cropping, and Food Security in Northern Mozambique: Research Methods.* Working Paper No. 22. (Maputo: Ministry of Agriculture, 1996).

The government-run Cotton Institute (IAM) provided the insecticides to companies to distribute amongst producers during the cotton campaign. However, Pitcher found, in her interviews with smallholders in the SODAN area, that 13 out of 33 had applied two or fewer treatments of insecticide when three or four treatments were recommended for cotton.[48] Some claimed not to have used insecticides because the agent who was supposed to have distributed them had sold them for a higher price to private farmers or demanded a chicken in return for them. Other producers maintained that inputs were delayed, jeopardizing their ability to control pests.[49] Thus corruption and favoritism to large producers partially accounted for low productivity in the smallholder sector.

In contrast, SODAN required small private farmers to cultivate a minimum of 10 ha of cotton in order to get access to resources (land, credit and inputs) and obtain a regular cash income (market and transport systems were geared largely to cotton production). Moreover, to be eligible for the Ibramugy input supply system, farmers had to have at least 20 ha under cotton cultivation. To meet their contract obligations, small private farmers hired tractors and cultivated cotton in a monoculture system. Company agents provided technical assistance for both their cotton and food crops, as well as insecticides for their cotton. However, these farmers reduced the fallow periods for their land and their shifting cultivation, as well as overcultivating their fields to meet the company production targets.

Cotton growers' farming practices have led to soil erosion, soil compaction, pest resistance and plant disease as well as soil infertility. Soil infertility, for example, was widespread among cotton growers, since they no longer left their fields to fallow. All cotton farmers complained of increased pest resistance to insecticides. With regard to small private farmers, the concessionary company supplied them with the same insecticides over a long period of time, which led to pest resistance building up. In contrast, for some smallholders, the price was too high. For other smallholders, the company provided the insecticides late and in insufficient quantities.

Yet other environmental changes were farmer group-specific. Soil erosion occurred largely on the small private cotton farms where cotton was farmed in a monocultural fashion, while it was less accentuated on the smallholder farms. Small private farmers observed soil erosion during the rainy season, as a result of inappropriate use of machinery and monoculture cultivation. Rented tractors typically arrived late to prepare the cotton fields. The exposure of plowed soils during the rainy season increased soil erosion. Furthermore, soil compaction was widespread on small private farms as a result of inappropriate plowing.

Small farmers were not entirely helpless in confronting environmental changes in the 1990s. They demonstrated their resilience and ingenuity by adapting different strategies to cope with their changing environment. These ranged from conservation techniques (such as mulching, minimum tillage and crop rotation) to sharing fields and combining production results in order to meet company requirements for services. Yet when the monopolistic and monopsonistic privileges of JVCs and private companies were combined with their strong connections with state power and political influence, the structural disadvantage of the producers stood out clearly.

---

[48] Pitcher, 'Recreating Colonialism', p. 66.
[49] MoA/MSU, 'A Socio-Economic Survey', p. 15.

Select Bibliography

Abel, N. and P. Blaikie, *Land Degradation, Stocking Rates and Conservation Policies in the Communal Rangelands of Botswana and Zimbabwe* (ODI Pastoral Development Network Paper 29a, May 1990).

Adams, M. E., 'Savanna environments' in W. A. Adams, A. S. Goudie and A. Orm (eds) *The Physical Geography of Africa* (Oxford: Oxford University Press, 1996).

Adams, W., *Green Development: Environment and Sustainability in the Third World* (London: Routledge, 1990).

Adjanohoun, E., *Végétation des savanes et des rochers découvertes en Côte d'Ivoire Centrale*, Mémoire ORSTOM 7 (Paris: ORSTOM, 1964).

Alemneh Dejene, *Environment, Famine and Politics in Ethiopia* (Boulder, CO, 1990).

Alemneh Dejene, *Peasants, Agrarian Socialism, and Rural Development in Ethiopia* (Boulder, CO: Westview Press, 1987).

Alexander, J., 'State, Peasantry and Resettlement in Zimbabwe', *Review of African Political Economy*, 21 (1994), pp. 325–45.

Allan, W. E., *The African Husbandman* (New York: Barnes & Noble Inc., 1965).

Allan, W.E., *Studies in African Land Usage in Northern Rhodesia*. Rhodes-Livingstone Papers no.15. (Cape Town: Oxford University Press, 1949).

Ambrose, S. H., 'Late Pleistocene human population bottlenecks, volcanic winter, and differentiation of modern humans', *Journal of Human Evolution*, 34 (1998), pp. 623–51.

Anderson, D. and R. Grove (eds), *Conservation in Africa: peoples, policies and practices* (Cambridge: Cambridge University Press, 1987).

Anderson, D., 'Cultivating pastoralists: ecology and economy among the Il Chamus of Baringo: 1840–1980', in Anderson and Johnson, *Ecology of Survival*, pp. 241–60.

Anderson, D., 'Depression, dust bowl, demography and drought: The colonial state and soil conservation in East Africa during the 1930s', *African Affairs*, 83 (1984), pp. 321–45.

Anderson, D. and R. Grove, 'Introduction: The scramble for Eden: past, present and future in African conservation', in Anderson and Grove, *Conservation in Africa*, pp. 1–12.

Anderson, D. and D. Johnson, *The Ecology of Survival. Case Studies from Northeast African History* (London/Boulder, CO: Lester Crook Academic Publishing and Westview Press, 1988).

Atwood, D., 'Land registration in Africa: the impact on agricultural production', *World Development*, 18, 5 (1990), pp. 659–71.

Aubréville, A., *Climats, forêts et désertification de l'Afrique tropicale* (Paris: Société d'Edition de Géographie Maritime et Coloniale, 1949).

Aubréville, A., *La Forêt Coloniale: Les forêts de l'Afrique occidentale française*, Annales de l'Académie des Sciences Coloniales, IX (Paris: Société d'Éditions Géographiques, Maritimes, et Coloniales, 1938).

Avenard, J.-M., 'La savane, conditions et mécanismes de la dégradation des paysages', in Jean-François Richard (ed.), *La Dégradation des Paysages en Afrique de l'Ouest* (Dakar: Presses Universitaires de Dakar, 1990), pp. 55–76.

Bahru Zewde, 'Forests and Forest Management in Wällo in Historical Perspective', *Journal of Ethiopian Studies*, XXXI, 1 (1998), pp. 87–121.

Baker, V. R., J. M. Bowler, Y. Enzel, and N. Lancaster, 'Late Quaternary palaeohydrology of arid and semi-arid regions', in K. J. Gregory, L. Starkel and V. R. Baker (eds), *Global Continental Palaeohydrology* (Chichester: Wiley, 1995), pp. 203–31.

Barnes, D. L. 'Problems and Prospects of Increased Pastoral Production in the Tribal Trust Lands', *Zambezia*, 6, 1 (1978), pp. 49–59.

Barnes, T. and J. Duncan (eds) *Writing Worlds: Discourse, Text and Metaphor in the Representation of Landscape* (London: Routledge, 1992).

Bartels, G., G. Perrier and B. Norton, 'The Applicability of the Carrying Capacity Concept in Africa: A Comment on the Thesis of de Leeuw and Tothill', in R. Cincotta, C. Gay and G. Perrier (eds), *New Concepts in International Rangeland Development: Theories and Applications* (Logan, UT: Department of Range Science, Utah State University, 1991), pp. 25–32.

Bassett, T. J. and D. Crummey (eds), *Land in African Agrarian Systems* (Madison: University of Wisconsin Press, 1993).

Bassett, T. J. and Z. Koli Bi, 'Environmental discourses and the Ivorian savanna', *Annals of the Association of American Geographers*, 90, 1 (2000), pp. 67–95.

Bassett, T. J. and Z. Koli Bi, 'Fulbe pastoralism and environmental change in northern Côte d'Ivoire', in M. DeBruijn and H. van Dijk (eds), *Pastoralism under Pressure? Fulbe societies confronting change in West Africa* (Amsterdam: Brill, 1999), pp. 139–59.

Bassett, T. J., 'The Uncaptured Corvée: Cotton in Côte d'Ivoire, 1912–1946', in A. Isaacman and R. Roberts (eds), *Cotton, Colonialism and Social History in Sub-Saharan Africa* (Portsmouth, NH and London: Heinemann and James Currey, 1995), pp. 247–67.

Bassett, T. J., 'Fulani Herd Movements', *The Geographical Review*, 76, 3 (1986), pp. 233–48.

Bassett, T. J., 'Hired Herders and Herd Management in Fulani Pastoralism (Northern Côte d'Ivoire)', *Cahiers d'Études Africaines*, Nos. 133–135 (1994), pp. 147–73.

Bassett, T. J., 'Introduction: the Land Question and Agricultural Transformation in Africa', in Bassett and Crummey (eds), *Land in African Agrarian Systems* (Madison: University of Wisconsin Press, 1993), pp. 3–31.

Bassett, T. J., 'Land Use Conflicts in Pastoral Development in Northern Côte d'Ivoire', in Bassett and Crummey (eds), *Land in African Agrarian Systems*, pp. 131–56.

Bassett, T. J., 'Mapping the Terrain of Tenure Reform: The Rural Land Holdings Project of Côte d'Ivoire', in J. Stone (ed.), *Maps and Africa* (Aberdeen: Aberdeen University African Studies Group, 1994), pp. 128–46.

Bassett, T. J., 'The Political Ecology of Peasant-Herder Conflicts in Northern Ivory Coast', *Annals of the Association of American Geographers*, 78, 3 (1988), pp. 453–72.

Batterbury, S., T. Forsyth and K. Thomson, 'Environmental Transformations in Developing Countries: Hybrid Research and Democratic Policy', *The Geographical Journal*, CLXIII (1997), pp. 126–32.

Bégué, L., 'Contribution à l'étude de la végétation forestière de la Haute-Côte d'Ivoire', *Bulletin du Comité d'Études Historiques et Scientifiques de l'Afrique Occidentale Française*, Série B, no. 4 (1937).

Behnke, R. and I. Scoones, 'Rethinking Range Ecology: Implications for Rangeland Management in Africa', in Behnke, Scoones and Kerven (eds), *Range Ecology at Disequilibrium*, pp. 1–30.

Behnke, R., I. Scoones and C. Kerven (eds), *Range Ecology at Disequilibrium: New Models of Natural Variability and Pastoral Adaptation in African Savannas* (London: Overseas Development Institute, 1993).

Beinart, W. and W. Coates, *Environment and History. The taming of nature in the USA and South Africa* (London and New York: Routledge, 1995).

Beinart, W., 'Soil Erosion, Conservation and Ideas about Development: A Southern African Exploration, 1900–1960', *Journal of Southern African Studies* XI, 1 (1984), pp. 52–83.

Belay Tegene, 'Indigenous Soil Knowledge and Fertility Management Practices of the South Wällo Highlands', *Journal of Ethiopian Studies*, XXXI, 1 (1998), pp. 123–58.

Belay Tegene, 'Potential and Limitations of an Indigenous Structural Soil Conservation Technology of Welo, Ethiopia', *Eastern Africa Social Science Research Review*, XIV, 1 (1998), pp. 1–18.

Bell, R.H.V., 'The effect of soil nutrient availability on community structure in African ecosystems', in Huntley and Walker (eds), *Ecology of Tropical Savannas*, pp. 193–216.

Bell, R., 'Conservation with a human face: conflict and reconciliation in African land use planning', in Anderson and Grove, *Conservation in Africa*, pp. 79–101.

Belsky, A.J., 'Spatial and temporal landscape patterns in arid and semi-arid African savannas', in L. Hansson, L. Fahrig, and G. Merriam (eds), *Mosaic Landscapes and Ecological Processes* (London: Chapman & Hall, 1995), pp. 31–56.

Berg, R., 'Foreign Aid in Africa: Here's the Answer – Is it Relevant to the Question?' in R. J. Berg and J.S.Whitaker (eds), *Strategies for African Development* (Berkeley: University of California Press, 1986), pp. 505–43.

Bergeret, A., 'Les Forestiers Coloniaux Français: Une doctrine et des politiques qui n'ont cessé de "rejeter de souche"', in Y. Chatelin and C. Bonneuil (eds), *Les Sciences Hors d'Occident au XXe Siècle*, Vol 3, *Nature et Environnement* (Paris: ORSTOM, 1995), pp. 59–74.

Bernard, L., M. Oualbadet, N. Ouattara, and R. Peltier, 'Parcs agroforestiers dans un terroir soudanien: Cas du village Dolékha au nord de la Côte d'Ivoire', *Bois et Forêts des Tropiques*, 244 (1995), pp. 25–42.

Berry, S., *No Condition is Permanent: The social dynamics of agrarian change in sub-Saharan Africa* (Madison: University of Wisconsin Press, 1993).

Beusekom, M. van, 'From Underpopulation to Overpopulation: French Perceptions of Population, Environment, and Agricultural Development in French Soudan (Mali), 1900–1960', *Environmental History*, 4, 2 (1999), pp. 198–219.

Biot, Y., P. Blaikie, C. Jackson and R. Palmer-Jones, *Rethinking Research on Land Degradation in Developing Countries* (Washington, DC: World Bank Discussion Paper 289, 1995).

Blackie, M., 'A Time to Listen: A Perspective on Agricultural Policy in Zimbabwe', *Zimbabwe Agricultural Journal*, 79 , 5 (1982), pp. 151–6.

Blaikie, P., 'Explanation and Policy in Land Degradation and Rehabilitation for Developing Countries', *Land Degradation and Rehabilitation*, 1 (1989), pp. 23–37.

Blaikie, P., 'A Review of Political Ecology: Issues, Epistemology and Analytical Narratives', *Zeitschrift für Wirtschaftsgeographie*, 43 (1999), pp. 131–47.

Blaikie, P., 'Changing Environments or Changing Views? A Political Ecology of Developing Countries', *Geography*, 80, 348 (1995), pp. 203–14.

Blaikie, P., *The Political Economy of Soil Erosion* (New York: Longman, 1985).

Bonkoungou, E. G., 'Inventaire et analyse biogéographique de la flore des galéries forestières de la Volta Noire en Haute Volta', *Notes et Documents Voltaïques*, 15, 1–2 (1984), pp. 64–83.

Bonneuil, C., 'Entre science et empire, entre botanique et agronomie: Auguste Chevalier, savant colonial', in Patrick Petitjean (ed.), *Les Sciences Hors d'Occident au XXe Siècle*, Vol 2, *Les Sciences Coloniales: Figures et Institutions* (Paris: ORSTOM, 1996), pp. 15–36.

Boserup, E., 'Environment, Population, and Technology in Primitive Societies', in Worster (ed.), *The Ends of the Earth*. pp. 23–38.

Boserup, E., *The Conditions of Agricultural Change. The Economics of Agrarian Change under Population Pressure* (New York: Aldine Publishing Co., 1965).

Botkin, D. B., *Discordant Harmonies: A New Ecology for the Twenty-first Century* (New York: Oxford University Press, 1990).

Bourlière, F. and M. Hadley, 'Present-day savannas: An overview', in F. Bourlière (ed.), *Ecosystems of the World 13: Tropical Savannas* (Amsterdam: Elsevier, 1983), pp. 1–18.

Boutillier, J. L., 'Les Structures Foncières en Haute-Volta', *Études Voltaïques*, Mémoire 5 (Ouagadougou: Centre IFAN–ORSTOM, 1964), pp. 5–181.

Boutrais, J., 'Éleveurs, bétail, et environnement', in C. Blanc-Pamard and J. Boutrais (eds), *À la Croisée des Parcours* (Paris: Editions ORSTOM, 1994), pp. 303–20.

Boutrais, J., *Hautes terres d'élevage au Cameroun* (Paris: ORSTOM, 2 vols., 1995).

Bratton, M., 'Farmer Organizations and Food Production in Zimbabwe', *World Development*, 14, 3 (1986), pp. 367–84.

Broch-Due, V. and R. A. Schroeder (eds), *Producing Nature and Poverty in Africa* (Stockholm: Nordiska Afrikainstitutet, 2000).

Brown, L., *Conservation for Survival: Ethiopia's Choice* (Addis Ababa, 1973).

Bruce, J., 'Do Indigenous Tenure Systems Constrain Agricultural Development?' in Bassett and Crummey, *Land in African Agrarian Systems*, pp. 35–56.

Bryant, R. and S. Bailey, *Third World Political Ecology* (London/New York: Routledge, 1997).

Bryant, R. L., 'Political Ecology: An Emerging Research Agenda in Third World Studies', *Political Geography*, 11, 1 (1992), pp. 12–36.

Bush, R. and L. Cliffe, 'Agrarian Policy in Migrant Labour Societies: Reform or Transformation in Zimbabwe?', *Review of African Political Economy*, 29 (1984), pp. 77–94.

Capron, J., *Communautés Villageoises Bwa: Mali, Haute Volta* (Paris: Institut d'Ethnologie, 1973).

Carson, R., *Silent Spring , With an Introduction by Vice President Al Gore* (Boston/New York: Houghton Mifflin Co., 1962, re-issue 1994).

Chambers, R., *Rural Development: Putting the Last First* (London: Longman, 1983).

Chauveau, J. P., P.-M. Bosc and M. Pescay, 'Le plan foncier rural en Côte d'Ivoire', in P. Lavigne Delville (ed.), *Quelles politiques foncières pour l'Afrique rurale? Réconcilier pratiques, légitimité et légalité* (Paris: Karthala/Coopération Française, 1998), pp. 553–82.

Chevalier, A., 'Mon exploration botanique du Soudan français', *Bulletin du Muséum d'Histoire Naturelle*, 5 (1900), pp. 248–53.

Cleaver, K. and G. Schreiber, *Reversing the Spiral: The Population, Agriculture, and Environment Nexus in sub-Saharan Africa* (Washington, DC: The World Bank, 1994).

Cole, M., *The Savannas: Biogeography and Biobotany* (New York: Academic Press, 1986).

Conte, C. A., 'The Forest becomes Desert: Forest Use and Environmental Change in Tanzania's West Usambara Mountains', *Land Degradation and Development*, X (1999), pp. 291–309.

Conway, D. and M. Hulme, 'Recent fluctuations in precipitation and runoff over the Nile sub-basins and their impact on Nile discharge', *Climatic Change*, 25 (1993), pp. 127–51.

Coulibaly, E., *Savoir et savoir-faire des anciens métallurgistes: Recherches préliminaires sur les procédés en sidérurgie directe dans le Bwamu (Burkina Faso-Mali)*, doctoral thesis, Université de Paris I, 1997.

Cronon, W., 'A Place for Stories: Nature, History, and Narrative', *The Journal of American*

History (1992), pp. 1347–76.

Cronon, W., *Nature's Metropolis. Chicago and the Great West* (New York: W. W. Norton, 1991).

Crummey, D., 'Deforestation in Wällo: Process or Illusion?' *Journal of Ethiopian Studies*, XXXVI, 1 (1998), pp. 1–41.

Crush, J., 'Imagining Development', in J. Crush (ed.), *Power of Development* (London: Routledge, 1995), pp. 1–23.

Curtin, P., *The Image of Africa: British Ideas and Action, 1780–1850* (Madison: University of Wisconsin Press, 1964), 2 Vols.

Dawit Wolde Giorgis, *Red Tears. War, Famine and Revolution in Ethiopia* (Trenton, NJ: Red Sea Press, 1989).

Delavignette, R., *Afrique Occidentale Française* (Paris, 1931).

Demeritt, D., 'Social Theory and the Reconstruction of Science in Geography', *Transactions of the Institute of British Geographers*, 21, 3 (1996), pp. 484–503.

Dessalegn Rahmato, *Famine and Survival Strategies. A Case Study from Northeast Ethiopia* (Uppsala: Scandinavian Institute of African Studies, 1991).

Dessalegn Rahmato, 'Environmentalism and Conservation in Wällo Before the Revolution', *Journal of Ethiopian Studies*, XXXI, 1 (1998), pp. 43–86.

Dessalegn Rahmato, 'The Unquiet Countryside: The Collapse of Socialism and Rural Agitations 1990–1991', in Abebe Zegeye and S. Pausewang (eds), *Ethiopia in Change* (London, 1994).

Devineau, J. L. and G. Serpentié, 'Paysage végétaux et systèmes agraires au Burkina Faso', in M. Pouget (ed.), *Caractérisation et suivi des milieux terrestres en régions arides et tropicales* (Paris: ORSTOM, 1991), pp. 373–83.

Diaz, H. F. and V. Markgraf (eds), *El Niño: Historical and Paleoclimatic Aspects of the Southern Oscillation* (Cambridge: Cambridge University Press, 1992).

Drinkwater, M., *The State and Agrarian Change in Zimbabwe's Communal Lands* (New York: St. Martins Press, 1991).

Dublin, H. T., 'Dynamics of the Serengeti-Mara Woodlands. An Historical Perspective', *Forest and Conservation History*, XXXV (1991), pp. 169–78.

Dumont, R., 'En Haute-Volta, une paysannerie à demi affamé', chapter 5 in *Paysans écrasés, terres massacrées* (Paris, 1978).

EC/FAO (European Commission/Food and Agriculture Organization), *Somalia: Livestock Export Market Study* (Nairobi: EC, 1995).

Eckholm, E. and L. R. Brown, 'Spreading Deserts—The Hand of Man', *World Watch Paper 13* (Washington, D.C., 1977).

Elias, E. and I. Scoones, 'Perspectives on soil fertility change: a case study from southern Ethiopia', *Land Degradation & Development*, 10 (1999), pp. 195–206.

Ellis, F., 'Household Strategies and Rural Livelihood Diversification', *Journal of Development Studies*, 35, 1 (1998), pp. 1–38.

Ellis, J. E. and D. M. Swift, 'Stability of African Pastoral Ecosystems: Alternate Paradigms and Implications for Development', *Journal of Range Management*, 41, 6 (1988), pp. 450–59.

Ellis, W., 'Africa's Stricken Sahel', *National Geographic Magazine*, 172, 2 (1987), pp. 140–79.

Escobar, A., 'After Nature: Steps to an Antiessentialist Political Ecology', *Current Anthropology*, 40, 1 (1999), pp. 1–30.

Escobar, A., 'Constructing Nature: Elements for a Post-Structuralist Political Ecology', in Peet and Watts (eds), *Liberation Ecologies*, pp. 46–68.

Escobar, A., *Encountering Development: The Making and Unmaking of the Third Word* (Princeton, NJ: Princeton University Press, 1995).

Fairhead, J. and M. Leach, 'Dessication and Domination: Science and Struggles over

Environment and Development in Colonial Guinea', *Journal of African History*, XLI (2000), pp. 35–54.

Fairhead, J. and M. Leach, 'Enriching the Landscape: Social History and the Management of Transition Ecology in the Forest-Savanna Mosaic of the Republic of Guinea', *Africa*, LXVI, 1 (1996), pp. 14–35.

Fairhead, J. and M. Leach, *Reframing Deforestation Global Analyses and Local Realities: Studies in West Africa* (London: Routledge, 1998).

Fairhead, J. and M. Leach, *Misreading the African Landscape: Society and Ecology in a Forest-Savanna Mosaic* (Cambridge: Cambridge University Press, 1996).

Falloux, F. and L. Talbot, *Crisis and Opportunity: Environment and Development in Africa* (London: Earthscan, 1993).

Feder, G. and R. Noronha, 'Land Rights and Agricultural Development in Sub-Saharan Africa', *World Bank Research Observer*, 2, 2 (1987) pp. 143–69.

Ferguson, J., *The Anti-Politics Machine: 'Development', Depoliticization, and Bureaucratic Power in Lesotho* (Minneapolis: University of Minnesota Press, 1990).

Fratkin, E., *Surviving Drought and Development* (Boulder, CO: Westview Press, 1991).

Frost, P. G. H., J.-C. Menaut, B. H. Walker, E. Medina, O. T. Solbrig and M. Swift, 'Responses of savannas to stress and disturbance', *Biology International* (Special Issue) 10, (1986), pp. 1–78.

Froude, M., 'Veld Management in the Victoria Province Tribal Areas', in *Rhodesia Agricultural Journal*, 71, 2, (1974), pp. 29–33.

Gandy, M., 'Crumbling Land: The Postmodernity Debate and the Analysis of Environmental Problems', *Progress in Human Geography*, 20, 1 (1996), pp. 23–40.

Gann, L. and M. Gelfand, *Huggins of Rhodesia: The man and his country* (London: Allen and Unwin, 1964).

Gartrell, B., 'Prelude to disaster: the case of Karamoja', in Anderson and Johnson, *Ecology of Survival*, pp. 193–217.

Gasse, F., 'Hydrological changes in the African tropics since the Last Glacial Maximum', *Quaternary Science Reviews*, 19 (2000), pp. 189–211.

Gasse, F., 'Water resources variability in tropical and subtropical Africa in the past', in *Water Resources Variability in Africa during the XXth Century*, IAHS Pub. No 252 (1998), pp. 97–105.

Gillon, D., 'The fire problem in tropical savannas', in F. Bourlière (ed.), *Ecosystems of the World 13: Tropical Savannas* (Amsterdam: Elsevier, 1983), pp. 617–41.

Glacken, C., *Traces on the Rhodian Shore* (Berkeley: University of California Press, 1967).

Glantz, M. H. (ed.), *Drought Follows the Plow* (Cambridge: Cambridge University Press, 1994).

Glantz, M. H., 'Drought and economic development in sub-Saharan Africa', pp. 37–58 in Michael H. Glantz (ed.), *Drought and Hunger in Africa: Denying Famine a Future* (Cambridge: Cambridge University Press, 1987).

Glantz, M. H., *Desertification: Environmental Degradation in and Around Arid Lands* (Boulder, CO: Westview Press, 1977).

Gobbins, K. and H. Prankerd, 'Communal Agriculture: A Study from Mashonaland West', *Zimbabwe Agricultural Journal*, 80, 4 (1983), pp.151–8.

Goldman, M., 'The Birth of a Discipline: Producing Authoritative Green Knowledge, World Bank Style', *Ethnography*, 2, 2 (2001), pp. 191–218.

Goodman, D., 'Agro-Food Studies in the "Age of Ecology": Nature, Corporeality, Bio-Politics', *Sociologia Ruralis* 39, 1, (1999), pp. 17–38.

Gowlett, J. A. J., 'Human adaptation and long-term climatic change in northeast Africa: An archaeological perspective', in Anderson and Johnson, *Ecology of Survival*, pp. 27–45.

Gray, L.C., 'Is Land Being Degraded: A Multi-scale Perspective on Landscape Change in

Southwestern Burkina Faso', *Land Degradation and Development*, 10, 4 (1999), pp. 327–34.

Greenberg , J. B. and T. K. Park, 'Political Ecology', *Journal of Political Ecology*, 1 (1994), pp. 1–12 (http://www.library.arizona.edu/ej/jpe/jpeweb.html).

Gritzner, J. A., *The West African Sahel: Human Agency and Environmental Change*, The University of Chicago Geography Research Paper No. 226 (1988).

Grove, R. and T. Falola, 'Chiefs, Boundaries, and Sacred Woodlands: Early Nationalism and the Defeat of Colonial Conservationism in the Gold Coast and Nigeria, 1870–1916', *African Economic History*, 24 (1996), pp. 1–23.

Grove, R., 'A Historical Review of Institutional and Conservationist Responses to Fears of Artificially Induced Global Climate Change: The Deforestation-Desiccation Discourse in Europe and the Colonial Context 1500–1940', in Y. Chatelin and C. Bonneuil (eds), *Les Sciences Hors d'Occident au XXᵉ Siècle*, Vol. 3, *Nature et Environnement* (Paris, ORSTOM, 1995), pp.155–74.

Grove, R., 'Early themes in African conservation: the Cape in the nineteenth century', in Anderson and Grove, *Conservation in Africa*, pp 21–39.

Grove, R., *Green Imperialism: Colonial Expansion, Tropical Island Edens and the Origins of Environmentalism, 1600–1860* (Cambridge: Cambridge University Press, 1995).

Grove, T. and A. Warren, 'Quaternary landforms and climate on the south side of the Sahara', *Geographical Journal*, 75 (1968), pp. 438–60.

Gupta, A. *Postcolonial Development: Agriculture in the Making of Modern India* (Durham, NC: Duke University Press, 1998).

Guyer, J. 'Diversity at Different Levels: Farm and Community in Western Nigeria', *Africa*, LXVI, 1 (1996), pp. 71–89.

Guyer, J. and P. Richards, 'The Invention of Biodiversity: Social Perspectives on the Management of Biological Variety in Africa', *Africa*, LXVI, 1 (1996), pp. 1–13.

Hall, S., 'The West and the Rest: Discourse and Power', in S. Hall *et al.*, *Modernity: An Introduction to Modern Societies* (Oxford: Blackwell, 1996).

Hare, K., 'The Making of Deserts: Climate, Ecology, and Society', *Economic Geography*, 53 (1977), pp. 332–45.

Harms, R., *Games Against Nature: An eco-cultural history of the Nunu of equatorial Africa* (Cambridge: Cambridge University Press, 1987).

Harris, W. C., *The Highlands of Ethiopia* (London: 3 vols., 1844).

Hassan, F. A., 'Historical Nile floods and their implications for climatic change', *Science*, 212 (1981), pp. 1142–5.

Hastings, J. R. and R. H. Turner, *The Changing Mile. An Ecological Study of Vegetation Change with Time in the Lower Mile of an Arid and Semiarid Region* (Tucson, AR: University of Arizona Press, 1965).

Helldén, U., 'Desertification—time for an assessment?' *Ambio*, 20, 8 (1991), pp. 372–83.

Hjort af Ornas, A. and M. A. Mohammed Salih (eds), *Ecology and Politics: Environmental Stress and Security in Africa* (Uppsala: Scandinavian Institute of African Studies, 1989).

Hjort, A., 'A Critique of "Ecological" Models of Pastoral Land Use', *Nomadic Peoples*, 10 (1982), pp. 11–27.

Hoben, A., 'The Cultural Construction of Environmental Policy: Paradigms and Politics in Ethiopia', in Leach and Mearns, *Lie of the Land*, pp. 186–208.

Hoffman, O., *Pratiques pastorales et dynamique du couvert végétale en pays Lobi (Nord est de la Côte d'Ivoire)*, Collection Travaux et Documents 189 (Paris: ORSTOM, 1985).

Homewood, K. and W. A. Rodgers, 'Pastoralism, conservation and the overgrazing controversy', in Anderson and Grove, *Conservation in Africa*, pp. 111–28.

Homewood, K. and W. A. Rodgers, *Maasailand Ecology: Pastoralist Development and Wildlife Conservation in Ngorongoro, Tanzania* (Cambridge: Cambridge University Press, 1991).

Huntley, B. J. and B. H. Walker (eds), *Ecology of Tropical Savannas* (Berlin: Springer-Verlag, 1982).

Hurni, H., 'Sustainable management of natural resources in African and Asian mountains', *Ambio*, 28, 5 (1999), pp. 382–9.

Ibo, G., 'La Politique Coloniale de Protection de la Nature en Côte d'Ivoire, (1900–1958)', *Revue Française d'Histoire d'Outre-Mer*, LXXX, 298 (1993), pp. 83–104.

Innes, J., 'Measuring Environmental Change', in D. L. Peterson and V. T. Parker (eds), *Ecological Scale: Theory and Applications* (New York: Columbia University Press, 1998), pp. 429–57.

Isaacman, A. and A. Chilundo, 'Peasants at Work: Forced-Cotton Cultivation in Northern Mozambique, 1938–1961', in A. Isaacman and R. Roberts (eds), *Cotton, Colonialism and Social History in Sub-Saharan Africa* (Portsmouth, NH: Heinemann, 1995), pp. 147–79.

Jackson, J. J., 'Some Observations on the Comparative Effects of Short Duration Grazing Systems and Continuous Grazing Systems on the Reproductive Performance of Ranch Cows', *Rhodesia Agricultural Journal,* 69, 5 (1972), pp. 95–102.

Jansson, K., M. Harris and A. Penrose, *The Ethiopian Famine* (London: Zed Press, 1987).

Johnson, D. L., S. H. Ambrose, T. J. Bassett, M. L. Bowen, D. E. Crummey, J. S. Isaacson, D. N. Johnson, P. Lamb, M. Şaul and A. E. Winter-Nelson, 'Meaning of Environmental Terms', *Journal of Environmental Quality*, XXVI, 3 (May–June, 1997), pp. 581–9.

Jones, B., 'Dessication and the West African colonies', *Geographical Journal*, XCI, 5 (1938), pp. 401–23.

Kabede Tato and H. Hurni (eds), *Soil Conservation for Survival* (Ankeny, 1992).

Kandeh, H. B. S. and P. Richards, 'Rural People as Conservationists: Querying Neo-Malthusian Assumptions about Biodiversity in Sierra Leone', *Africa*, LXVI, 1 (1996), pp. 90–103.

Keeley, J. and I. Scoones, 'Knowledge, power and politics: the environmental policy-making process in Ethiopia', *Journal of Modern African Studies*, XXXVIII, 1 (2000), pp. 89–120.

Kepe, T. and I. Scoones, 'Creating Grasslands: Social Institutions and Environmental Change in Mkambati Area, South Africa', *Human Ecology*, 27, 1 (1999), pp. 29–53.

Kessler J. and H. Breman, 'The Potential of Agroforestry to Increase Primary Production in the Sahelian and Sudanian Zones of West Africa', *Agroforestry Systems*, 13 (1991), pp. 41–62.

Kowal, J.M. and A.H. Kassam, *Agricultural Ecology of Savanna: A Study of West Africa* (Oxford: Clarendon Press, 1978).

Lamprey, H. and H. Yussuf, 'Pastoral and Desert Encroachment in Northern Kenya', *Ambio*, 10, 2 (1981).

Le Roux, P., A. Stubbs and P. Donnelly, 'Problems and Prospects of Increasing Beef Production in the Tribal Trust Lands', *Zambezia*, 6, 1 (1978), pp. 37–48.

Leach, M. and Robin Mearns (eds), *The Lie of the Land: Challenging Received Wisdom on the African Environment* (Oxford/Portsmouth, NH: James Currey/Heinemann, 1996).

Leeuw, P. N. de and J.C. Tothill, *The Concept of Rangeland Carrying Capacity in Sub-Saharan Africa – Myth or Reality?* (London: ODI Pastoral Development Network Paper 29b, 1990).

Little, P. D. and M. M. Horowitz (eds), *Lands at Risk in the Third World: Local-Level Perspectives* (Boulder, CO: Westview Press, 1987).

Little, P. D., 'Pastoralism, Biodiversity, and the Shaping of the Savanna Landscapes in East Africa', *Africa*, LXVI, 1 (1996), pp. 37–51.

Little, P. D., 'Social Differentiation and Pastoralist Sedentarization in Northern Kenya', *Africa*, 55 (1985), pp. 242–61.

Little, P. D., 'The Link between Local Participation and Improved Conservation: a Review of Issues and Experiences', in D. Western and R. M. Wright (eds), *Natural Connections:*

Perspectives in Community-based Conservation (Washington, D. C.: The Island Press, 1994), pp. 347–72.

Little, P. D., 'Traders, Brokers, and Market "Crisis" in Southern Somalia', *Africa*, 62, 1 (1992), pp. 94–124.

Little, P., 'The Social Context of Land Degradation ('Desertification') in Dry Regions', in L. Arizpe, P. Stone and D. Major (eds), *Population and Environment: Rethinking the Debate* (Boulder, CO: Westview Press, 1994), pp. 209–51.

Lord Hailey, *An African Survey. A Study of Problems Arising in Africa South of the Sahara* (London: Oxford University Press, 1938).

Low, D. A. and A. Smith, *History of East Africa* (Oxford: Clarendon Press, 3 vols., 1963, 1968, 1976).

Maack, P. A., '"We Don't Want Terraces!" Protest and Identity under the Uluguru Land Usage Scheme', in Maddox, *et al. Custodians of the Land*, pp. 152–69.

Mackenzie, F., 'Selective Silence: A Feminist Encounter with Environmental Discourse in Colonial Africa', in J. Crush (ed.), *Power in Development* (London: Routledge, 1995), pp. 100–12.

MacKenzie, J. M., 'Chivalry, social Darwinism and ritualised killing: the hunting ethos in Central Africa up to 1914', in Anderson and Grove, *Conservation in Africa*, pp. 41–61.

Maddox, G., J. Giblin and I. Kimambo (eds), *Custodians of the Land. Ecology and Culture in the History of Tanzania* (London/Nairobi/Dar es Salaam/Athens, OH: James Currey/Mkuki na Nyota/E.A.E.P/Ohio University Press, 1996).

Mainguet, M., *Desertification: Natural Background and Human Mismanagement* (Berlin: Springer-Verlag, 2nd edn, 1994).

Mashiringwani, N. A., 'The Present Nutrient Status of the Soils in the Communal Farming Areas of Zimbabwe', in *Zimbabwe Agricultural Journal*, 80, 2 (1983), pp. 73–5.

Matiza, T. and S. Crafter (eds), *Wetlands Ecology and Priorities for Conservation in Zimbabwe* (IUCN, 1994).

Matlon, P., 'Indigenous Land Use Systems and Investments in Soil Fertility in Burkina Faso', in J. Bruce and S. Migot-Adholla (eds) *Searching for Land Tenure Security in Africa* (Dubuque, IA: Kendall Hunt, 1994), pp. 41–70.

McCann, J., 'Ethiopia', in Michael Glantz (ed.) *Drought Follows the Plow* (Cambridge: Cambridge University Press, 1994), pp. 103–15.

McCann, J., *Green Land, Brown Land, Black Land. An Environmental History of Africa, 1800–1900* (Portsmouth, NH/Oxford: Heinemann/James Currey, 1999).

McCracken, J., 'Colonialism, capitalism and ecological crisis in Malawi: a reassessment', in Anderson and Grove, *Conservation in Africa*, pp. 63–77.

McGregor, J., 'Conservation, Control and Ecological Change: The Politics and Ecology of Colonial Conservation in Shurugwi, Zimbabwe', *Environment and History*, 1, 3, (1995), pp. 257–79.

McMillan, D. E., *Sahel Visions: Planned Settlement and River Blindness Control in Burkina Faso* (Tucson: University of Arizona Press, 1995).

Menaut, J.-C. and J. Cesar, 'The structure and dynamics of a West African savanna', in Huntley and Walker (eds), *Ecology of Tropical Savannas*, pp. 80–100.

Mesfin Wolde-Mariam, *Suffering Under God's Environment: A Vertical Study of the Predicament of Peasants in North-Central Ethiopia* (Berne: African Mountains Association and Geographica Bernensia 1991).

Menaut, J.-C., 'The vegetation of African savannas', in F. Bourlière (ed.), *Ecosystems of the World 13: Tropical Savannas* (Amsterdam: Elsevier, 1983), pp. 109–49.

Migot-Adholla, S., P. Hazell, B. Blarel and F. Place. 'Indigenous Land Rights Systems in Sub-Saharan Africa: A Constraint on Productivity?' *The World Bank Economic Review*, 5, 1 (1991), pp. 155–75.

Miller, C. and H. Rothman, *Out of the Woods. Essays in Environmental History* (Pittsburgh: University of Pittsburgh Press, 1997).

Millington, A., 'Environmental degradation, soil conservation and agricultural policies in Sierra Leone, 1895–1984', in Anderson and Grove, *Conservation in Africa*, pp. 229–48.

Miracle, M., *Agriculture in the Congo Basin; Tradition and change in African rural economies* (Madison: University of Wisconsin Press, 1967).

Mohamed Salih, M. A. and Shibru Tedla (eds), *Environmental Planning, Policies and Politics in Eastern and Southern Africa* (London/New York: MacMillan Press/St. Martin's Press, 1999).

Mortimore, M., *Roots in the African Dust. Sustaining the Drylands* (Cambridge: Cambridge University Press, 1998).

Mortimore, M., *Adapting to Drought: Farmers, famines and desertification in West Africa* (Cambridge: Cambridge University Press, 1989).

Moyo, S., P. Robinson, Y. Katerere, St. Stevenson and D. Gumbo, *Zimbabwe's Environmental Dilemma: Balancing Resource Inequities* (Harare, 1991).

Mtetwa, R., 'Myth or Reality: The "Cattle Complex" in South East Africa, with Special Reference to Rhodesia', *Zambezia*, 6, 1 (1978), pp. 23–35.

Munro, W., 'Building the Post-Colonial State: Villagization and Resource Management in Zimbabwe', *Politics and Society*, 23, 1 (March 1995), pp. 107–140.

Myers, G., 'Competitive Rights, Competitive Claims: Land Access in Mozambique', *Journal of Southern African Studies*, 20, 4, (1994), pp. 603–33.

Netting, R. McC., *Cultural Ecology* (Menlo Park, CA: Cummings Publishing Co., 1977).

Netting, R. McC., *Hill Farmers of Nigeria: Cultural Ecology of the Kofyar of the Jos Plateau* (Seattle: University of Washington Press, 1968).

Netting, R. McC., *Smallholders, Householders: Farm Families and the Ecology of Intensive, Sustainable Agriculture* (Stanford, CA: Stanford University Press, 1993).

Neumann, R., *Imposing Wilderness. Struggles over Livelihoods and Nature Preservation in Africa* (Berkeley: University of California Press, 1998).

Niamer-Fuller, M., *Managing Mobility in African Rangelands: The legitimization of transhumance* (London: Intermediate Technology Publications, 1999).

Nichol, J. E., 'Geomorphological evidence and Pleistocene refugia in Africa', *Geographical Journal*, 165 (1999), pp. 79–89.

Nicholson, S. E., 'The Methodology of Historical Climate Reconstruction and its Application to Africa', *Journal of African History*, XX, 1 (1979), pp. 31–49.

Nicol, C., *From the Roof of Africa* (London: Hodder and Stoughton, 1971).

Nsiah-Gyabah, K., *Environmental Degradation and Desertification in Ghana: A Study of the Upper West Region* (Aldershot: Avebury Studies in Green Research, 1994), pp. 105–37.

O'Leary, M., 'Drought and Change Among Northern Kenya Nomadic Pastoralists: The Case of the Rendille and Gabra', in Gísli Pálsson (ed.), *From Water to World-making: African Models and Arid Lands* (Uppsala: Scandinavian Institute of African Studies, 1990), pp. 151–74.

O'Leary, M., 'Ecological Villains or Economic Victims: The Case of the Rendille of Northern Kenya', *Desertification Control Bulletin*, 11 (1984), pp. 17–21.

O'Leary, M., *The Economics of Pastoralism in Northern Kenya: The Rendille and the Gabra* (Nairobi: UNESCO, 1985).

Oba, G., N. C. Stenseth and W. J. Lusigi, 'New perspectives on sustainable grazing management in arid zones of sub-Saharan Africa', *BioScience*, 50, 1 (2000), pp. 35–51.

Okoth-Ogendo, H. W. O., 'Some issues of theory in the study of tenure relations in African agriculture', *Africa*, 59, 1 (1989), pp. 6–17.

Oliver, J., 'Nature and Nurture in Animal Production: A Heretical View', *Zambezia*, 7, 2 (1979), pp. 125–8.

Ouadba, J.-M., 'Note sur les caractéristiques de la végétation ligneuse et herbacée d'une jachère protégée en zone soudanienne dégradée', in C. Floret and G. Serpantié (eds) *La jachère en Afrique de l'Ouest* (Paris: ORSTOM, 1993), pp. 331–40.

Ouedraogo, R., J.-P. Sawadogo, V. Stam and T. Thiombiano, 'Tenure, Agricultural Practices and Land Productivity in Burkina Faso: Some Recent Empirical Results', *Land Use Policy*, 13, 3 (1996), pp. 229–32.

Peet, R and M. Watts, 'Introduction: Development Theory and Environment in an Age of Market Triumphalism', *Economic Geography*, 69, 3 (1993), pp. 227–53.

Peet, R. and M. Watts, *Liberation Ecologies: Environment, development, social movements* (Routledge: London, 1996).

Peluso, N., 'Coercing Conservation: The Politics of State Resource Control', in R. Lipschutz and K. Conca (eds), *The State and Social Power in Global Environmental Politics* (New York: Columbia University Press, 1993), pp. 46–70.

Person, Y., *Samori: Une révolution dyula* (Dakar: IFAN, 2 vols., 1968).

Peters, P., '"Who's local here?" The Politics of Participation in Development', *Cultural Survival Quarterly*, (Fall 1994), pp. 22–5.

Petit, S., *Environnement, conduite des troupeaux et usage de l'arbre chez les agropasteurs de l'ouest burkinabe*, doctoral thesis, 2 vols., Université d'Orléans, 2000.

Pitcher, M. A., 'Conflict and Cooperation: Gendered Roles and Responsibilities within Cotton Households in Northern Mozambique', *African Studies Review*, 39, 3 (1996), pp. 81–112.

Pitcher, M. A., 'Disruption Without Transformation: Agrarian Relations and Livelihoods in Nampula Province, Mozambique, 1975–1995', *Journal of Southern African Studies*, 24, 1 (1998), pp. 115–40.

Pitcher, M. A., 'Recreating Colonialism or Reconstructing the State: Privatisation and Politics in Mozambique', *Journal of Southern African Studies*, 22 (1996), pp. 49–74.

Platteau, J. P., 'Does Africa need land reform?' in Toulmin and Quan (eds), *Evolving Land Rights, Policy and Tenure Reform in Africa*, pp. 51–74.

Platteau, J. P., 'The evolutionary theory of land rights as applied to sub-Saharan Africa: A critical assessment', *Development and Change*, 27 (1996), pp. 29–86.

Poiani, K., B. Richter, M. Anderson and H. Richter, 'Biodiversity Conservation at Multiple Scales: Functional Sites, Landscapes, and Networks', *BioScience*, L, 2 (2000), pp. 133–46.

Prudencio, C., 'Ring Management of Soils and Crops in the West African Semi-arid Tropics: The Case of the Mossi Farming Season in Burkina Faso', *Agriculture, Ecosystems and Environment*, 47 (1993), pp. 237–64.

Ranger, T., *Peasant Consciousness and Guerrilla War* (Oxford/Berkeley: James Currey, University of California Press, 1985).

Ranger, T., *Voices from the Rocks. Nature, Culture and History in the Matopos Hills of Zimbabwe* (Oxford/Bloomington, IN: James Currey/Indiana University Press, 1999).

Ranger, T., *Revolt in Southern Rhodesia, 1896–97; a study in African resistance* (Evanston, IL: Northwestern University Press, 1967).

Rasmusson, E. M. and J. M. Wallace, 'Meteorological aspects of the El Niño/Southern Oscillation', *Science*, 222 (1983), pp. 1195–1202.

Rattray, J., R. Cormack and R. Staples, 'The Vlei Areas of S. Rhodesia and their Uses', *Rhodesia Agricultural Journal*, 50, 6 (November–December 1953), pp. 465–83.

Reid, R., R. Kruska, N. Muthui, A. Taye, S. Wotton, C. Wilson and W. Mulatu, 'Land-use and land-cover dynamics in response to changes in climatic, biological and socio-political forces: the case of southwestern Ethiopia', *Landscape Ecology*, 15 (2000), pp. 339–55.

Ribot, J. C., 'Theorizing Access: Forest Profits along Senegal's Charcoal Commodity Chain', *Development and Change*, 29 (1998), pp. 307–41.

Richards, P., *Indigenous Agricultural Revolution* (London: Hutchinson, 1985).

Riou, G., *Savanes, l'herbe, l'arbre et l'homme en terres tropicales* (Paris: Masson Armand Colin, 1995).

Roe, E., '"Development Narratives" or making the best of blueprint development', *World Development*, XIX, 4 (1991), pp. 287–300.

Roe, E., L. Hutsinger and K. Labnow, 'High-Reliability Pastoralism Versus Risk-Averse Pastoralism', *Journal of Arid Environments*, 39 (1998), pp. 39–55.

Sauer, C., *Land and Life* (Berkeley: University of California Press, 1963).

Şaul, M., 'Farm Production in Bare, Burkina Faso: The Technical and Cultural Framework of Diversity', in G. Dupré (ed.), *Savoirs paysans et développement* (Paris, Karthala, 1991), pp. 301–29.

Şaul, M., 'Land Custom in Bare: Agnatic Corporation and Rural Capitalism in Western Burkina', in Bassett and Crummey (eds), *Land in African Agrarian Systems*, pp 75–100.

Şaul, M., 'Money and Land Tenure as Factors in Farm Size Differentiation in Burkina Faso', in R. E. Downs and S. P. Reyna (eds), *Land and Society in Contemporary Africa* (Hanover, NH: 1988), pp. 243–79.

Savonnet, G., 'Un Système de Culture Perfectionné, Pratiqué par les Bwabas: Bobo-Oulé de la Région de Houndé (Haute-Volta)', *Bulletin de l'I.F.A.N.*, XXI, sér. B, 3–4 (1959), pp. 425–58.

Savonnet-Guyot, C., *État et Sociétés du Burkina: Essai sur le politique Africain* (Paris: Éditions Karthala, 1986).

Scholes, R. J. and B. H. Walker, *An African Savanna: Synthesis of the Nylsvley study* (Cambridge: Cambridge University Press, 1993).

Schwartz, A., 'Brève histoire de la culture du coton au Burkina Faso', *Découvertes du Burkina*, 1 (1993), pp. 207–37.

Scoones, I., 'Patch Use by Cattle in Dryland Zimbabwe: Farmer Knowledge and Ecological Theory', in B. Cousins (ed). *People, Land and Livestock* (Mt. Pleasant, Zimbabwe: Centre for Applied Social Sciences, 1989), pp. 227–309.

Scoones, I., 'The dynamics of soil fertility change: historical perspectives on environmental transformation from Zimbabwe', *Geographical Journal*, 163 (1997), pp. 161–9.

Scoones, I., 'Coping with Drought: Responses of Herders and Livestock in Constrasting Savanna Environments in Southern Zimbabwe', *Human Ecology*, 20, 3 (1992), pp. 31–52.

Scoones, I., *Living with Uncertainty: New directions in pastoral development in Africa* (London: Intermediate Technology Publications, 1995).

Scoones, I., *Patch Use by Cattle in Dryland Zimbabwe: Farmer Knowledge and Ecological Theory* (London: ODI Pastoral Development Network Paper 28b, 1989).

Scoones, I., 'The Dynamics of Soil Fertility Change: Historical Perspectives on Environmental Transformation from Zimbabwe', *The Geographical Journal*, 162, 2 (1997), pp. 161–69.

Scott, J., *Seeing like a State:How certain schemes to improve the human condition have failed* (New Haven, CT: Yale University Press, 1998).

Sebsebe Demissew, 'A Study of the Vegetation and Floristic Composition of Southern Wällo, Ethiopia', *Journal of Ethiopian Studies*, XXXI, 1 (1998), pp. 159–92.

Simpson, H. J., M. A Cane, S. K. Lin, S. E. Zebiak and A. L. Herczeg, 'Forecasting annual discharge of River Murray, Australia, from a geophysical model of ENSO', *Journal of Climate*, 6, 5 (1993), pp. 386–90.

Skinner, E. P. 'The Changing Status of the "Emperor of the Mossi" under Colonial Rule and since Independence', in M. Crowder and O. Ikime (eds), *West African Chiefs under Colonial Rule and Independence* (New York and Ile-Ife, 1970).

Sobania, N., 'Pastoralist migration and colonial policy: a case study from northern Kenya', in Anderson and Johnson, *Ecology of Survival*, pp. 219–39.

Solbrig, O. T., E. Medina and J. F. Silva (eds), *Biodiversity and Savanna Ecosystem Processes,*

Ecological Studies V. 121 (Berlin: Springer, 1996).

Spear, T., 'Struggles for the Land. The Political and Moral Economies of Land on Mount Meru', in Maddox et al., Custodians of the Land, pp. 213–40.

Spooner, B., 'Desertification: The Historical Significance', in R. Huss-Ashmore and Solomon H. Katz (eds), African Food Systems in Crisis. Part One: Microperspectives (New York: Gordon and Breech Science Publishers, 1989), pp. 111–62.

Sprugel, D. G., 'Disturbance, equilibrium, and environmental variability; What is 'natural' vegetation in a changing environment?', Biological Conservation, 58 (1991), pp. 1–18.

Ståhl, M., 'Environmental Degradation and Political Constraints in Ethiopia', Disasters, 14, 2 (1990).

Stähli, P., 'Changes in Settlement and Land Use in Simen, Ethiopia, especially from 1954 to 1975', in B. Misserli and K. Aerni (eds), Simen Mountains–Ethiopia, Vol. 1, Cartography and its Application for Geographical and Ecological Problems (Bern: Geographisches Institut der Universität Bern, 1978).

Stamp, L. D., 'The southern margin of the Sahara: Comments on some recent studies on the question of dessication in West Africa', Geographical Review, XXX, 2 (1940), pp. 297–300.

Stebbing, E. P., 'The encroaching Sahara: The threat to the West African colonies', Geographical Journal, DXXXV, 5 (1935), pp. 506–24.

Steele, M., 'The Economic Function of African-owned Cattle in Rhodesia', Zambezia, 9, 2 (1981), pp. 29–48.

Steinhart, E., 'Hunters, Poachers and Gamekeepers: Towards a Social History of Hunting in Colonial Kenya', Journal of African History, 30, 2 (1989), pp. 247–64.

Stevens, S. 'The Legacy of Yellowstone', in S. Stevens (ed.), Conservation through Survival: Indigenous Peoples and Protected Areas (Washington, DC: Island Press, 1997), pp. 13–32.

Stocking, M., 'Soil Conservation Policy in Colonial Africa', Agricultural History, 59, 2 (1985), pp. 148–61.

Stocking, M., 'Relationship of Agricultural History and Settlement to Severe Soil Erosion in Rhodesia', Zambezia, 6, 2 (1978), pp. 129–45.

Stocking, M., 'Soil Erosion: Breaking New Ground', in Leach and Mearns, The Lie of the Land , pp. 140–54.

Sutter, J. W., 'Cattle and Inequality: Herd Size Differences and Pastoral Production Among the Fulani of Northeastern Senegal', Africa, 57 (1987), pp. 196–217.

Swain, A., 'Ethiopia, the Sudan and Egypt: The Nile River dispute', Journal of Modern African Studies, 35, 4 (1997), pp. 674–94.

Swift, J., 'Desertification: Narratives, Winners and Losers', in Leach and Mearns, Lie of the Land, pp. 73–90.

Talbot, M. R. and M. A. J. Williams, 'Cyclic alluvial fan sedimentation on the flanks of fixed dunes, Janjari, central Niger', Catena, 6 (1979), pp. 43–62.

Talbot, M. R. and M. A. J. Williams, 'Erosion of fixed dunes in the Sahel, central Niger', Earth Surface Processes, 3 (1978), pp. 107–13.

Talbot, M. R., 'Environmental responses to climatic change in the West African Sahel over the past 20,000 years', in M.A.J.Williams and H. Faure (eds), The Sahara and the Nile. Quaternary environments and prehistoric occupation in northern Africa (Rotterdam: Balkema, 1980), pp. 37–62.

Taylor, P. J. and F. W. Buttel, 'How Do We Know We Have Global Environmental Problems? Science and the Globalization of Environmental Discourse', Geoforum, 23, 3 (1992), pp. 405–16.

Teilhard de Chardin, P. and P. Lamare, 'Le canon de l'Aouache et le volcan Fantal', Mémoire Société Géologique de France, 14 (1930), pp. 13–20.

Tersiguel, P., Le pari du tracteur: La modernisation de l'agriculture cotonnière au Burkina Faso

(Paris: ORSTOM, 1995).

Thomas, D. S. G. and N. J. Middleton, *Desertification: Exploding the Myth* (Chichester: Wiley, 1994).

Thomas, W. L. (ed.), *Man's Role in Changing the Face of the Earth* (Chicago: University of Chicago Press, 1956).

Tiffen, M., M. Mortimore and F. Gichuki (eds), *More People, Less Erosion: Environmental Recovery in Kenya* (Chichester, UK: John Wiley and Sons for the Overseas Development Institute, 1994).

Timberlake, L., *Africa in Crisis: The Causes, the Cures of Environmental Bankruptcy* (London: Earthscan, 1985).

Toulmin, C. and J. Quan (eds), *Evolving Land Rights, Policy, and Tenure in Africa* (London: DFID/IIED/NRI, 2000).

Tucker, C. and H. D. Newcomb, 'Expansion and Contraction of the Sahara Desert from 1980 to 1990', *Science*, 253 (1991), pp. 299–301.

Turner, B. L., G. Hyden and R. W. Kates (eds), *Population Growth and Agricultural Change in Africa* (Gainesville: University Press of Florida, 1993).

Turton, D., 'The Mursi and National Park development in the Lower Omo River', in Anderson and Grove, *Conservation in Africa*, pp. 169–86.

UNCED (United Nations Conference on Environment and Development), *Earth Summit Agenda 21: Programme of Action for Sustainable Development* (New York: United Nations Environment Programme, 1992).

UNCOD (United Nations Conference on Desertification), *Report of the United Nations Conference on Desertification, Nairobi, 29 August–9 September 1977* (Nairobi: United Nations Environment Programme, 1977).

UNEP (United Nations Environment Programme), *Status of Desertification and Implementation of the UN Plan of Action to Combat Desertification* (Nairobi: UNEP, 1991).

UNEP, *Report of the Executive Director. Status of Desertification and Implementation of the United Nations Plan of Action to Combat Desertification* (Nairobi: 1992).

UNEP, *Report of the United Nations Conference on Desertification, 29 August–9 September 1977* (Nairobi: UNEP, 1977).

UNEP, *World Atlas of Desertification*, Editorial commentary by N. Middleton and D.S.G. Thomas (London: Arnold, 1997).

Vayda, A. P. and B. B. Walters, 'Against Political Ecology', *Human Ecology*, 27, 1 (1999), pp. 167–79.

Vines, A., *Renamo: Terrorism in Mozambique* (London: James Currey, 1991).

Vrba, E. S., G. H. Denton, T. C. Partridge and L. H. Burckle (eds), *Paleoclimate and Evolution, with Emphasis on Human Origins* (New Haven, CT: Yale University Press. 1995).

Warren, A. and C. T. Agnew, *An Assessment of Desertification and Land Degradation in Arid and Semi-Arid Areas* (London: International Institute for Environment and Development, 1988).

Warren, D. M., L. J. Slikkerveer and D. Brokensha (eds), *The Cultural Dimension of Development: Indigenous Knowledge Systems* (London: Intermediate Technology Publications, 1995).

Westoby, M., B. Walker and I. Noy-Meir, 'Opportunistic Management for Rangelands Not at Equilibrium', *Journal of Range Management*, 42, 4 (1989), pp. 266–74.

Whetton, P., D. Adamson, and M. A. J. Williams, 'Rainfall and river flow variability in Africa, Australia and East Asia linked to El Niño-Southern Oscillation events', in P. Bishop (ed.), *Lessons for Human Survival: Nature's Record from the Quaternary. Geological Society of Australia Symposium Proceedings*, 1 (1990), pp. 71–82.

Whitlow, R. 'Vlei Cultivation in Zimbabwe: Reflections on the Past', *Zimbabwe Agricultural Journal*, 80, 3 (May–June 1983), pp. 123–5.

Whitlow, R., 'Research on Dambos in Zimbabwe', *Zimbabwe Agricultural Journal*, 82, 2 (March–April 1985), pp. 59–66.

Wiens, J. A. 'Landscape Mosaics and Ecological Theory', in L. Hansson, L. Fahrig and G. Merriam (eds), *Mosaic Landscapes and Ecological Processes* (London, Chapman and Hall, 1995), pp. 1–26.

Wilde, J. C. de, *Experiences with Agricultural Development in Tropical Africa*, Vol. 1, *The Synthesis* (Baltimore: Johns Hopkins University Press, 1967).

Wilgen, B. W. van, C.S. Everson and W. S. W. Trollope, 'Fire Management in Southern Africa: Some Examples of Current Objectives, Practices, and Problems', in J. G. Goldhammer (ed.), *Fire in the Tropical Biota: Ecosystem Processes and Global Challenges* (Berlin: Springer-Verlag, 1990), pp. 179–215.

Williams, M. A. J., 'Recent Tectonically-Induced Gully Erosion at K'one, Metahara-Wolenchiti Area, Ethiopian Rift', *SINET Ethiopian Journal of Science*, 4 (1) (1981), pp. 1–11.

Williams, M. A. J., P. M. Bishop, and F. M. Dakin, 'Late Quaternary lake levels in southern Afar and the adjacent Ethiopian Rift', *Nature*, 267 (1977), pp. 690–93.

Williams, M. A. J., 'Soil salinity in the west central Gezira, Republic of the Sudan', *Soil Science*, 105, 6 (1968), pp. 451–64.

Williams, M. A. J. and D. A. Adamson (eds), *A Land Between Two Niles: Quaternary Geology and Biology of the Central Sudan* (Rotterdam: Balkema, 1982).

Williams, M. A. J. and F. M. Williams, 'Evolution of the Nile Basin', in M. A. J. Williams and H. Faure (eds), *The Sahara and the Nile* (Rotterdam: Balkema, 1980), pp. 207–224.

Williams, M. A. J. and Robert C. Balling Jr., *Interactions of Desertification and Climate* (London: Arnold for the World Meteorological Organization and the United Nations Environmental Program, 1996).

Williams, M. A. J., D. Adamson, B. Cock and R. McEvedy, 'Late Quaternary environments in the White Nile region, Sudan', *Global and Planetary Change*, 26, 1–3 (2000), pp. 305–16.

Williams, M. A. J., D. Dunkerley, P. De Deckker, P. Kershaw, and J. Chappell, *Quaternary Environments* (London: Arnold, 1998).

Williams, M. A. J., F. M. Williams, and P. M. Bishop, 'Late Quaternary history of Lake Besaka, Ethiopia', *Palaeoecology of Africa*, 13 (1980), pp. 93–104.

Wilson, K. '"Water Used to be Scattered in the Landscape": Local Understandings of Soil Erosion and Land Use Planning in Southern Zimbabwe', *Environment and History*, 1, 3 (1995), pp. 281–96.

Winter-Nelson, A., 'Rural Taxation in Ethiopia, 1981–1989: A Policy Analysis Matrix Assessment for Net Consumers and Net Producers', *Food Policy*, 22 (1997), pp. 419–32.

Wolf, E., 'Ownership and Political Ecology', *Anthropological Quarterly*, 45 (1972), pp. 201–5.

Wolf, E. R., *Europe and the People Without History* (Berkeley: University of California Press, 1982).

Wood, A., 'Zambia's Soil Conservation Heritage: A Review of Policies and Attitudes towards Soil Conservation from Colonial Times to the Present', in Kebede Tato and Hurni, *Soil Conservation*, pp. 156–71.

World Bank, *Zimbabwe: Land Subsector Study* (Washington, DC: World Bank, 1986).

Worster, D. (ed.), *The Ends of the Earth* (Cambridge: Cambridge University Press, 1988).

Worster, D., 'Transformations of the Earth: Toward an Agroecological Perspective in History', *Journal of American History*, 76, 4 (1990), pp. 1087–1110.

Worster, D., *Dust Bowl. The Southern Plains in the 1930s* (Oxford/New York: Oxford University Press, 1979).

Yeraswork Admassie, *Twenty Years to Nowhere. Property Rights, Land and Conservation in Ethiopia* (Lawrenceville, NJ: The Red Sea Press, 1997).

Zielinski, G. A., P. A. Mayewski, L. D. Meeker, S. Whitlow and M. S. Twickler, 'Potential atmospheric impact of the Toba mega-eruption – 71,000 years ago', *Geophysical Research Letters*, 23, 8 (1996), pp. 837–40.

Zimmerer, K. S. and K.R. Young (eds), *Nature's Geography: New Lessons for Conservation in Developing Countries* (Madison: University of Wisconsin Press, 1998).

Index

Abaina, Raimundo 232
Aberra Alemu 46
access, to resources 14-15, 20, 29-30, 70, 72,
    74, 79, 113, 114, 118, 163, 165, 166, 168,
    171, 173, 175-7, 183, 184, 194, 204, 209,
    246
Adhana Haile Adhana 95
adjustment, structural 70, 139, 231
afforestation 19, 119, 205, 206, 214, 216, 218
agriculture 5, 7, 13, 28, 29, 59-60, 64, 70,
    73-90 passim, 95, 99-101, 124-6, 134-60
    passim, 171, 179, 184-7, 194, 211, 212,
    227-47; commercial 155, 185, 232-3;
    extensification 29, 59, 77-8; intensification
    70, 77, 82, 85, 89, 90, 127, 137; near
    field/far field 76, 86; plantations 27, 126,
    129, 139, 148, 154-60 passim; shifting 76,
    241; slash-and-burn 66
agroforestry 73, 90, 124, 131, 133, 134, 213
aid 16, 28, 29, 65, 89, 139, 195, 217; food
    168, 205, 211-15, 219-20
Allan, William 196
Alvord, E.D. 185
Amazon 10, 96
Anderson, David 2-4 passim
animal traction 74, 76, 136 see also draft
ANOVA 87
Antoine, Tamini 81
Antonio, Sabonete 242
Araùjo, August 245
Armac'äho 97, 108, 113, 116, 117
Asfaw, Sheikh Mohammed 114
Aubréville, Auguste 11-12
Australia 163

Aylieff, John 215

Bailey, Sinead 29
banditry 124
Bare 142, 145-6, 148, 150
Barnes, T. 8
Bassett, Thomas J. 1-30, 53-71
beans 99, 240-3 passim; locust 131, 133, 134,
    159
beef 20, 184, 186, 187
Bégué, L. 125
Behnke, R.H. 196
Belgians 15
Berry, Sara 19
Besaka, Lake 45-9, 52
Besse, Kahoun 81
Binger, L.G. 125
biodiversity 17, 54, 91, 163
biomass 56-60 passim, 82
Blaikie, Piers 23
Bobo-Dioulasso 125, 127, 128, 130, 135, 156
Bognounou, Ouétian 121-60
borrowing 182, 202; land 79-80, 157-8
Borru Peasant Association 96, 98-101, 107,
    110, 111, 115, 116
Boserup, Ester 20, 94, 95
Bosshard, W.C. 206
Bougoula 142, 144, 150-2
Boutillier, J.L. 79
Bowen, Merle L. 23, 225-47
Breitenbach, F. von 207
Britain 11, 13, 15, 125, 184-92
Brown, Leslie 206-7
Bryant, Raymond 29

bunding 15, 17, 88, 96, 205, 209, 214, 217-24 *passim*
Burkina Faso 18-20 *passim*, 25, 26, 58, 72-90, 121-60; Sourou project 139; Tagouara plateau 123, 150-4, 156, 158-9; Volta Valley project 139-40
burning 27, 53-8 *passim*, 70, 164
bush 60, 66, 69, 146, 150, 156, 198; savanna 77, 141
Bwa 74-82, 85-9 *passim*, 146

camels 173, 174
Canada 205
Cape Colony 4
capitalism 8, 184, 185
Capron, J. 75
carrying capacity 17, 26, 162-4, 178-9, 181, 183, 186, 195-201 *passim*, 212
Carson, Rachel 16
cashew trees 19, 148, 230
cassava 230, 241-4 *passim*
cattle 26, 59, 64, 66, 126, 140-1, 160, 173-5, 178-80, 182-202 *passim*
Catholic Relief Services 17
Celestino, Manuel 230
centralization 184, 185, 194
cereals 99-101 *passim*, 126, 137-8
César, J. 59
charcoal 129
ch'at 101, 102, 113
Chauveau, Jean-Pierre 55
Chevalier, August 125
Chilundo, Arlindo 225-47
China 36
Cliffe, Lionel 183
climate 2, 5, 7, 10, 32, 33, 51, 56, 97, 98, 162, 163, 179, 198, 202
cocoa 160
coffee 27, 101, 102, 113, 160
collectivization 209, 212, 217, 228-30
colonial period 4, 5, 11, 13-15, 121, 124-36, 140, 159, 160, 184-92
commercialization 124, 133, 160, 183
commons, tragedy of 71, 194
compensation 66, 69, 139
conflicts 123, 171, 173, 201-3 *passim*; farmer/herder 66, 69- 70, 141, 164; land 80, 158
Congo 15
conservation 3, 15, 17, 29, 55, 68, 70, 74, 88, 124, 125, 159, 184, 187, 191, 240 *see also* Ethiopia; World Conservation Strategy 194
co-operatives 87, 216, 228
corruption 247
Côte d'Ivoire 12, 15, 18-20 *passim*, 24, 25,

53-71, 127; CIDT 59; High Commissioner of Savanna Region 53-4; land degradation in 53-5, 66, 69-70; NEAP 54, 65, 68; Plan Foncier Rural 69-70; PNTGER project 70; SODEPRA 58
cotton 20, 25, 38, 59, 75, 76, 87, 126, 134-8 *passim*, 146, 158- 60, 182, 227-47 *passim*
Cousins, Ben 201
credit 59, 118, 141, 231, 240, 244, 247
Cronon, William 8
crop rotation 15, 81, 194, 221, 230, 242-3, 246-7
Crummey, Donald 1-30, 91-120
cultural factors 2, 20, 165, 187

dam building 50-1, 140
Dämbeya 97, 108, 113-14, 116-17
debts 87, 138, 229
decentralization policy 69
decolonization 155
deforestation 5, 9-12 *passim*, 16, 19, 24, 25, 32, 54, 121 *see also* Ethiopia
degradation, land 2, 3, 8-9, 11, 17-18, 25, 26, 33, 52, 72, 73, 89, 163, 164, 167-70 *passim*, 177 *see also* Côte d'Ivoire; Ethiopia; Mozambique; Zimbabwe
descent groups 79, 81
desertification 9-10, 15, 25, 32, 33, 53, 68, 164, 167, 168, 199; UN Convention to Combat 9, 33; *World Atlas of* 33
Dessalegn Rahmato 17, 19, 23-5 *passim*, 205-24
dessication 5, 11, 16, 25, 33, 38
destocking 15, 181, 184, 189, 190, 195
devaluation, currency 229
Dimikuy 74, 77-9 *passim*, 81, 86-9 *passim*
disease 123-4; livestock 70; plant 234, 242
disequilibrium approach 18, 162, 164, 166, 177, 199
Dohoun 74, 77-9 *passim*, 82, 86-9 *passim*
donors 24, 28, 29, 53, 69, 70, 118, 138, 195, 202, 205, 206, 211, 212, 215, 221
draft 182-4, 189
drainage ditches/canals 88, 101, 209, 217, 220, 221, 224, 245
drought 3, 6, 10, 15, 16, 32-6 *passim*, 46, 48, 51, 58, 76, 162-4 *passim*, 173, 179, 197, 211, 214
Dumont, René 155
Duncan, J. 8
Dya, Sanou 159
Dzimba, Governor Gaspar 230

education 86; higher 139

Egypt 36
Elias, E. 40-1
elites 28, 70, 94-6 passim, 154
El Niño Southern Oscillation 33, 51
employment 139, 212, 218, 220
enclosures 20, 55, 96, 119, 169, 214, 219, 220
environmentalism 205-24; peasant 209-10, 224; state 208-9, policy 210-24
equilibrium approach 6, 7, 17, 163, 164, 217
Eritrea 13, 213
erosion, soil 3, 10, 21, 23-5 passim, 32, 66, 88, 185, 186, 192 see also Ethiopia; Mozambique
Ethiopia 12, 13, 16-20 passim, 24, 29, 32, 38-47, 91-120, 205-24; conservation policy 17, 201, 205-22, National Strategy 94-5; deforestation in 12, 24, 25, 38-41, 52, 91, 94-6 passim, 108, 119, 206-7, 209, 211, 212, 219, 220; Derg 96, 116, 119, 205, 209, 211-13, 216-24 passim; EFAP 207; EHRS 16, 207, 211, 223; Gondär 91, 93, 96, 97, 108, 113; land degradation in 38-41, 91, 94, 95, 208, 210-12, 220, 222, 223; National Action Plan/Programme to Combat Desertification 95, 96; Project 2488 214-19; Relief and Rehabilitation Commission 46, 95; SCRP 16, 217, 222, 223; soil erosion in 17, 32, 39-45, 95, 119, 207, 209, 211, 212, 220-3 passim; Tigray 213; Wällo 25, 91-120, 209, 212-21 passim
Ethiopian Rift 41-50 passim, 52
eucalyptus 17, 19, 66, 101, 109, 114, 116, 120
European Union 205; Commission 214
expatriate advisers 27, 138, 205-8 passim, 211-12, 221-2
exports 124, 126, 135-6, 160, 210
expropriation 139, 140
extension services 24, 200, 244, 245

Fairhead, James 3, 10, 12, 13
fallows 25, 39-40, 72, 78, 81, 82, 124, 137, 140, 141, 150, 159, 160, 234, 241-7 passim
famine 3, 15-18 passim, 24, 25, 32, 38, 91, 93, 95, 108, 119, 211, 214
fans, alluvial 49-50, 52
FAO 206, 207
farmers 1, 5, 11, 15-19 passim, 53, 65, 66, 69-70, 72-120 passim, 134, 137, 138, 141, 156-9, 182-4, 187-90, 193-202, 208-12, 216- 28, 230-47 passim; contract 227, 231, 234, 244, 246, 247
fertilizer use 19, 28, 86-90, 136-8 passim
fire 6, 7, 23, 25, 53-9, 164; suppression 57, 68; timing 57, 59

fish 140
floods 32, 36, 41, 45-51 passim, 174
Food for Work schemes 212-15, 217, 219, 221, 224
forests, classified 129; fringing 123-4, 140
France 10-11, 13, 15, 69, 124-36 passim
fruit trees 19, 24, 25, 99, 113, 125-6, 152, 154-60
fuelwood 19, 66, 107, 114, 119, 128, 129, 160, 212, 219

Gabbra 168
game 14, 60-1, 64, 71
geomorphology 1, 21, 31-52
Gerba Peasant Association 96, 98-101, 107, 110, 111, 115
Germany 15
gésho 101, 102, 113
gestion des terroirs 69, 72
Gikuchi, Francis 3, 22
global warming 36
Gold Coast 127
government, local 209, 229, 232
grasses 56-7, 59, 66, 141, 160; 171, 200
Gray, Leslie C. 19, 26, 29, 72-90, 140
grazing 14, 15, 23, 25, 56-9 passim, 64, 66, 119, 120, 141, 171-5 passim, 184, 186, 191-5, 198, 200-2 passim, 210, 216; rotational 66, 70, 184, 195, 200
groundnuts 182, 230, 234, 240-4 passim
Grove, Richard 3, 4
Guinea 3
Gwobeya Peasant Association 96, 98-101, 107, 110, 111, 115
herbicides 19
herders 1, 11, 15, 17-19 passim, 26, 59, 65, 66, 69, 70, 140, 141, 160, 163-4, 168, 171-6
heritage, national 210
Herweg, Karl 222
Hoben, A. 41, 208
hunting 7, 13, 53, 60-6 passim
Hurni, Hans 16, 39, 40, 207, 222-4 passim
hydro-electric power 50-2 passim, 140, 171

ideology 184-5, 208
income 101, 160; non-farm 99, 118, 246
India 36
Indonesia 36
inheritance, land 80, 81
innovation 18, 20-1, 94-6, 115-18, 185
inputs, use of 74, 89, 124, 227, 240, 243-7 passim see also individual entries
insecticides 136, 227, 228, 231, 235, 240, 243-7 passim

insecurity 168
intercropping 28, 227, 241, 243
International Red Cross 12, 17; Union for
    Conservation of Nature 12
interviews 23, 58, 60-5 *passim*, 93, 96-8,
    102-3, 220
investment, agricultural 18, 54, 72-4, 80, 84-
    90, 118, 119, 209, 215
iron work 126, 128, 160
irrigation 9, 47, 48, 50-2, 139, 140, 221

Jackson, J. J. 200
Johnson, Douglas H. 3
joint ventures 227, 231, 232, 243, 246, 247
João Ferreira dos Santos 231
Journal of Ethiopian Studies 93
Joyce, Major F. 221

Käbbädä Hamza 112
Kankalaba 142, 144, 150, 152-4
kapok 131, 133
Kassaw Bäqqälä 112
Kebede Tato 222
Kenya 14, 15, 19, 20, 29, 166-71, 173, 177;
    Machakos 3, 118; Marsabit, IPAL project
    in 25, 166-71, 176, 177
knowledge, local 1, 12-13, 18-21, 24, 26-30
    passim, 71, 111, 117, 181, 198, 201, 203,
    219-23 *passim*
Koli Bi, Zueli 12, 19, 23, 25, 53-71
K'one volcanic complex 41-5

labor 74, 89, 111, 124, 146, 154, 165, 168,
    175, 176, 183, 229, 240; forced 127, 213
Lamprey, Hugh 167
land 14-15, 20, 69, 72-90, 111, 118-19, 154-
    9, 165, 168, 192-3, 225, 247; legislation
    54-5, 68-70; reform 54, 68-70, 118, 209,
    210, 225; rights 19, 20, 26, 54-5, 72-4,
    78-81, 85, 90, 148, 154, 157; scarcity 72-
    4 *passim*, 77, 80-2, 85, 89, 90, 99, 119,
    138, 154, 158-9, 183, 184, 192, 203,
    232; tenure 55, 72, 73, 80, 89, 118, 119,
    168, 182, 193, 194, 216; use 25, 31-52,
    57-60 *passim*, 68-70, 74, 79, 141-54,
    163, 169, 185, 186, 194-5, 209, 211,
    233-45
Landolphia 126, 129-30, 133, 159
landscape change 21-4, 30; 'walks' 23
Laurent, P. J. 80
Leach, Melissa 3, 10, 12, 13
lines, stone 76, 88
Little, Peter D. 17-18, 25, 26, 29, 161-77
livestock/raising 7, 13, 17, 26, 29, 57-9, 70,
    74, 99, 171-6, 178-202; pressure 178, 179,
    187
Logan, W.E.M. 206
MacArthur, John D. and Catherine T. Foun-
    dation 93, 97
Mackenzie, Fiona 180
maize 75, 76, 99, 136-8 *passim*, 150, 182,
    241-3 *passim*
Mali 58
Malthus, Robert 8-9, 20, 91, 94, 208
Mammo Seyem 213
management, resource 165, 178-204, 209
Mängesté Täklé 108, 113
mangoes 19, 125, 148, 150, 155-9
manure 19, 73, 74, 76, 82, 86, 87, 89, 90,
    141, 182, 184, 186
market gardening 101, 115, 117, 118
markets 19-20, 25, 118, 246, 247; cattle 187-
    9; game 60-11, 64; land 55, 194
Massua, Jacinto 235
Mathieu, P. 80
Maugini, Armando 97, 107
Mavala, Armando 245
McCann, James 12, 24, 40, 41
meat 182; bush 61
Mengistu, Haile Mariam 38
Menilek, Emperor 116
Merera Ejjeta 96, 97
migration 20, 58, 72-82, 85-9 *passim*, 127,
    140, 146, 148, 150, 168; herd 173-5, 201
milk 182
millet 76, 87, 137, 150, 182, 230, 241-4
    passim
modernization 28, 55, 194
Mombeshora, B. 193
monetization 126
mono-culture 234, 245, 247
Monteuil, L.-P. 125
Mortimore, Michael 3, 10, 22
Mose/Mossi 74-82, 85-90 *passim*, 146, 150,
    155
Mozambique 20, 225-47; EEAN 228, 229,
    231; Frelimo 225, 228-47; Ibramugy 234,
    244, 247; land degradation in 225, 233,
    234, 240, 241, 246; Namialo 227-47;
    Renamo 225, 230; SODAN 227, 231,
    232, 235, 240, 244, 246, 247; soil erosion
    in 230, 233, 234, 245, 247
Muacamisa, Afonso 243
multinational corporations 227
Munro, William A. 17-18, 23, 26, 29, 178-
    204

Naba Kougri 155

Namibia 13
nationalism 15
nationalization 20, 119, 228
NGOs 12, 15-16, 24, 28, 53, 65, 72, 94, 118,
    195, 201-5 *passim*, 214, 215
Nicol, Clive 39, 40
Niger 49-50, 52, 127
Niger delta 51, 127
Nigeria 57
Nile 32, 34, 36, 38, 40, 51
Northern Rhodesia 196
nutritional status 212-13
Nyasaland 15

offtake 179, 181, 186-9 *passim*, 192, 195, 200
O'Leary, Michael 168
orchards 25, 124, 148, 150, 152, 154-60
Otieno, Joseph 19
Ouadba, Jean-Marie 121-60
Ouattara, Tiona 19, 23, 25, 53-71
over-cropping 9, 16, 184, 207
overgrazing 9, 16, 27, 41, 49, 50, 164, 195,
    196, 200
overstocking 17, 178, 179, 181, 184
ownership 208-9; forest 210, 216, 217; land
    69, 80, 81, 119; livestock 164, 165, 173-5,
    182, 183
oxen 59, 76, 99, 141

Pacheleque, José 235
paddocking 184, 191, 194, 195, 200-3 *passim*
palm, *ban* 152, 160
Park, Mungo 5
Parkia biglobosa 88, 131-4 *passim*, 160
parks, national 14, 54, 66, 163, 210
pastoralists 5, 15, 17, 53, 58-9, 161-77
patronage 29, 183, 191
peanuts 126, 134-5, 146, 157
Peet, Richard 24
Péni 128, 142, 144, 149, 150, 156, 157
Perrot, Emile 125
pesticides 19, 28, 244
pests 227, 234, 235, 241-3 *passim*, 247
photographs, aerial 3, 22, 23, 60, 65, 69, 77,
    90, 141, 144; matched 3, 22, 23, 25, 96,
    103, 107
Pitcher, M.A. 247
plans/planning 29, 65-70 *passim*, 185-6
plows/plowing 59, 75, 76, 89, 99, 128, 138,
    141, 184, 221
poaching 54, 66
political ecology 2, 29, 71, 161-2, 164-70,
    174-7 *passim*
political factors 28, 29, 70-1, 192, 194, 229

population 39, 125, 182; density 97, 140, 159,
    163, 192, 212; growth 4, 9, 20-1, 24, 72,
    74-7, 90, 91, 94, 95, 107, 119, 137, 161,
    208; pressure 179, 182, 183, 187, 211, 212
poverty 110, 111, 168, 171, 192, 213
prices 136, 187; producer 20, 23, 99, 228,
    240, 244
privatization 20, 54-5, 68, 71, 73, 225, 227,
    232, 245-6
productivity 7, 91, 118, 163, 173, 180, 182,
    185, 195, 203, 229, 234
Prudencio, C. 76, 86
public works projects 126-8, 213-15 *passim*

questionnaires 24, 227

race factors 184, 185
railroad 127-8
rainfall, 6, 7, 10, 33, 34, 41, 49, 51, 58, 75,
    98, 123, 162, 173, 195, 198, 234
ranches 20, 169
rangelands 17-18, 20, 26, 66, 70, 161, 163,
    164, 168, 169, 179
Ranger, Terence 184
Raogo, Korbeogo 82
re-afforestation 41, 213, 214, 216
registration, land 69, 70, 72
religious sanctions 209
remote sensing 3, 22, 23
Rendille 168
research 30, 124-6, 138, 161-71 *passim*
reserves 163; forest 15; game 14, 171; native
    14, 186-93 *passim*
resettlement 17, 96, 99, 118, 139, 209
revenue, state 210-11
rice 241
rights, land 19, 20, 27, 68, 201; livestock 190,
    194
risk 179, 180, 203
river flow 33, 34, 36-8, 50-2
roads 127, 128, 154, 213
Rostow, Walter 28
royalties 210
rubber 129-30, 133, 159

Sahara 5, 10, 15, 33
Sahel 10, 15, 33, 49-50, 52, 96; Club du 16
sales, cattle 187-9 *passim*; game 61-2, 64
Samori 130
Sara 74, 77-81, 85-90 *passim*, 128; -Hantiaye
    142-4, 146, 148
Şaul, Mahir 19, 23, 121-60
savanna 5-6; bush 69, 77, 141; Guinea 11, 56,
    75; Sudan 75, orchard 123, 129

savannaization 53, 54, 65, 68, 69
Savonnet, G. 75-6
Scholes, R.J. 7
Scoones, I. 40-1, 196, 198-9
Scott, James 27
sea surface temperature 33-4, 51
sedentarization 20, 168, 171
sedimentation 36-8, 41, 52
Selassie, Haile 38, 116, 117, 119
Senegal 130, 134
Serno, G. 234
sesame 99, 126, 134, 135, 241, 243
settlers, white 4, 14, 184, 185, 199
shea nuts/butter 130-1, 133, 159
SIDA 205, 214
simulium fly 123
Sobania, N. 168
Society for Preservation of Wild Fauna of the
    Empire 14
soil, compaction 227, 235, 247; fertility 40-1,
    52, 72-90 *passim*, 229, 230, 234, 241, 245;
    loss 25, 33, 38-41, 52, 203, 207, 212, 223,
    234, 243; management 15, 19-20, 29, 72-
    90; regeneration 23, 81
Somalia 29, 32, 50-2, 171-7
sorghum 76, 87, 99, 136, 137, 146, 150, 241,
    243
Soumousso 142, 144, 148, 150
South Africa 7, 14, 166
Southern Oscillation Index 36
Southern Rhodesia 14 *see also* Zimbabwe
springs, hot 48, 52
Ståhl, Michael 95, 207
state farms 228-31 *passim*
Stebbing, E.P. 16
Stocking, Michael 223
stocking rates 163, 186, 189-90, 192, 194-8
    passim, 201
Street, Alayne 50
subsidies 138, 229
Sudan 32, 36, 38, 51-2
sugar cane 139
Sulula 97, 98, 108
surveys, household 65, 69, 96
Sutcliffe, J.P. 223
Swain, E.H.F. 206

T'abisa 97, 98, 107
Talbot, Michael 49
Tanganyika 15
Tärräfäch Färräda 114
taxation 118, 126, 130, 140, 210
technology 19, 23-5 *passim*, 28, 91, 94, 184,
    212, 214, 220, 222, 229

t'ef 99
terracing 15, 17, 75, 76, 101, 152, 205, 209,
    214, 217-19 *passim*, 221, 245
Tersiguel, P. 75
Tiffen, Mary 3, 22
timber 123, 128, 210
Timberlake, Lloyd 9-10
Tique, César A. 225-47
T'is Aba Lima 97, 98, 107
Toba volcano, Sumatra 31
tobacco 19, 182
tourism 66, 210
tractors 89, 136, 227, 234-5, 244, 246, 247
trade 99, 125, 126, 133
transport 99, 127-8, 155, 160, 182, 184, 247
tree, density 86-90, 109; felling 7, 10, 210,
    219; /grass ratios 6-7, 56-7, 60, 66;
    -planting 16-19 *passim*, 54, 56, 66, 68, 80,
    96, 101-3, 107, 109-19 *passim*, 126, 210,
    213, 214, 216
tsetse fly 123, 173, 175

unemployment 139
United Nations 9-10; Conference on
    Environment and Development 33, 65;
    Environmental Program 12; UNESCO
    167, 206
United States 135; Dust Bowl 4, 15; PL480
    213; USAID 12

Vayda, Andrew P. 177
vegetation, climax 17, 163; transects 22, 23;
    wild 121-60
villages, communal 228-30
villagization 209, 229
volcano-tectonic factors 41-5, 48-9, 52

Walker, B.H. 7
Walters, Bradley B. 177
war 124-5, 171, 173, 225, 230; liberation 192;
    World War I 125, 135; World War II 130
water resources 7, 15, 32, 114, 171, 209, 212,
    213
Watts, Michael 24
wealth 74, 86, 89, 110
Westoby, M. 197
wetlands 171, 199-201
wildlife 22, 55, 58, 60-4, 66, 69, 71, 127,
    206-7, 210; World Wildlife Fund 12
Williams, Martin 21, 24, 25, 29, 31-52
Winter-Nelson, Alex 12, 17, 19, 23-5 *passim*,
    91-120
Wolf, Eric 162, 165
women 133, 139, 155, 168; female-headed

households 111, 168, 201
woodcutting 114, 117, 128
woodlots 216, 217, 219
Wood A. 207
World Bank 12, 16, 55, 65, 68-70 *passim*,
  205, 206, 214
World Food Programme 205, 214, 215, 218-
  20 *passim*
Wylde, A.B. 221

Yemane Zecharias 48
yields 19, 99, 119, 136, 137, 212, 222-3, 229,
  234, 241-3 *passim*

Zibita, Kahoun 82
Zimbabwe 26, 29, 41, 178-204; Agritex 194,
  201; Communal Lands 178-9, 181-3, 192-
  204; land degradation in 178, 184-6
  passim, 193, 197-200 *passim*, 203; National
  Land Use Programme 194; Native Land
  Husbandry Act 184, 186; Natural
  Resources Act 199- 200; UDI 191, 192;
  VIDCOs 194-5, 202; Water Act 199-200

Printed and bound by CPI Group (UK) Ltd, Croydon, CR0 4YY

27/10/2024

14580347-0005